Semiclassical Methods in Molecular Scattering and Spectroscopy

NATO ADVANCED STUDY INSTITUTES SERIES

Proceedings of the Advanced Study Institute Programme, which aims at the dissemination of advanced knowledge and the formation of contacts among scientists from different countries

The series is published by an international board of publishers in conjunction with NATO Scientific Affairs Division

A	Life Sciences	Plenum Publishing Corporation
B	Physics	London and New York
C	Mathematical and Physical Sciences	D. Reidel Publishing Company Dordrecht, Boston and London
D	Behavioural and Social Sciences	Sijthoff & Noordhoff International Publishers
E	Applied Sciences	Alphen aan den Rijn and Germantown U.S.A.

Series C – Mathematical and Physical Sciences

Volume 53 – Semiclassical Methods in Molecular Scattering and Spectroscopy

Semiclassical Methods in Molecular Scattering and Spectroscopy

Proceedings of the NATO ASI held in Cambridge, England, in September 1979

Edited by

M. S. CHILD
University of Oxford, Theoretical Chemistry Department, Oxford, England.

D. Reidel Publishing Company

Dordrecht : Holland / Boston : U.S.A. / London : England

Published in cooperation with NATO Scientific Affairs Division

Library of Congress Cataloging in Publication Data

CIP

NATO Advanced Study Institute, Cambridge, England, 1979.
Semiclassical methods in molecular scattering and spectroscopy.

(NATO advanced study institute series: Series C, Mathematical and physical sciences ; v. 53)
Includes bibliographies and indexes.
1. Molecular spectra-Congresses. I. Child, M. S. II. Title. III. Series.

QD96.M65.N37 1979 539'.6 80-10382

ISBN-13: 978-94-009-8998-6 e-ISBN-13: 978-94-009-8996-2
DOI: 10.1007/978-94-009-8996-2

Published by D. Reidel Publishing Company
P. O. Box 17, 3300 AA Dordrecht, Holland

Sold and distributed in the U.S.A. and Canada
by Kluwer Boston Inc., Lincoln Building,
160 Old Derby Street, Hingham, MA 02043, U.S.A.

In all other countries, sold and distributed
by Kluwer Academic Publishers Group,
P.O. Box 322, 3300 AH Dordrecht, Holland

D. Reidel Publishing Company is a member of the Kluwer Group

TABLE OF CONTENTS

PREFACE

The enclosed proceedings of the NATO Advanced Study Institute held in September 1979 are intended to provide a comprehensive introduction to the development of semiclassical methods for molecular scattering and spectroscopy over the past twenty years.

Progress in the scattering field may be taken to date from the famous paper by Ford and Wheeler in 1959, and coincidentally the RKR technique for diatomic spectroscopy was rediscovered at roughly the same time. Subsequent developments have emphasised the insight to be gained by combining classical and quantum mechanical ideas in a wide variety of more complicated problems. Three main types of development may be recognised. First advances have occurred in mathematical technique, both in the development of higher order JWKB and phase integral expansions as discussed in chapters 1 and 3, and also in the use of uniform approximations to remove the spurious singularities inherent in the primitive JWKB approach. The latter are introduced in the context of elastic scattering in Chapter 2, and later recur in other applications in chapters 4 and 5. A second development has been to extend semiclassical ideas to tunneling and curve-crossing situations. Here spectroscopic applications of this type are covered in chapter 4, and scattering situations are treated under the heading of non-adiabatic transitions in chapter 6. Finally major progress has been made over the past ten years in the treatment of non-separable situations. Applications to inelastic scattering are covered in chapters 5 and 7; extensions to reactive situations are also outlined. The final chapter covers quantisation procedures for the bound states of non-separable situations.

The level of presentation is intended as that of a graduate student in chemical physics, Problems are provided at the end of each chapter so that readers may test their comprehension of this fascinating but possibly unfamiliar field.

M.S. Child

LIST OF AUTHORS

M.S. Child, Theoretical Chemistry Department, 1 South Parks Road, Oxford, OX1 3TG, U.K.

J.N.L. Connor, Department of Chemistry, University of Manchester, Manchester, M13 9PL, U.K.

A.S. Dickinson, Department of Atomic Physics, University of Newcastle, Newcastle-upon-Tyne, NE1 7RU, U.K.

N. Froman, Institute of Theoretical Physics, University of Uppsala, Uppsala, Sweden.

Thomas F. George, Department of Chemistry, University of Rochester, Rochester, New York, U.S.A.

N.C. Handy, University Chemical Laboratory, Lensfield Road, Cambridge, U.K.

Kai S. Lam, Department of Chemistry, University of Rochester, Rochester, New York, U.S.A.

Robert J. Le Roy, Guelph-Waterloo Centre for graduate work in Chemistry, University of Waterloo, Waterloo, Ontario, Canada, N2L 3G1

SEMICLASSICAL AND HIGHER-ORDER APPROXIMATIONS: PROPERTIES. SOLUTION OF CONNECTION PROBLEMS

Nanny Fröman

Institute of theoretical physics, University of Uppsala, Uppsala, Sweden

Abstract. 1. Introduction. 2. First-order JWKB-approximation. 3. Method for solving connection problems. 4. Anti-Stokes' and Stokes' lines. 5. Properties of the JWKB-approximation in the complex plane. Connection formulas for a simple turning point and for a barrier. 6. Bohr-Sommerfeld quantization condition. 7. Phase shifts. 8. Normalization. 9. Expectation values and matrix elements. 10. Higher-order approximations. 11. Exercises.

1. INTRODUCTION

Only a few quantal problems are exactly soluble by analytical methods. In practice one has almost always to resort to an appropriate approximation method or to numerical integration of the Schrödinger equation. Among the great variety of approximate methods used in quantum mechanics, one may discern three different main types: perturbation methods, variational methods and asymptotic methods. The topic of the present lecture course falls into this last category.

The significance of an asymptotic treatment is not only to provide a convenient means for approximate calculations of details, but often it also has a fundamental physical significance. In our present context, the asymptotic treatment of the Schrödinger equation in its zeroth order yields the classical Hamilton-Jacobi equation, in its first order it yields the semi-classical approximation, and in higher orders it yields, in many important situations, a full account of quantal details, the physical quantities being obtained with a degree of accuracy hardly obtainable by any other means.

It is beyond the scope of the present lectures to enter in-

M. S. Child (ed.), Semiclassical Methods in Molecular Scattering and Spectroscopy, 1–44.

to any account of the extensive literature that exists on the theory of asymptotic approximations. A great number of references can be found in Heading's book [1] and in more recent review articles [2,3]. My aim is to present the basic properties of the semi-classical approximation and the higher-order approximations with a minimum of mathematical and technical details. To achieve this goal I will not, in the first place, aim at generality but rather try to bring out the essential points as simply as possible. Thus, we shall in the main part of the lectures be concerned with the use of the first-order JWKB-approximation, or the semi--classical approximation. The higher-order approximations which, by the way, will not be obtained from the JWKB-expansion but from a more convenient expansion, are brought in only after we have become familiar with the first-order approximation. To generalize the result of a certain application from the first-order JWKB--approximation to the kind of higher-order approximations we shall use, is, however, a rather straightforward matter, when one knows how to do it. The procedure will be described in Sec. 10.

2. FIRST-ORDER JWKB-APPROXIMATION

2.1. The JWKB-expansion

We consider the Schrödinger equation

$$d^2\psi/dz^2 + Q^2(z)\,\psi = 0\ , \tag{2.1}$$

where, with obvious notations,

$$Q^2(z) = 2m[E - V(z)]/\hbar^2\ . \tag{2.2}$$

The function $V(z)$ may be the actual physical potential or an effective potential. Thus, if we are concerned with the radial Schrödinger equation, $V(z)$ is assumed to include also the centrifugal term. According to the standard procedure for obtaining the JWKB-expansion we look for a solution of the form

$$\psi = \exp\{i\ S/\hbar\}\ . \tag{2.3}$$

Substitution into the Schrödinger equation yields for dS/dz the nonlinear Riccati equation

$$-\frac{i\hbar}{2m}\frac{d^2S}{dz^2} + \frac{1}{2m}\left(\frac{dS}{dz}\right)^2 + V(z) = E\ . \tag{2.4}$$

In the limit $\hbar \to 0$ this equation reduces to the classical Hamilton-Jacobi equation, which on integration immediately yields as the zeroth-order approximation the classical action

$$S_0 = \pm \int^z \{2m[E - V(z)]\}^{\frac{1}{2}} \, dz \ . \tag{2.5}$$

The first-order and higher-order JWKB-approximations are obtained by assuming a series expansion in powers of $\hbar$

$$S = S_0 + \hbar\, S_1 + \hbar^2\, S_2 + \dots \ , \tag{2.6}$$

substituting it into (2.4) and equating coefficients of each power of $\hbar$. In that way one obtains

$$- i \frac{d^2 S_0}{dz^2} + 2 \frac{dS_0}{dz} \frac{dS_1}{dz} = 0 \ , \tag{2.7a}$$

$$- i \frac{d^2 S_1}{dz^2} + \left(\frac{dS_1}{dz}\right)^2 + 2 \frac{dS_0}{dz} \frac{dS_2}{dz} = 0 \ , \quad \text{etc.} \tag{2.7b}$$

From (2.7a) combined with (2.5) it follows that

$$S_1 = i \ln\{2m[E - V(z)]\}^{\frac{1}{4}} \ . \tag{2.8}$$

Thus, the first-order JWKB-functions are

$$\psi \approx \left(\frac{2m}{\hbar^2} (E-V(z)) \right)^{-\frac{1}{4}} \exp\left\{ \pm \frac{i}{\hbar} \int^z [2m(E-V(z))]^{\frac{1}{2}} \, dz \right\}$$

$$= Q^{-\frac{1}{2}}(z) \exp\left\{ \pm i \int^z Q \, dz \right\} . \tag{2.9}$$

One may expect these functions to satisfy the Schrödinger equation "locally" very well if

$$|\hbar^2 \, dS_2/dz| \ll |\hbar \, dS_1/dz| \ll |dS_0/dz| \ . \tag{2.10}$$

The last inequality yields, with due regard to the definition (2.2) of Q^2 ,

$$\frac{1}{2} \left| \frac{Q'}{Q^2} \right| \ll 1 \ . \tag{2.11a}$$

Since $Q = 2\pi/\lambda$, where λ is the local de Broglie wavelength, we can write (2.11a) alternatively as

$$|d\lambda/dz| \ll 4\pi \ . \tag{2.11b}$$

This condition implies that the potential must change slowly over a distance comparable to the wavelength. It is immediately clear that the condition (2.11) will be violated whenever E equals $V(z)$, which is the case at the classical turning points, or when $V(z)$ fluctuates rapidly. When we consider the "global" problem, i.e., when we require a certain linear combination of the functions (2.9) to represent one and the same exact solution in a certain region of the complex z-plane, the condition (2.11) is not sufficient, as will be seen below.

2.2. Connection problems

In physical applications it is usually necessary to know the behaviour of one and the same solution not only locally in the neighbourhood of a certain point but over extended regions of the independent variable. One then encounters difficulties even if the condition (2.11) is excellently fulfilled at every point in the region of interest. We shall examine the situation closer.

Let us for convenience introduce the notations $f_1(z)$ and $f_2(z)$ for the JWKB-functions in the right-hand member of (2.9). For future needs we also take into account the possibility of so called modifications, a well-known example of which is the replacement of $l(l+1)$ by $(l+\frac{1}{2})^2$ in radial problems, and write $Q_{mod}(z)$ instead of $Q(z)$ in the JWKB-functions, i.e.,

$$f_1(z) = Q_{mod}^{-\frac{1}{2}}(z)\exp\{+iw(z)\} \tag{2.12a}$$

$$f_2(z) = Q_{mod}^{-\frac{1}{2}}(z)\exp\{-iw(z)\}, \tag{2.12b}$$

where

$$w(z) = \int^{z} Q_{mod}(z)\,dz . \tag{2.13}$$

For the present discussion we consider for simplicity the unmodified approximations (2.9). Suppose that $Q^2(z)$ has a zero on the real axis at $z = x'$ (see Fig. 1), so that x' is a classical turning point with a classically forbidden region to the left and a classically allowed region to the right. In the neighbourhood of the classical turning point x' the functions $f_1(x)$ and $f_2(x)$, are no longer approximate solutions of the differential equation. The zeros of $Q^2(x)$ are regular points for the differential equation, which implies that the exact solutions $\psi(x)$ are regular there. The approximate solutions $f_1(x)$ and $f_2(x)$, however, become infinite at these zeros. There is thus a certain region in the neighbourhood of each classical turning point in which the approximation breaks down. Furthermore, if a given solution $\psi(x)$ is represented approximately by a certain linear combination of $f_1(x)$ and $f_2(x)$ on one side of the turning point, then it has to be represented by another linear combination on the other side of the turning point. This is a typical reflection of the so called Stokes' phenomenon, which is well known in the theory of asymptotic expansions.

More generally, if we allow the independent variable to take values in the complex z-plane, the same linear combination may be used within a certain region of the z-plane, whereas in other regions different linear combinations have to be used to represent the same solution ψ of the differential equation. The problem of determining which linear combinations of approximate solutions that represent the same solution ψ in different regions is the so called connection problem or global problem. It is deeper than the local problem of finding approximate solutions considered in paragraph 2.1.

Connection formulas for tracing first-order JWKB-solutions across a classical turning point on the real axis were first derived by Gans [5], and then independently by Jeffreys [6] and by Kramers [7]. Brillouin [8], Wentzel [9] and Kramers [7] in 1926 introduced the method in quantum mechanics. Over the years there has been very much dispute and discussion in the literature as to the precise meaning and properties of these connection formulas. After having been subject to rigorous analysis and numerical studies they are now fully understood. It is, however, not only the connection problem at a classical turning point that is intricate. Caution has to be exercised whenever one moves along a path where $|\exp\{iw(z)\}|$ increases or decreases rapidly.

To summarize: Even if one knows a set of linearly independent functions which satisfy a given differential equation approximately, and even to a very high degree of accuracy, it may still be a difficult matter to use these solutions for solving problems encountered in theoretical physics. The boundary conditions imposed by the physical problem require a special solution to be represented approximately. However, it may occur that a linear combination of approximate solutions, satisfying a boundary condition, not at all represents the exact solution approximately, when one proceeds away from the boundary.

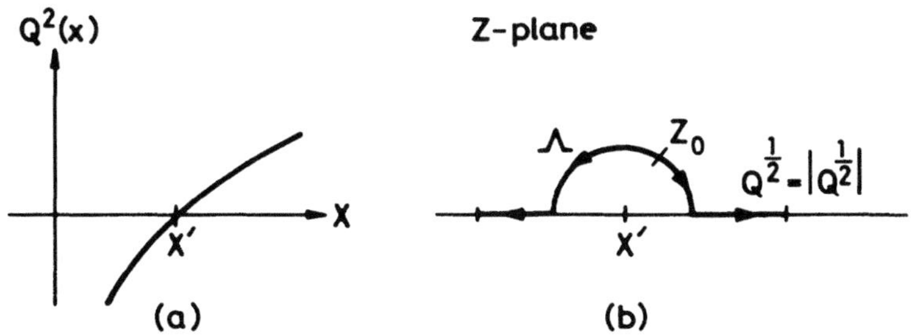

Fig. 1. a) Qualitative behaviour of Q^2 at a classical turning point.

b) Path Λ for tracing solutions in the complex plane. Arrows indicate directions in which $|\exp\{iw(z)\}|$ increases.

For the solution of the connection problems at a classical turning point, say x' , there are two essentially different approaches: one amounts to finding the form of the solution as one proceeds through the turning point, the other amounts to avoiding the turning point by tracing the solution in the complex plane well away from the turning point.

The technique most frequently met with in text books falls into the first of these two categories. Its idea is briefly as follows.

In the neighbourhood of the turning point x' under consideration, $Q^2(z)$ is approximated by the linear term in the Taylor expansion around $z = x'$, and hence the differential equation (2.1) becomes

$$d^2\psi/dz^2 + c^2(z-x')\,\psi = 0 \, , \tag{2.14}$$

where c^2 is the derivative of $Q^2(z)$ at $z = x'$. This equation can be solved exactly in terms of Bessel functions of the order $\pm 1/3$, the general solution being

$$A\,(z-x')^{\frac{1}{2}} J_{\frac{1}{3}}\left(\tfrac{2}{3}c(z-x')^{\frac{3}{2}}\right) + B\,(z-x')^{\frac{1}{2}} J_{-\frac{1}{3}}\left(\tfrac{2}{3}c(z-x')^{\frac{3}{2}}\right) \, , \tag{2.15}$$

where A and B are constants. For sufficiently small values of $|z-x'|$ the general solution of (2.1) is approximated by (2.15), and for sufficiently large values of $|z-x'|$ it is approximated by a linear combination of the JWKB-functions f_1 and f_2 defined by (2.12a,b). Well away from the turning point, the Bessel functions are approximated by well-known asymptotic expressions and (2.15) becomes a linear combination of the functions f_1 and f_2 , solving (2.14) approximately. Thus for equation (2.14), the Stokes' phenomenon can be mastered, since the exact solution is known, and hence the appropriate linear combination of asymptotic solutions on either side of the turning point can be found directly. The philosophy of the JWKB-procedure for solving the connection problems for the approximate solutions of the original equation is to assume that there is an overlapping region, in which the two conditions one is imposing are both valid, namely a region in which at the same time $|z-x'|$ is sufficiently small, in order that the linear expansion of Q^2 is justified, and sufficiently large, in order that the asymptotic expressions for the Bessel functions, i.e., the JWKB-solutions of (2.14), are valid. In this region one can then fit a certain JWKB-solution of the original differential equation (2.1) to a Bessel function solution of (2.14). In that way, JWKB-solutions of (2.1) are continued through the turning point via Bessel function representations.

Another method for obtaining solutions valid through the turning point is due to Langer [10] and is based on the idea of using an approximating equation, the solutions of which are approximate solutions of the original differential equation through the turning point as well as further away from it in the asymptotic regions. The solutions (2.15) are valid approximations to the solutions of the original equation only where the original equation is well approximated by (2.14). The functional form of (2.15), however, suggests a possibly useful form for a uniform approximation. Thus, Langer found that the expression

$$A\ Q^{-\frac{1}{2}}(z)\ w^{\frac{1}{2}}\ J_{\frac{1}{3}}(w) + B\ Q^{-\frac{1}{2}}(z)\ w^{\frac{1}{2}}\ J_{-\frac{1}{3}}(w)\ , \qquad (2.16)$$

with

$$w = \int_{x'}^{z} Q(z)\ dz$$

satisfies

$$d^2\psi/dz^2 + (Q^2(z) + \eta(z))\ \psi = 0\ , \qquad (2.17)$$

where $\eta(z)$ can be shown to be small, for small as well as large values of $|z-x'|$. Thus (2.16) is an approximate solution of (2.1) valid uniformly through the turning point and further away from it into the region where the asymptotic expressions approach the JWKB-solutions of (2.1).

The other, decisively different, approach for solving connection problems was initiated by Zwaan [11]. In 1929 Zwaan derived one of the connection formulas by letting the independent variable take complex values and proceeding with the solution in the complex plane, circumventing the classical turning point by a path proceeding qualitatively as Λ in Fig. 1. The change of the coefficients of the initial linear combination was determined from simple qualitative rules known from the study of Stokes' phenomenon. Kemble [12] adopted Zwaan's idea and took the first step towards a quantitative treatment. In 1965 the method was put on a rigorous basis [13].

An advantage of the method for tracing solutions, based on the expansion of $Q^2(x)$ around the turning point, and of the Langer method, is that one obtains an approximate solution valid through the turning point. However, if the behaviour of the solution in the neighbourhood of the turning point is not of interest, which is the case for wide classes of physical applications, it is preferable to stay away from the turning point. Not only does one avoid introjecting the problem of finding a solution which is not needed, but more important - it is only by tracing solutions at a distance from the turning points that one can understand and exploit fully the accuracy obtainable by asymptotic methods. For instance, on the basis of uniform approximations along the real axis one can hardly understand, why the JWKB-quantization condition for the harmonic oscillator yields exact eigenvalues, while it is easy to prove this fact by tracing solutions in the complex plane far away from the turning points.

This lecture course will be based on the rigorous method for tracing solutions in the complex plane [13]. We shall learn a rule of thumb enabling us to derive e.g. quantization conditions and connection formulas at a classical turning point by simple qualitative arguments. We can apply the rule with confidence, knowing that we have a rigorous treatment in the background, which also provides upper bounds for the errors involved in our final results. I shall briefly review the rigorous method and derive the rule of thumb quantitatively. Barrier transmission provides an example of a problem which requires a detailed mathematical treatment for a full understanding and, accordingly, I will, to some extent, resort to the formal mathematical treatment in discussing the connection formula for a barrier. For the rest, I will try to avoid technical and mathematical detail as much as possible.

3. METHOD FOR SOLVING CONNECTION PROBLEMS

We shall give in a condensed form the main ideas of the rigorous method on which our treatment will be based.

In the Schrödinger equation (2.1), the function $Q^2(z)$ is assumed to be an analytic function of the complex variable z in the region considered. We introduce the row vector

$$\underline{f}(z) = (f_1(z),\ f_2(z)) \ , \tag{3.1}$$

where $f_1(z)$ and $f_2(z)$ are the JWKB-functions defined by (2.12a,b) with (2.13), and we assume the complex plane to be cut in such a way that the function $Q^{\frac{1}{2}}_{\text{mod}}(z)$ is single-valued. We also introduce a column vector

$$\underline{a}(z) = \begin{pmatrix} a_1(z) \\ a_2(z) \end{pmatrix} \tag{3.2}$$

where $a_1(z)$ and $a_2(z)$ are functions of z, uniquely determined by the requirements

$$\psi = a_1(z)\, f_1(z) + a_2(z)\, f_2(z) = \underline{f}(z)\, \underline{a}(z) \tag{3.3a}$$

$$\frac{d\psi}{dz} = a_1(z)\,\frac{df_1(z)}{dz} + a_2(z)\,\frac{df_2(z)}{dz} = \frac{d\underline{f}(z)}{dz}\,\underline{a}(z) \ . \tag{3.3b}$$

We notice that the derivate of ψ is obtained by differentiating eq. (3.3a), treating $a_1(z)$ and $a_2(z)$ <u>formally</u> as constants. This one achieves by imposing the condition

$$\frac{da_1(z)}{dz}\, f_1(z) + \frac{da_2(z)}{dz}\, f_2(z) = 0 \tag{3.3c}$$

on the functions $a_1(z)$ and $a_2(z)$. Substituting (3.3a) into the Schrödinger equation (2.1) and using (3.3c), we obtain a system of two coupled differential equations of the first order for $a_1(z)$ and $a_2(z)$, which we write as follows

$$da_1/dz = \tfrac{1}{2}i\ \varepsilon(z)\ Q_{mod}(z)\ [a_1 + \exp\{-2iw(z)\}a_2] \tag{3.4a}$$

$$da_2/dz = -\tfrac{1}{2}i\ \varepsilon(z)\ Q_{mod}(z)\ [\exp\{2iw(z)\}\ a_1 + a_2] \tag{3.4b}$$

where

$$\varepsilon = \frac{Q^2 - Q^2_{mod}}{Q^2_{mod}} + Q_{mod}^{-\frac{3}{2}} \frac{d^2}{dz^2} Q_{mod}^{-\frac{1}{2}} . \tag{3.5}$$

We realize that if ε is identically equal to zero, it follows from (3.4a,b) that a_1 and a_2 are constants, and hence f_1 and f_2 are exact solutions of the differential equation (2.1). If ε is small in a certain region of the z-plane, then a_1 and a_2 are approximately constant in the neighbourhood of any point z in that region, and the functions f_1 and f_2 satisfy the differential equation approximately, i.e., f_1 and f_2 solve the local problem in that region. We should note that the condition $|\varepsilon| \ll 1$ is a more precise condition than (2.11a) for the local validity of the JWKB-approximation.

The solution of the system of first-order differential equations (3.4a,b) can be obtained in closed form by means of an iteration procedure. The result can be described as follows. Given an arbitrary column vector $\underline{a}(z_0)$ at a given point z_0, we have for $\underline{a}(z)$ the following formula

$$\underline{a}(z) = \underline{F}(z,z_0)\ \underline{a}(z_0) , \tag{3.6}$$

where $\underline{F}(z,z_0)$ is a matrix, the elements of which are given by convergent series. Useful estimates of the elements of $\underline{F}(z,z_0)$ have been derived in [13] on the assumption that the points z and z_0 can be connected by a path Λ in the complex plane, along which $|\exp\{iw(z)\}|$ increases monotonically from z_0 to z. These so called basic estimates can be written as follows

$$|F_{11}(z,z_0) - 1| \leq \tfrac{1}{2}\mu + \ldots \tag{3.7a}$$

$$|F_{12}(z,z_0)| \leq |\exp\{-2iw(z_0)\}|\ (\tfrac{1}{2}\mu + \ldots) \tag{3.7b}$$

$$|F_{21}(z,z_0)| \leq |\exp\{2iw(z)\}|(\tfrac{1}{2}\mu + \ldots) \tag{3.7c}$$

$$|F_{22}(z,z_0) - 1| \leq \tfrac{1}{2}\mu + |\exp\{2i[w(z) - w(z_0)]\}|\ (\tfrac{1}{4}\mu^2 + \ldots) \tag{3.7d}$$

where

$$\mu = \int_{z_0}^{z} |\varepsilon(z_1)\, Q_{mod}(z_1)\, dz_1| \tag{3.8}$$

with the integration performed along the path of monotonicity.

The matrix $\underline{F}(z,z_0)$ satisfies the following general relations

$$\frac{\partial}{\partial z} \det \underline{F}(z,z_0) = 0; \quad \det F(z,z_0) = 1 \tag{3.9}$$

$$\underline{F}(z_0,z) = [\underline{F}(z,z_0)]^{-1} = \begin{pmatrix} F_{22}(z,z_0) & -F_{12}(z,z_0) \\ -F_{21}(z,z_0) & F_{11}(z,z_0) \end{pmatrix} \tag{3.10}$$

$$\underline{F}(z,z_0) = \underline{F}(z,z_1)\, \underline{F}(z_1,z_0)\ . \tag{3.11}$$

General properties of the function Q^2 in the original differential equation (2.1) are reflected in properties of the matrix $\underline{F}$. Thus, symmetry relations between elements of $\underline{F}$ exist in the special case that $Q^2(z)$ is real on the real axis and the points z_0 and z both lie on this axis. If $Q^2(-z) = Q^2(z)$, or if $Q^2(z)$ is periodic with respect to z, further symmetry relations exist.

The relation (3.6) displays the connection problem in a condensed mathematical form. In physical applications the general procedure is to derive an exact expression for the physical quantity or the relation of interest and then omit "small" quantities, expressed in terms of elements of the matrix $\underline{F}$. In general situations when the points z_0 and z cannot be joined by a path along which $|\exp\{iw(z)\}|$ is monotonic, one has to divide the path chosen into monotonic parts, use the relations (3.9)-(3.11), and possibly other relations existing, to bring the expression for the quantity of interest into a convenient form, and then use the basic estimates (3.7a-d) for obtaining upper bounds for the terms neglected.

The μ-integral (3.8) plays an important role as a measure of the "goodness" of the JWKB-approximation. For instance, if the points z_0 and z both lie <u>in a classically allowed region</u> (of the real axis) where w is real and $\mu \ll 1$, the matrix $\underline{F}(z,z_0)$ is approximately a unit matrix, and both a_1 and a_2 are approximately constant. On the other hand, <u>in a classically forbidden region</u> (of the real axis) the quantity $|\exp\{iw\}|$ increases or decreases very rapidly, and some of the elements of the matrix $\underline{F}$ become very large. Therefore, at least one of the coefficients a_1 and a_2 changes violently. In order that a_1 or a_2 be approximately constant it is necessary, as already mentioned, that the quantity ε is small. As we have now seen, this is, however, not at all a sufficient condition. The rather delicate properties of the JWKB-approximation along a path where $|\exp\{iw(z)\}|$ increases or decreases rapidly have been analyzed in detail in [14] and will here be considered in Sec. 5.2.

4. ANTI-STOKES' AND STOKES' LINES

When tracing a solution in the complex plane, one must know qualitatively the properties of monotonicity of $|\exp\{iw(z)\}|$ along the path. These properties are mapped in the pattern of anti-Stokes' lines in the complex plane. If one knows the picture of anti-Stokes' and Stokes' lines in the neighbourhood of the transition points which must be taken into account, it is in general not difficult to get the whole overall pattern.

Anti-Stokes' and Stokes' lines are defined by the conditions that

$$dw = Q_{mod}(z)dz \quad \text{is} \quad \begin{cases} \text{real along an anti-Stokes' line} & (4.1a) \\ \text{purely imaginary along a Stokes' line.} & (4.1b) \end{cases}$$

Anti-Stokes' lines are level lines for constant absolute value and Stokes' lines are level lines for constant phase of $\exp\{iw(z)\}$.

Putting $z = x + iy$, and $\arg Q_{mod} = \chi$, we easily find from (4.1a,b) that

$$dy/dx = - \operatorname{tg}\chi \quad \text{for an anti-Stokes' line,} \tag{4.2a}$$

$$dy/dx = 1/\operatorname{tg}\chi \quad \text{for a Stokes' line.} \tag{4.2b}$$

Thus, at ordinary points, where Q_{mod} is neither zero nor singular, the anti-Stokes' and Stokes' lines are orthogonal to each other.

Assuming Q^2_{mod} to be a linear function of z in the neighbourhood of a simple zero, say z', we easily find that three anti-Stokes' lines emerge from z', forming angles $2\pi/3$ with each other. Correspondingly, there emerge three Stokes' lines, bisecting the angles between the emerging anti-Stokes' lines. The whole pattern of anti-Stokes' and Stokes' lines in the region around a simple zero of $Q^2_{mod}(z)$ is shown in Fig. 2a and a path Λ circumventing the zero is depicted. At the point z_0, the absolute value of $\exp\{iw(z)\}$ has a minimum or a maximum, depending on the choice of sign of $Q_{mod}(z)$. Anticipating our discussion of higher-order approximations in Sec. 10, we show in Fig. 2b the pattern of Stokes' and anti-Stokes' lines around z', when the third-order phase-integral approximation is used. In the immediate neighbourhood of z' there is a spectacular difference between Figs. 2a and 2b. However, well away from z', on the path Λ, the patterns are very similar, the effect of the use of the higher-order approximation being only a slight distortion of Stokes' and anti-Stokes' lines.

Fig. 2c shows the pattern of anti-Stokes' and Stokes' lines around a simple pole of $Q_{mod}^2(z)$, and Fig. 2d shows the pattern pertaining to a real potential barrier.

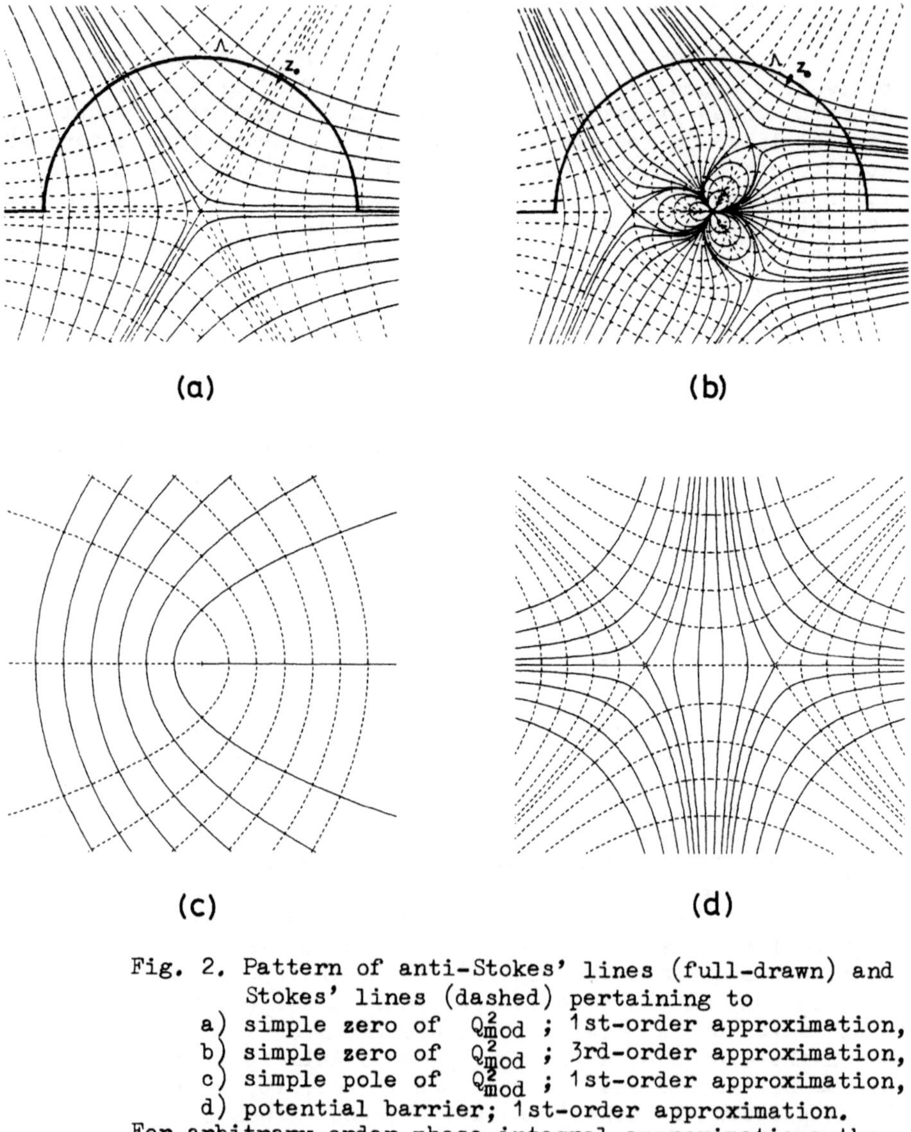

Fig. 2. Pattern of anti-Stokes' lines (full-drawn) and Stokes' lines (dashed) pertaining to
a) simple zero of Q_{mod}^2 ; 1st-order approximation,
b) simple zero of Q_{mod}^2 ; 3rd-order approximation,
c) simple pole of Q_{mod}^2 ; 1st-order approximation,
d) potential barrier; 1st-order approximation.
For arbitrary-order phase-integral approximations the anti-Stokes' and Stokes' lines are defined by (4.1a,b) with Q_{mod} replaced by the function $q(z)$ given by (10.6) - (10.10).

5. PROPERTIES OF THE JWKB-APPROXIMATION IN THE COMPLEX PLANE

It is very important to give the connection problem a precise meaning. Lack of clearness in that respect may give rise to misconceptions, which have not failed to show themselves in the literature. With regard to the needs for the solution of physical problems, the natural formulation which conforms to relation (3.6) and which we shall adopt, is the following: given a linear combination of JWKB-functions representing a certain exact solution in one part of the complex plane, then what combination is appropriate for approximating that solution in other parts of the plane? We shall answer this question for some important situations.

5.1. Anti-Stokes' line

Let us consider the situation that both z and z_0 lie on an anti-Stokes' line and also, for simplicity, that the lower limit of integration in the integral defining $w(z)$ lies on the same anti-Stokes' line. We then have $|\exp\{iw(z)\}| = |\exp\{iw(z_0)\}| = 1$, and hence from the basic estimates (3.7) we conclude that if $\mu(z_0,z) << 1$, the matrix $\underline{F}$ is approximately a unit matrix, and both $a_1(z)$ and $a_2(z)$ are approximately constant. If we have fitted, approximately or exactly, the appropriate linear combination of JWKB-functions to a given wave function $\psi(z)$ at a point, say z_0 , on an anti-Stokes' line, then the same linear combination is a good approximation of $\psi(z)$ along the whole anti-Stokes' line, in either direction from z_0 , as long as we do not approach zeros or singularities of $Q^2_{mod}(z)$, such that the condition $\mu << 1$ becomes violated. In particular, if the potential is real on the real axis, the classically allowed regions are anti-Stokes' lines and the JWKB-approximation has the simple properties just mentioned.

5.2. Arbitrary path along which $|\exp\{iw(z)\}|$ is monotonic in the strict sense. Stokes' line

As soon as one does not move along an anti-Stokes' line, the properties of the JWKB-approximation become more complicated, and caution must be exercised even in an interval for which the μ-integral is small. If one moves away from a point, say z_0 , in an arbitrary direction in the complex plane, not coinciding with an anti-Stokes' line, one of the absolute values $|\exp\{\pm iw(z)\}|$ is increasing while the other is decreasing, the change being most rapid in the direction of the Stokes' line through z_0 . Thus, if $|\exp\{iw(z)\}|$ is the increasing exponential along the path considered, we realize from the basic estimates (3.7) that, since $|\exp\{iw(z)\}|$ soon becomes very large, the condition $\mu << 1$ no longer ensures that the matrix $\underline{F}$ be approximately a unit matrix.

Let us assume that the $\underline{a}$-vector at the point z_0 is

$$\underline{a}(z_0) = \begin{pmatrix} C \\ D \end{pmatrix} . \tag{5.1}$$

The solution $\psi(z)$ and its derivative at $z = z_0$ are given by the general formulas (3.3a,b), i.e.,

$$\psi(z_0) = \underline{f}(z_0)\ \underline{a}(z_0) = Q_{mod}^{-\frac{1}{2}}(z_0) \Big(C \exp\{iw(z_0)\} + + D \exp\{-\ iw(z_0)\} \Big). \tag{5.2}$$

Moving from z_0 on a path along which $|\exp\{iw(z)\}|$ increases, we have for the a-coefficients at a point z , according to (3.6) and (5.1),

$$|a_1(z) - a_1(z_0)| \leqslant |F_{11}(z,z_0) - 1|\ |C| + |F_{12}(z,z_0)|\ |D| \tag{5.3a}$$

$$|a_2(z) - a_2(z_0)| \leqslant |F_{21}(z,z_0)|\ |C| + |F_{22}(z,z_0) - 1|\ |D|. \tag{5.3b}$$

From these inequalities and the basic estimates (3.7) we can draw the following conclusions as to the behaviour of $a_1(z)$ and $a_2(z)$.

If

$$|D|\ |\exp\{-\ iw(z_0)\}| \lesssim |C|\ |\exp\{iw(z_0)\}| , \tag{5.4}$$

i.e., if $|D\ f_2(z_0)|$ is of less or at the most the same order of magnitude as $|C\ f_1(z_0)|$, then from (5.3a) we find that $a_1(z)$ is approximately constant when we move away from z_0 along a line on which $|\exp\{iw(z)\}|$ increases. Thus

$$a_1(z) \approx a_1(z_0) = C , \tag{5.5}$$

the relative error being at the most of the order of magnitude μ . However, the coefficient $a_2(z)$ differs considerably from $a_2(z_0)$ unless z is close to z_0 , the order of the relative change being in general larger than $\mu \exp\{2i[w(z) - w(z_0)]\}$ as is seen from (5.3b) and (3.7c,d). The whole term $a_2(z)\ f_2(z)$, however, is insignificant compared to $a_1(z)\ f_1(z)$, unless z lies close to z_0 . Thus, the fact that $a_2(z)$ changes very much when the point z moves away from z_0 , does not spoil our possibility of using the JWKB-approximation to represent $\psi(z)$ along a path from z_0 towards increasing values of $|\exp\{iw(z)\}|$, as long as $\mu << 1$. In fact, according to (2.12a,b), (3.3a), (3.6) and (5.1) we obtain

$$\psi(z) = Q_{mod}^{-\frac{1}{2}}(z) \exp\{iw(z)\} \Big([F_{11}(z,z_0) + F_{21}(z,z_0) \exp\{-2iw(z)\}]\ C + [F_{12}(z,z_0) + F_{22}(z,z_0) \exp\{-2iw(z)\}]\ D \Big) . \tag{5.6}$$

From the estimates (3.7) and the condition (5.4) we obtain

$$\left| \frac{\psi(z)}{C\; Q_{mod}^{-\frac{1}{2}}(z)\; \exp\{iw(z)\}} - 1 \right| \lesssim \tfrac{3}{2}\mu \; +$$

$$+ \; |\exp\{-2i[w(z) - w(z_0)]\}|\; (1+\tfrac{1}{2}\mu) \; , \qquad (5.7)$$

and thus $\psi(z)$ is given by

$$\psi(z) \approx C\; f_1(z) = C\; Q_{mod}^{-\frac{1}{2}}(z)\; \exp\{iw(z)\} \qquad (5.8)$$

with a relative error at the most of the order of magnitude μ, if the condition (5.4) is fulfilled and if $|\exp\{-2i[w(z) - w(z_0)]\}|$ is less than or at the most of the order of magnitude μ, which means that z must not lie too close to z_0. If z lies close to z_0, the term $D\; f_2(z)$ is significant and must be retained, but far away from z_0 it is insignificant.

On the other hand, if the condition (5.4) is not fulfilled, i.e., if

$$|D\; \exp\{-\, iw(z_0)\}| >> |C\; \exp\{iw(z_0)\}| \; , \qquad (5.9)$$

we see from (5.3a,b) and the basic estimates that both $a_1(z)$ and $a_2(z)$, in general, differ considerably from $a_1(z_0)$ and $a_2(z_0)$, respectively. Therefore, in this case we cannot use the JWKB-approximation, unless z lies close to z_0.

If, instead, we move on a path along which $|\exp\{-iw(z)\}|$ increases monotonically, the coefficients $a_1(z)$ and $a_2(z)$ interchange their properties as compared to the previous case, and the discussion of the usefulness of the JWKB-approximation can be persued with the same arguments as above. We find that we can use the JWKB-approximation to represent $\psi(z)$ only if

$$|C\; \exp\{iw(z_0)\}| \lesssim |D\; \exp\{-iw(z_0)\}| \; . \qquad (5.10)$$

If this condition is fulfilled, we find that

$$\psi(z) \approx D\; f_2(z) = D\; Q_{mod}^{-\frac{1}{2}}(z)\; \exp\{-iw(z)\} \qquad (5.11)$$

with a relative error at the most of the order of magnitude $\mu(z_0,z)$, if z does not lie close to z_0. In the latter case the term $C\; f_1(z)$ is significant and must be retained.

Our results as to the properties and validity of the JWKB--approximation along a path on which the quantities $|\exp\{\pm iw(z)\}|$ increase or decrease (as the case may be) rapidly, can be summarized as a thumb-rule: If at the point z_0, the a-coefficients are given by (5.1) and hence $\psi(z_0)$ by (5.2), and if both terms in (5.2) are of the same order of magnitude at $z = z_0$, then both of the conditions (5.4) and (5.10) are fulfilled, and the JWKB-

-approximation can be used to represent $\psi(z)$ along a certain path in the complex plane in both directions from z_0 , as long as $\mu(z_0,z) << 1$. As one moves away from the point z_0 , keeping C and D constant, and of course replacing z_0 by the moving coordinate z , both of the terms in (5.2) are at first significant, but very soon the term containing the decreasing exponential becomes insignificant, and then $\psi(z)$ is approximately given by (5.8) in the direction of increasing values of $|\exp\{iw(z)\}|$ and by (5.11) in the direction of increasing values of $|\exp\{-iw(z)\}|$. If, however, the absolute value of one of the terms in (5.2) is much larger than the absolute value of the other term, then only one of the conditions (5.4) and (5.10) is fulfilled, and we can use the JWKB-approximation to represent $\psi(z)$ only in the direction in which the exponential factor of the term, dominating at the point z_0 , increases.

For the special case that one moves along a Stokes' line, the real part of $w(z)$ is constant and hence $\exp\{\pm iw(z)\}$ are non-oscillatory, and we may say in short, albeit less precisely: On a Stokes' line one must not proceed with the JWKB-approximation in the direction of decreasing absolute values of $\psi(z)$.

Anticipating our discussion in the sections below, we mention that the properties of the JWKB-approximation along a Stokes' line obviously imply that the connection formulas be one-directional.

I would like to emphasize that the restrictions on the usefulness of approximate solutions discussed above are not specific for the Schrödinger equation and the JWKB-approximation, but appear whenever one is dealing with approximate solutions of differential equations.

5.3. Connection formulas at a simple zero of $Q^2_{mod}(z)$

Using the rules formulated in Secs. 5.1 and 5.2 for tracing a linear combination of JWKB-functions, representing a certain exact solution ψ , along a path on which $|\exp\{iw(z)\}|$ increases monotonically or is constant, and using also the single-valuedness of the wave function, we can now obtain connection formulas at a simple zero of $Q^2_{mod}(z)$.

We assume that the complex plane is cut along an anti-Stokes' line emerging from the zero, as shown in Fig. 3, and we examine the connection from this anti-Stokes' line and to its continuation as a Stokes' line on the other side of the zero, and vice versa. Suppose that for points z_3 on the Stokes' line in question a solution ψ is represented by

$$\psi(z_3) \approx B\ f_2(z_3) = B\ Q^{-\frac{1}{2}}_{mod}(z_3)\ \exp\{-iw(z_3)\}\ , \qquad (5.12)$$

where B is a constant. This representation is valid in the direction of increasing values of $|\exp\{-iw(z)\}|$, i.e., along the Stokes' line in the direction towards the transition point z', and along Λ_+ and Λ_- up to z_1 and z_5, respectively. We next assume that at points z_0 and z_6 on the upper and lower edge, respectively, of the cut along the anti-Stokes' line, the solution ψ is represented by the linear combinations

$$\psi(z_0) = A_+ f_1(z_0) + B_+ f_2(z_0) \tag{5.13a}$$

and

$$\psi(z_6) = A_- f_1(z_6) + B_- f_2(z_6) , \tag{5.13b}$$

respectively. When we proceed with the solution (5.13a) along Λ_+, the term $A_+ f_1$ will soon die out and become insignificant and only the term $B_+ f_2$ will survive. Hence, when we arrive at z_1, the two solutions (5.12) and (5.13a) fit together, if $B_+ = B$. Analogously, if we proceed with the solution (5.13b) along Λ_- to z_5, it fits to the solution (5.12), if $B_- = B$. Thus, we have

$$B_+ = B_- = B. \tag{5.14}$$

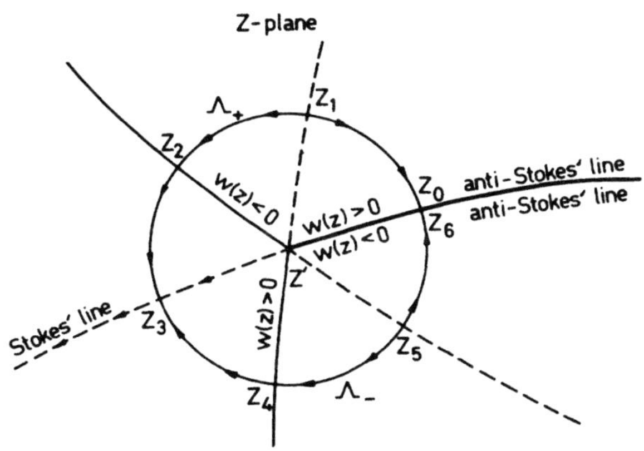

Fig. 3. Anti-Stokes' and Stokes' lines emerging from a simple zero z' of Q^2_{mod}. Arrows indicate directions in which $|\exp\{iw(z)\}|$ increases.

Since the true solution is single-valued, when we trace it one turn around the transition point z', the expressions (5.13a) and (5.13b) shall be identical. If $w(z') = 0$, $w(z)$ has opposite signs on the upper and lower edge of the cut, and the function f_1 on the upper edge corresponds to f_2 on the lower edge, and vice versa. Taking also into account the fact that the phase of Q^2_{mod} changes by $i\pi/2$, when we move one turn in the positive direction around the zero, we obtain, with due regard also to (5.14),

$$A_+ = -iB_- = -iB \; ; \; A_- = +iB_+ = +iB \, . \tag{5.15}$$

Thus, the solution (5.12), increasing along the Stokes' line in the direction towards the transition point, is, with the area around the transition point excluded, represented by (5.13a) with $A_+ = -iB$, $B_+ = B$, or by (5.13b) with $A_- = +iB$, $B_- = B$ on the anti-Stokes' line, continuing the Stokes' line in question on the other side of the transition point.

We can write our result as the following one-directional connection formula from the Stokes' line to the upper edge of the cut along the anti-Stokes' line

$$B\, Q_{mod}^{-\frac{1}{2}}(z) \exp\{-iw(z)\} \rightarrow$$
$$\rightarrow e^{-i\frac{1}{4}\pi} B\, Q_{mod}^{-\frac{1}{2}}(z) \left(e^{i[w(z)-\frac{1}{4}\pi]} + e^{-i[w(z)-\frac{1}{4}\pi]} \right) , \tag{5.16}$$

where $|\exp\{-iw(z)\}|$ increases towards the transition point, z', and $w(z') = 0$, and where the arrow indicates the direction in which inferences can be drawn.

In particular, we may let the transition point be a classical turning point, say x' in Fig. 1. Choosing $B = \exp\{i\pi/4\}$, we then obtain the well known connection formula

$$|Q_{mod}^{-\frac{1}{2}}(x)| \exp\{-|w(x)|\} \rightarrow 2|Q_{mod}^{-\frac{1}{2}}(x)| \cos[|w(x)|-\tfrac{1}{4}\pi] \, , \tag{5.17}$$

which we have now written in a form independent of the choice of phase for the square root of Q_{mod}.

It is easy to understand intuitively the one-directedness of (5.16) and (5.17). For instance, even a very small departure of the phase in the right-hand member of (5.17) from the exact value causes the exponentially increasing term to appear in the classically forbidden region, which completely destroys the agreement with the left-hand member of (5.17).

It should be noted that for the derivation of (5.17) one may, instead of utilizing the single-valuedness of the wave function, exploit the fact that for a real wave function the coefficients A_+ and B_+ must be complex conjugates, if $Q^2(x)$ is real.

We next assume that a certain solution ψ is represented on the anti-Stokes' line by a known linear combination of f_1 and f_2, which, on the upper edge of the cut, we write as

$$\psi(z) \approx C\, f_1(z) + D\, f_2(z) \, . \tag{5.18}$$

With similar, but more intricate, arguments, as used above, we arrive at the result that the solution ψ on the Stokes' line is represented by $(C + iD)\, f_1(z)$. Assuming that $w(z) = 0$ at the transition point, we obtain the connection formula

$$(C+iD)\ Q_{mod}^{-\frac{1}{2}}(z)\ \exp\{iw(z)\} \leftarrow$$

$$\leftarrow Q_{mod}^{-\frac{1}{2}}(z)\ [C\ \exp\{iw(z)\} + D\ \exp\{-iw(z)\}] \tag{5.19}$$

for tracing the solution (5.18) from the upper edge of the cut along the anti-Stokes' line to the Stokes' line.

In particular, we now let the transition point be x' in Fig. 1, assume $Q^2(x)$ to be real, and choose

$$C = \tfrac{1}{2}\exp\{i(\gamma-\tfrac{1}{4}\pi)\}\ ,\quad D = \tfrac{1}{2}\exp\{-i(\gamma-\tfrac{1}{4}\pi)\}\ , \tag{5.20}$$

where γ is an unspecified real phase. We then obtain the connection formula

$$\sin\gamma|Q_{mod}^{-\frac{1}{2}}(x)|\ \exp\{|w(x)|\} \leftarrow |Q_{mod}^{-\frac{1}{2}}(x)|\ \cos[|w(x)|+\gamma-\tfrac{1}{4}\pi]\ , \tag{5.21}$$

which is valid provided γ is not close to zero or a multiple of π ; cf. (5.17). Rigorous derivations of the connection formulas (5.17) and (5.21) with upper bounds for the errors involved can be found in [13]; more detailed comments are given in [14].

5.4. Connection formula for a real potential barrier

For energies far below the top of a thick potential barrier, the classical turning points are well separated. Thus, for an incoming plane wave, being partly reflected and partly transmitted, the transmission coefficient can be obtained by tracing the transmitted wave backwards through the barrier with the aid of the connection formulas (5.19) and (5.17), respectively, at the classical turning points. The reflection coefficient is then obtained from the fact that the sum of the reflection and the transmission coefficient is unity. However, if the wave function given on one side of the barrier, is a standing wave, the problem of determining the wave function on the other side of the barrier becomes very delicate for a wave penetrating a thick barrier, when one is close to a resonance, and for a full understanding one must resort to a rigorous analysis. Furthermore, when the energy increases, the classical turning points approach each other and coincide at the top of the barrier. By further increase of the energy, beyond the top of the barrier, there appear two complex conjugate transition points instead of the classical turning points. It is obvious that the connection formulas (5.17) and (5.21) are not applicable for energies in the neighbourhood of the top of the barrier. It also turns out that there is a certain quantity σ (see below) in the phase of the wave function that cannot be determined within the framework of the JWKB-approximation or the higher-order phase-integral approximations described in Sec. 10. As concerns the first-order JWKB-approximation, this difficulty was overcome by the use of exactly soluble model barriers, such as a parabolic barrier [15-18] and an inverted Morse potential [19]; see also [20]. A rigorous analysis of the barrier transmission problem with the aid of the matrix $\underline{F}$ and the arbitrary-order phase-integral approximations to be described in Sec. 10, was performed in [21]. Here we shall only quote the final connection formula and discuss the delicate points in the light of the rigorous treatment.

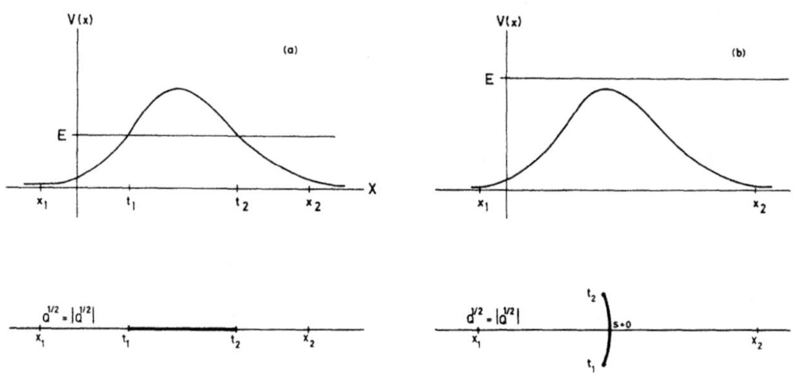

Fig. 4. Qualitative behaviour of $V(x)$. Heavy lines indicate cuts along Stokes' lines joining the transition points t_1 and t_2 for a sub-barrier energy (a) and a super-barrier energy (b). In the latter case the point where the cut crosses the real axis is called s . The values of $Q^{\frac{1}{2}}_{mod}$ given refer to the first Riemann sheet.

If the wave function at the point x_1 to the left of the barrier is (see Fig. 4)

$$\psi(x_1) \approx |Q^{-\frac{1}{2}}_{mod}(x_1)| \cos[|w_1(x_1)| - \tfrac{1}{4}\pi + \gamma_1] , \qquad (5.22)$$

where

$$w_1(x_1) = \mathrm{Re}\int_{t_1}^{x_1} Q_{mod}(z)\, dz =$$

$$= \begin{cases} \displaystyle\int_{t_1}^{x_1} Q_{mod}(x)\, dx , & \text{sub-barrier case,} \qquad (5.23a) \\ \displaystyle\int_{s-0}^{x_1} Q_{mod}(x)\, dx , & \text{super-barrier case,} \qquad (5.23b) \end{cases}$$

and γ_1 is an arbitrary constant, then at the point x_2 on the other side of the barrier we have

$$\psi(x_2) \approx \Omega_1 \, |Q_{mod}^{-\frac{1}{2}}(x_2)| \cos[|w_2(x_2)| - \tfrac{1}{4}\pi + \varphi_1 - \sigma] \quad (5.24)$$

with the following definitions of the occurring quantities:

$$w_2(x_2) = \mathrm{Re} \int_{t_2}^{x_2} Q_{mod}(z)\, dz = \begin{cases} \displaystyle\int_{t_2}^{x_2} Q_{mod}(x)\, dx & \text{sub-barrier case,} \quad (5.25a) \\ \displaystyle\int_{s+0}^{x_2} Q_{mod}(x)\, dx & \text{super-barrier case,} \quad (5.25b) \end{cases}$$

$$\frac{1}{\Omega_1} = \frac{e^{K}\left[1 + (1 + e^{-2K})^{\frac{1}{2}}\right]}{\left\{1 + \left(\dfrac{\sin(\gamma_1 - \gamma_1^{(0)})}{\frac{1}{2}(1 + e^{-2K})^{\frac{1}{4}} - \frac{1}{2}(1 + e^{-2K})^{-\frac{1}{4}}}\right)^2\right\}^{\frac{1}{2}}}, \quad (5.26)$$

$$\varphi_1 \approx \mathrm{arctg}\left(\frac{(1 + e^{-2K})^{\frac{1}{4}} - (1 + e^{-2K})^{-\frac{1}{4}}}{(1 + e^{-2K})^{\frac{1}{4}} + (1 + e^{-2K})^{-\frac{1}{4}}} \mathrm{tg}[\tfrac{1}{2}\pi - (\gamma_1 - \gamma_1^{(0)})]\right) \quad (5.27)$$

with arctg chosen such that φ_1 lies in the same quadrant as $\frac{1}{2}\pi - (\gamma_1 - \gamma_1^{(0)})$,

$$\gamma_1^{(0)} = \tfrac{1}{2} \arg F_{22} - \sigma , \quad (5.28a)$$

$$\sigma = \tfrac{1}{2} (\arg F_{12} - \tfrac{1}{2}\pi) , \quad (5.29)$$

where F_{22} and F_{12} are elements of the matrix $\underline{F}(x_1,x_2)$,

$$\gamma_1^{(0)} \approx - \sigma , \quad \text{near the top of the barrier,} \quad (5.28b)$$

$$K = \pm \left| \int_{t_1}^{t_2} Q_{mod}(z)\, dz \right| , \quad \begin{array}{l} + \text{ sign, sub-barrier case} \\ - \text{ sign, super-barrier case.} \end{array} \quad (5.30)$$

The exact expressions (5.28a) and (5.29) for $\gamma_1^{(0)}$ and σ, respectively, are retained for reasons, which will be explained presently. For the first-order approximation we have

$$\sigma \approx \frac{1}{2}\left\{ - \arg \Gamma\left(\frac{1}{2} + i\,\frac{K}{\pi}\right) + \frac{K}{\pi} \ln \left|\frac{K}{\pi}\right| - \frac{K}{\pi} \right\}, \quad (5.31)$$

where Γ is the gamma function. The third- and fifth-order expressions for σ can be found in [22]. The formula (5.31) and the corresponding higher-order formulas for σ have been obtained by means of a comparison equation method for connecting arbitrary-order phase-integral solutions on opposite sides of a barrier which close to the top has approximately parabolic shape. For energies close to the top of the barrier, we can put $\gamma_1^{(0)} \approx -\sigma$ and the 'parabolic' connection formulas thus obtained can be used in both directions. The connection formulas at a single turning point are not applicable at all, since the two turning points involved lie close together.

From (5.26) it is obvious that, if $\exp\{2K\} \gg 1$ and if we imagine that the phase γ_1 increases continuously, the quantity $1/\Omega_1$ passes through very sharp maxima and the phase φ_1 changes steeply by π, when γ_1 is equal to $\gamma_1^{(0)}$, or differs from it by an integer multiple of π.

If the region to the left of the barrier lies inside some atomic system, and the region to the right of the barrier lies outside this system, the maxima of $1/\Omega_1$ correspond to resonances for the physical system. Orbiting resonances in molecular scattering provide examples of this type of resonance.

Although the quantity $\gamma_1^{(0)}$, defined by (5.28a) with (5.29), is very small well below the top of the barrier, it is still dangerous to neglect it, or to use the formula $\gamma_1^{(0)} \approx -\sigma$, close to a sharp resonance, since, due to the largeness of the factor $\exp\{K\}$ in (5.26) and (5.27), those approximations will in general imply that the resulting formulas for $1/\Omega_1$ and φ_1 will yield a displacement of the resonance from its actual position by a very large number of half-widths, unless phase-integral approximations of a rather high order are used. This fact is illustrated for an exactly soluble model in [23]. However, it is in general not necessary to know the numerical value of the small quantity $\gamma_1^{(0)}$ but only to be aware of this small correction quantity, which is only slightly energy-dependent. The appearance of the quantity $\gamma_1^{(0)}$ in the expressions (5.26) and (5.27) for $1/\Omega_1$ and φ_1, respectively, is, for energies well below the top of the barrier, closely related to the one-directional nature of the connection formulas at the first and second turning points. This one-directedness thus reflects itself, in the present context, as the above-mentioned displacement of the sharp resonance.

Thus, the formulas (5.22) - (5.31) give a detailed description of the wave function in the presence of a potential barrier. The shape of the resonance curve is completely determined by those formulas. In principle so also is its position, but since the approximate expressions $\gamma_1^{(0)} \approx -\sigma$, or possibly $\gamma_1^{(0)} = 0$, are results of the omission of quantities, which, although small compared to unity, are in general large compared to the half-width of the resonance when $\exp\{2K\} \gg 1$, the position of very narrow resonances, when determined from those approximate formulas, may be displaced by several powers of ten times the half-width. If higher-order phase-integral approximations are used, the quantity $\gamma_1^{(0)}$ diminishes, and, by proceeding to sufficiently high orders, one may achieve that the true position even of a narrow resonance is obtained to within a half-width, if $\gamma_1^{(0)}$ is neglected. Further comments on this point can be found in [24].

In the first-order approximation we have $K > 0$ for energies below the top of the barrier, $K = 0$ at the top, and $K < 0$ for energies above the top of the barrier. The formula for the transmission coefficient of a real barrier, valid for both sub-barrier and super-barrier energies, is (see eq. (80.b) in [21])

$$T = (1 + \exp\{2K\})^{-1} \qquad (5.32)$$

Hence, in the first-order approximation $T \leq \frac{1}{2}$ for sub-barrier energies. However, if higher-order approximations are used, the quantity K may become negative also for energies below the top, and it turns out that the transmission coefficient may be obtained with the aid of higher-order approximations, also when the first-order approximation is not applicable at all. An example is displayed in Table I.

Table I. The transmission coefficient T for a real symmetrical Eckart-Epstein barrier [25], given as a function of the energy. $E/V_{max} = 1.0$ corresponds to the top of the barrier.

E/V_{max}	first order appr.	third order appr.	exact value
0.4	0.240	0.603	0.576
0.6	0.330	0.7032	0.7034
0.8	0.418	0.775	0.786
1.0	0.500	0.828	0.841
1.2	0.574	0.867	0.880
1.4	0.640	0.895	0.907

6. BOHR-SOMMERFELD QUANTIZATION CONDITION

Let us consider a single-well potential and assume that, for the function $Q^2_{mod}(z)$, the zeros off the real axis as well as the singularities lie far away from the classically allowed region of the real axis. Cutting the complex z-plane as shown in Fig. 5, we choose $Q_{mod}(z)$ to be positive on the upper lip of the cut in the classically allowed region.

The exact wavefunction ψ, corresponding to an eigenvalue of the energy, must tend to zero in the classically forbidden regions on either side of the classically allowed region. We can use this fact and apply the connection formula (5.17) for tracing solutions bounded to the left and to the right, respectively, from the classically forbidden regions into the classically allowed region. The requirement that those solutions fit continuously to each other in the classically allowed region yields the Bohr-Sommerfeld quantization condition.

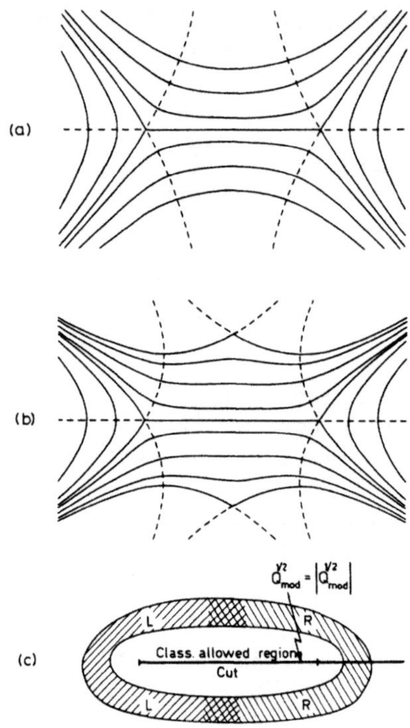

Fig. 5. Anti-Stokes' lines (full-drawn) and Stokes' lines (dashed) in the case of a harmonic oscillator (a) and a quartic oscillator (b). Fig. (c) shows the regions L (indicated by ////) and R (indicated by \\\\).

In the procedure just described one utilizes the approximate wavefunction in the classically allowed region obtained with the aid of the appropriate connection formula. We shall, however, give also the following alternative proof in which we stay away from the classically allowed region and use only the main rule of thumb, formulated in Sec. 5.2, for tracing a solution in the complex plane.

Whether the bottom of the potential is approximately parabolic or not, it follows from the general behaviour of the anti--Stokes' lines in the complex plane (see Fig. 5) that, if one excludes a certain region around the classically allowed region and also a more remote region containing the further zeros and singularities of $Q^2_{mod}(z)$, which may possibly exist, and considers a certain band encircling the classically allowed region, there is in this band a region L to the left in which any point z can be reached by starting far out to the left on the real axis, where the wave function is proportional to $f_2(z)$, and moving on a path along which $|\exp\{-iw(z)\}|$ increases monotonically, and there is a region R to the right in which any point z can be reached by starting far out to the right of the turning point, where the wave function is proportional to $f_2(z)$, and moving on a path along which $|\exp\{-iw(z)\}|$ increases monotonically, and for L and R there is a common region above as well as below the real axis. The requirement that those solutions fit continuously to each other in the part of the band, where L and R overlap, yields the result that

$$f_2(z) = Q_{mod}^{-\frac{1}{2}}(z) \exp\{-iw(z)\} \tag{6.1}$$

represents a bound state wave function in the whole band consisting of the regions L and R. Since the exact wave function is single-valued, its asymptotic expression (6.1) must be single-valued when z moves, for example, one turn around both classical turning points. However, by such a circulation, $\exp\{-iw(z)\}$ changes, due to the fact that Q_{mod} has opposite signs on the lower and upper edge of the cut between t_1 and t_2, and, furthermore, $Q_{mod}^{\frac{1}{2}}$ changes its sign. The requirement that $f_2(z)$ be single-valued, when one moves from a point $x_2 + i0$, in the right-hand classically forbidden region on the upper edge of the cut, around the two zeros of $Q_{mod}^2(z)$ to the point $x_2 - i0$, yields

$$i[w(x_2+i0) - w(x_2-i0)] = i\pi + 2in\pi ,$$

i.e.,

$$\int_{t_1}^{t_2} Q_{mod}(x)\, dx = (n+\tfrac{1}{2})\pi , \qquad n = 0,1,2, \ldots , \tag{6.2}$$

which is the Bohr-Sommerfeld quantization condition [9,7,26,27].

For the harmonic oscillator, the classical turning points are the only transition points, since there are no other zeros or singularities of $Q^2(z)$ in the complex plane. Then we can, by letting the radius of a circle, on which $f_2(z)$ represents the bound state solution, tend to infinity, achieve the result that $f_2(z)$ approximates the exact wave function with as great an accuracy as desired. Hence the Bohr-Sommerfeld quantization condition is exact for the harmonic oscillator.

For complicated potentials with several potential wells, the tracing of solutions in the complex plane becomes difficult. The rigorous F-matrix treatment is most useful in the respect that one sees precisely where the quantities appear, which by omission yield errors in the final formula. One may also use the connection formula (5.17) for a simple turning point and the connection formula for a barrier given in Secs. 5.3 and 5.4, respectively.

The rigorous F-matrix method [13] has been used by Rosenzweig and Krieger [28] to prove the exact validity of the JWKB quantization condition for a number of one-dimensional and radial potentials, without solving the Schrödinger equation or calculating the eigenvalues. For one of those potentials no analytic solutions to the Schrödinger equation are known.

In Table II, results obtained by the use of the arbitrary-order phase-integral quantization condition (11) in [22] for a symmetric double-oscillator are compared with numerically obtained exact results [29].

Table II. Energy eigenvalues for a double-well potential (harmonic oscillator with a Gaussian barrier). Barrier top at 9.0

First order		Third order		Exact
σ negl.	σ incl.	σ negl.	σ incl.	values
3.09	3.085	3.076	3.0756	3.07540
3.10	3.087	3.080	3.0788	3.07851
5.16	5.149	5.140	5.1382	5.13830
5.19	5.173	5.166	5.1643	5.16437
7.02	6.99	6.976	6.97143	6.97140
7.14	7.11	7.104	7.09844	7.09840
8.64	8.60	8.52	8.57625	8.57621
8.98	8.98	8.8	8.96793	8.96788
10.11	10.14	10.127	10.1224	10.1223
10.80	10.83	10.822	10.8242	10.8242
11.823	11.84	11.828	11.8298	11.8298
12.688	12.705	12.694	12.6956	12.6955

7. PHASE SHIFT

For spherically symmetric potentials the Schrödinger equation can be separated in spherical polar coordinates. The method of partial wave expansion for solving the scattering problem then reduces to the problem of finding asymptotic solutions of the radial Schrödinger equation

$$d^2u/dr^2 + Q^2(r)\, u = 0\ , \tag{7.1}$$

where, with standard notations,

$$Q^2(r) = k^2 - U(r) - l(l+1)/r^2 \, . \tag{7.2}$$

If $\lim r^2 U(r) = 0$ when $r \to 0$, eq. (7.1) has two linearly independent solutions which at the origin behave as r raised to $l+1$ and $-l$, respectively. The first-order unmodified JWKB-functions, however, behave as r raised to the powers $\frac{1}{2} \pm \sqrt{l(l+1)}$. Examining the μ-integral, we find that it diverges at the origin for unmodified approximations, but converges for modified approximations with

$$Q^2_{mod}(l,k;r) = k^2 - U(r) - (l+\tfrac{1}{2})^2/r^2 \, . \tag{7.3}$$

Hence, the μ-integral converges if, in the JWKB-functions, $l(l+1)$ is replaced by $(l+\frac{1}{2})^2$. This is the Kramers-Kemble-Langer modification. It is easily seen that the JWKB-functions will display the correct behaviour at the origin after this replacement.

If the potential tends to zero faster than $1/r$, when $r \to \infty$, the asymptotic behaviour of u is

$$u \sim \sin[kr - \tfrac{1}{2}l\pi + \delta_l(k)] \, , \tag{7.4}$$

where $\delta_l(k)$ is the phase shift, defined such that it is zero for $U(r) = 0$.

We remark that if the potential is more singular than $1/r^2$ at the origin, the JWKB-functions have the correct behaviour at $r = 0$ both when they are unmodified and when they are modified according to (7.3). However, it is convenient to use the modification (7.3) also in that case, since the JWKB phase shift then becomes exactly zero when $U(r) = 0$.

7.1. One-turning-point problem. Simple scattering

In the simple situation that there is only one classical turning point r_0, and if we choose the wave function to be real and positive immediately to the right of the origin, the connection formula (5.17) yields the asymptotic form

$$u \sim |Q_{mod}^{-\frac{1}{2}}(r)| \sin\left(\int_{r_0}^{r} |Q_{mod}(r)|\, dr + \tfrac{1}{4}\pi]\right) , \tag{7.5}$$

except for a positive constant factor. For a free particle an analogous expression with $U(r) = 0$ is valid. For the phase shift we thus obtain

$$\delta_l(k) = \lim_{R\to\infty} \left\{ \int_{r_0}^{R} |Q_{mod}(r)|\, dr - \int_{(l+\frac{1}{2})/k}^{R} |k^2 - (l+\tfrac{1}{2})^2/r^2|^{\frac{1}{2}}\, dr \right\} =$$

$$= \int_{r_0}^{\infty} [|Q_{mod}(r)| - k]\, dr - kr_0 + \tfrac{1}{2}(l+\tfrac{1}{2})\pi \, . \tag{7.6}$$

In Table III results obtained for a Lennard-Jones potential are displayed. The figures given are expected to be significant.

Table III. Asymptotic phase shift δ_l for a Lennard--Jones potential. Parameter values and energy the same as in [36].

Order of approximation	l = 50	l = 150
1	-11.63	28.8
3	-11.6356 361	28.7530 0
5	-11.6356 3615 3	28.7529 958

7.2. Three-turning-point problem. Resonance scattering

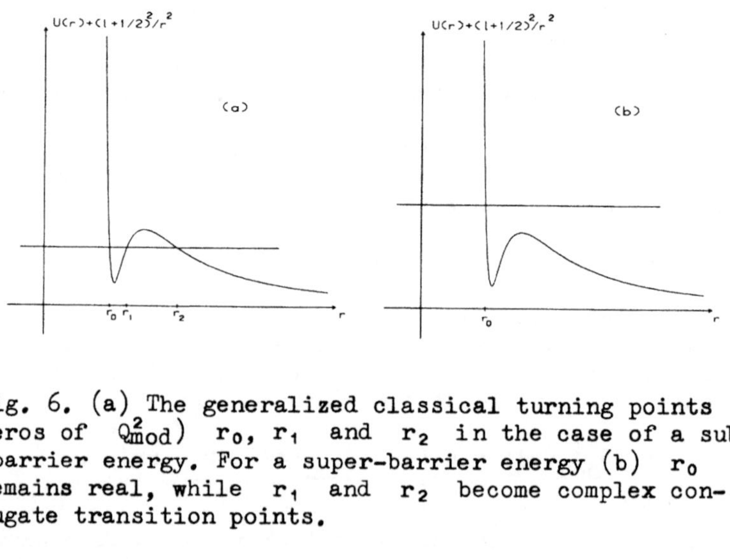

Fig. 6. (a) The generalized classical turning points (zeros of Q^2_{mod}) r_0, r_1 and r_2 in the case of a sub--barrier energy. For a super-barrier energy (b) r_0 remains real, while r_1 and r_2 become complex conjugate transition points.

For the situation that there is a potential barrier (see Fig. 6), we are faced with a three-turning-point problem. The connection formula (5.17) yields the wave function, to the left of the barrier as

$$u \approx |Q_{mod}^{-\frac{1}{2}}(r)| \cos\left(\int_{r_0}^{r} |Q_{mod}(r)|\, dr - \tfrac{1}{4}\pi\right) . \qquad (7.7)$$

Taking into account both the case of sub-barrier and of super--barrier energy, we rewrite (7.7) in accordance with (5.22), obtaining

$$u \approx |Q_{mod}^{-\frac{1}{2}}(r)| \cos[|w_1(r)| - \tfrac{1}{4}\pi + \tfrac{1}{2}\pi - L(k)] \tag{7.8}$$

where $w_1(r)$ is defined by (5.23a,b), with t_1 and x_1 replaced by r_1 and r, respectively,

$$\gamma_1 = \tfrac{1}{2}\pi - L(k) \tag{7.9}$$

and

$$L(k) = \mathrm{Re}\int_{r_0}^{r_1} Q_{mod}(r)\,dr = \tag{7.10}$$

$$= \begin{cases} \displaystyle\int_{r_0}^{r_1} Q_{mod}(r)\,dr\,, & \text{sub-barrier case,} \quad (7.11a) \\ \displaystyle\int_{r_0}^{s-0} Q_{mod}(r)\,dr\,, & \text{super-barrier case.} \quad (7.11b) \end{cases}$$

Inserting the expression (7.9) for γ_1 into the asymptotic expression for u obtained according to (5.24) - (5.30), we easily derive the following expression for the phase shift

$$\delta_l(k) = \lim_{R\to\infty} [|w_2(R)| - kR] + \tfrac{1}{2}(l+\tfrac{1}{2})\pi - \sigma +$$

$$+ \mathrm{arctg}\left(\frac{(1+e^{-2K})^{\frac{1}{4}} - (1+e^{-2K})^{-\frac{1}{4}}}{(1+e^{-2K})^{\frac{1}{4}} + (1+e^{-2K})^{-\frac{1}{4}}}\, \mathrm{tg}(L+\gamma_1^{(0)}) \right), \tag{7.12}$$

where $w_2(R)$ is defined according to (5.25a,b). We recognize, on the first line of this formula, the same expression as the phase-shift (7.6), pertaining to the one-turning-point problem, but with r_0 replaced by the outermost turning point r_2 for the sub-barrier case and by $s+0$ for the super-barrier case, and with the additional term $-\sigma$. This part of the phase shift represents the slowly varying background scattering, while the term arctg (...) represents the resonance part of the phase-shift, which changes steeply by π when the energy passes through a narrow resonance. The resonances occur when $L + \gamma_1^{(0)}$ is equal to $(n+\frac{1}{2})\pi$, where n is an integer. If the amplitude outside the barrier is asymptotically equal to unity, the amplitude to the left of the barrier is $|k/Q_{mod}(r)|^{\frac{1}{2}}\,\Omega_1^{-1}$, and hence goes through sharp maxima simultaneously as the phase shift changes rapidly by π.

The half-width Γ , on the energy scale, of a resonance peak, located at $E = E_n$, is given by the following formula, derived in [24] (see also references given therein)

$$\Gamma \approx \frac{2\hbar\ [(1 + e^{-2K})^{+\frac{1}{4}} - (1 + e^{-2K})^{-\frac{1}{4}}]}{|T - 2\hbar\ (d\sigma/dE)_{E=E_n}|} , \tag{7.13}$$

where

$$T = 2\hbar \left(\frac{\partial L}{\partial E}\right)_{E=E_n} = \frac{2m}{\hbar} \mathrm{Re}\left(\int_{r_0}^{r_1} \frac{dr}{Q_{mod}(r)}\right)_{E=E_n} . \tag{7.14}$$

In the case of sub-barrier penetration, T is the classical period of a complete oscillation of the particle back and forth between the classical turning points r_0 and r_1 . The quantity $(d\sigma/dE)_{E=E_n}$ in (7.13) is very important when the energy is close to the top of the barrier, but is otherwise negligible. In fact, at the top of the barrier the denominator in the right-hand side of (7.13) is finite, while T is infinite. For the first-order approximation we have [24]

$$\frac{d\sigma}{dE} = \frac{1}{2\pi}\left(1.9635 - 2.1036\ (2K/\pi)^2 + 2.0090\ (2K/\pi)^4 + \ldots + \ln|K/\pi|\right) dK/dE$$

where

$$\frac{dK}{dE} = i\ \frac{m}{\hbar^2} \int_{r_1}^{r_2} \frac{dr}{Q_{mod}(r)} . \tag{7.15}$$

8. NORMALIZATION

An exact formula for the normalization integral of the wave function for a bound state in a one-dimensional potential well of arbitrary shape was derived by Furry [30]. The formula was later simplified by Yngve [31], who gave it in the following form

$$\int [\psi(z)]^2 \frac{\partial}{\partial E} Q^2(z,E)\ dz = \left(\frac{\partial}{\partial E}\ (\psi_1\psi_2' - \psi_2\psi_1')\right)_{E=E_n} , \tag{8.1}$$

where ψ_1 is a solution bounded at the left-hand endpoint and ψ_2 a solution bounded at the right-hand endpoint of the interval for the independent variable. Furthermore, $\psi(z) = \psi_1(z) = \psi_2(z)$ when E equals the eigenvalue E_n. Since the right-hand member of (8.1) is the derivative with respect to the energy of the Wronskian of the two solutions ψ_1 and ψ_2 , it is independent of z .

We now specialize to the case of a single-well potential and use for ψ_1 and ψ_2 the JWKB-approximation, choosing the left-hand turning point as the lower limit in the integral defining $w(z)$. If we put $\psi_1 \approx C_1 f_2$ and $\psi_2 \approx C_2 f_2$ in the classically forbidden regions to the left and to the right, respectively, and use the connection formula (5.17) for tracing those solutions into the classically allowed region, the fitting of those solutions to each other, when $E = E_n$, yields $|C_1| = |C_2|$ and also, as described previously, the Bohr-Sommerfeld quantization condition, which we now, for convenience, write as

$$L(E) = (n+\tfrac{1}{2})\pi , \qquad n = 0, 1, 2, \dots . \tag{8.2}$$

Evaluating the right-hand member of (8.1) at an arbitrary point in the classically allowed region, using the JWKB-functions obtained as just described and putting $|C_1| = |C_2| = |C|$, we find that for the normalized wave function

$$\frac{1}{|C|^2} \approx 2 \left| \frac{\hbar^2}{m} \frac{\partial}{\partial E} L(E) \right|_{E=E_n} . \tag{8.3}$$

When $L(E)$ is given by the first-order expression in the left-hand member of (6.2) with $Q_{mod} = Q$ and use is made of (2.2), the formula (8.3) becomes

$$\frac{1}{|C|^2} = \left(\frac{2\hbar^2}{m}\right)^{\frac{1}{2}} \int_{t_1}^{t_2} \frac{dz}{[E - V(z)]^{\frac{1}{2}}} . \tag{8.4}$$

For the calculation of the energy derivative of higher-order expressions for $L(E)$, we refer to [32] and [33].

The probability density at the origin for an s-state, which is related to the normalization of the wave function, is of interest in various physical problems. It can be expressed exactly in terms of the expectation value of $dV(r)/dr$. The expectation value formula derived below is based on the use of the quantization condition, which is in general "more accurate" than the corresponding wave function. Hence it is in general advantageous to calculate the probability density from the expectation value formula for $dV(r)/dr$ instead of using the normalized wave function directly. Explicit formulas for different cases are given in [34].

9. EXPECTATION VALUES AND MATRIX ELEMENTS

9.1. Exact, general formula for expectation values

We shall start by deriving an exact formula for the expectation value of a multiplicative operator with respect to a bound state in a general potential well.

We let the function $Q^2(z)$ in the differential equation (2.1) be

$$Q^2(z,E,\kappa) = \frac{2m}{\hbar^2}\,[E - V(z,\kappa)] \,. \tag{9.1}$$

Here $V(z,\kappa)$ is an auxiliary potential defined by

$$V(z,\kappa) = V(z) - \kappa\, f(z) \,, \tag{9.2}$$

where $V(z)$ is the effective potential pertaining to the physical problem we are interested in, κ is an arbitrarily small parameter, and $f(z)$ is the unspecified function of z for which the expectation value is to be calculated. The meaning of the other notations introduced above is obvious. Earlier papers exploiting the idea of introducing an auxiliary potential as described above are mentioned in [32].

For the eigenstate $\psi(z,E_\nu) = \psi(z,E_\nu,0)$ with quantum number ν pertaining to the potential (9.1) with $\kappa = 0$, we shall now consider the calculation of the expectation value of the function $f(z)$.

According to the Hellman-Feynman theorem we have

$$\langle \partial V(z,\kappa)/\partial\kappa \rangle_\nu = \partial E_\nu(\kappa)/\partial\kappa \tag{9.3}$$

which, combined with (9.2), yields

$$\langle f(z) \rangle_\nu = - \partial E_\nu(\kappa)/\partial\kappa \,. \tag{9.4}$$

We now assume the quantization condition for the auxiliary potential $V(z,\kappa)$ to be known in the implicit form

$$L(E,\kappa,\nu) = 0 \,, \qquad \nu = 0,\ 1,\ 2,\ \ldots \,. \tag{9.5}$$

For each value of the quantum number ν the relation (9.5) determines the corresponding energy eigenvalue $E_\nu(\kappa)$. Differentiating (9.5) with respect to κ, while the quantum number ν is kept constant, we obtain

$$\left(\frac{\partial L(E,\kappa,\nu)}{\partial\kappa} + \frac{\partial L(E,\kappa,\nu)}{\partial E}\,\frac{\partial E_\nu(\kappa)}{\partial\kappa}\right)_{E=E_\nu} = 0 \,,$$

i.e.,

$$\partial E_\nu(\kappa)/\partial\kappa = - \left(\frac{\partial L(E,\kappa,\nu)/\partial\kappa}{\partial L(E,\kappa,\nu)/\partial E}\right)_{E=E_\nu(\kappa)} \,. \tag{9.6}$$

Inserting (9.6) into (9.4) and letting κ tend to zero, which means that the auxiliary potential (9.2) becomes our true effective potential, we obtain the general formula

$$\langle f(z)\rangle_\nu = \left(\frac{\partial L(E,\kappa\ \nu)/\partial\kappa}{\partial L(E,\kappa\ \nu)/\partial E}\right)_{\kappa=0,\ E=E_\nu(0)} \tag{9.7}$$

for the expectation value of $f(z)$ with respect to the eigenstate with quantum number ν in the general potential $V(z) = V(z,0)$.

So far no approximation has been introduced. Provided the quantization condition (9.5) is an exact relation, one therefore obtains from (9.7) the exact expectation value of $f(z)$.

9.2. Phase-integral formulas for expectation values and off--diagonal matrix elements

9.2.1. Bound states. If we insert into (9.7) for $L(E,\kappa,\nu)$ the corresponding phase-integral expression for a certain problem considered, we obtain a phase-integral formula for the calculation of quantal expectation values.

We now apply formula (9.7) to a single-well potential. We allow for the possibility of an arbitrary modification, in which case the Bohr-Sommerfeld quantization condition yields

$$L(E,\kappa,\nu) = \int_{t_1}^{t_2} Q_{mod}(z,E,\kappa)\,dz - (\nu+\tfrac{1}{2})\pi . \tag{9.8}$$

According to (9.1) with (9.2) and our assumption concerning a possible modification, we have (if $Q^2-Q^2_{mod}$ is independent of κ and E)

$$\frac{\partial Q_{mod}}{\partial\kappa} = \frac{1}{2}\frac{1}{Q_{mod}}\frac{\partial Q^2_{mod}}{\partial\kappa} = \frac{m}{\hbar^2}\frac{f(z)}{Q_{mod}} , \tag{9.9a}$$

$$\frac{\partial Q_{mod}}{\partial E} = \frac{1}{2}\frac{1}{Q_{mod}}\frac{\partial Q^2_{mod}}{\partial E} = \frac{m}{\hbar^2}\frac{1}{Q_{mod}} . \tag{9.9b}$$

Inserting (9.8) into (9.7), we obtain, with due regard to (9.9a, b),

$$\langle f(z)\rangle_\nu = \int_{t_1}^{t_2} f(z)\frac{dz}{Q_{mod}(z)}\bigg/\int_{t_1}^{t_2}\frac{dz}{Q_{mod}(z)} . \tag{9.10}$$

If $Q_{mod} \equiv Q$ this formula is identical to the formula for the classical time average. When encountered in the literature, it is usually derived by inserting the first-order JWKB-approximation for the wave function in the integral defining the expectation value, and then simplifying the resulting expression by neglecting contributions from the classically forbidden regions, and replacing the square of the cosine in the classically allowed region by its average $\frac{1}{2}$. In view of those approximations one expects a rather rough formula, and the very high accuracy of the formula has puzzled many authors.

The derivation given above shows that one may, quite generally, expect the usefulness of the phase-integral formula for expectation values obtained from (9.7) to be about the same as that of the phase-integral quantization condition inserted, and the latter is known to be very accurate for important classes of physical problems. Illustrative examples of results obtained by using the expectation value formula obtained from (9.7) by inserting the phase-integral quantization condition for a symmetric double-oscillator are shown in Table IV.

Table IV. Expectation values of x^2 for a particle in a symmetric double-well potential calculated for the lowest four levels with even parity. The levels 0+ and 1+ are below, and the levels 2+ and 3+ are above the top of the barrier.

Order of approx.	$\langle x^2\rangle_{0+}$	$\langle x^2\rangle_{1+}$	$\langle x^2\rangle_{2+}$	$\langle x^2\rangle_{3+}$
1	1.961	1.130	0.975	1.510
3	1.95810	1.13142	0.9786	1.5071
5	1.95805	1.13146	0.97841	1.50773
Exact	1.95800	1.13145	0.97844	1.50677

We shall next generalize the single-well formula (9.10) into a formula, applying also to off-diagonal matrix elements, by a simple heuristic argument [35]. Let us consider the matrix element of $f(z)$ between states with energies E_ν and $E_{\nu'}$ and assume that E_ν is the ground state energy. We rewrite the definition of an off-diagonal matrix element as

$$\int \psi(z,E_\nu)\; f(z)\; \psi(z,E_{\nu'})\; dz = \int \psi(z,E_\nu)\, \frac{f(z)\psi(z,E_{\nu'})}{\psi(z,E_\nu)}\, \psi(z,E_\nu)\; dz \tag{9.11}$$

and apply (9.10) to obtain the expectation value of $f(z)\psi(z,E_{\nu'})/\psi(z,E_\nu)$, which is a regular function of z on the real axis, since the ground state eigenfunction $\psi(z,E_\nu)$ has no zeros there. In that way, we obtain a formula which, as was done in [35], we generalize to apply to arbitrary states, without a full proof, and write as follows

$$\langle \nu | f(z) | \nu' \rangle = \frac{\displaystyle\int_\Gamma f(z) \frac{\exp\{\pm i[w(z,E_\nu) - w(z,E_{\nu'})]\}}{Q_{mod}^{\frac{1}{2}}(z,E_\nu)\, Q_{mod}^{\frac{1}{2}}(z,E_{\nu'})} dz}{\left(\displaystyle\int_\Gamma \frac{dz}{Q_{mod}(z,E_\nu)}\right)^{\frac{1}{2}} \left(\displaystyle\int_\Gamma \frac{dz}{Q_{mod}(z,E_{\nu'})}\right)^{\frac{1}{2}}} . \qquad (9.12)$$

The lower limit of integration in the integrals defining $w(z,E_\nu)$ and $w(z,E_{\nu'})$ is chosen as the left-hand turning point corresponding to the energies E_ν and $E_{\nu'}$, respectively. The contour Γ shall enclose the four turning points involved. The derivation sketched above yields the plus sign in the exponent, when $E_\nu < E_{\nu'}$, but one can, by qualitative arguments, realize that the minus sign should be used. Thus, with our choice of phase for Q_{mod}, one should choose the plus sign in the exponent occurring in the numerator of (9.12) if $E_\nu > E_{\nu'}$, but the minus sign if $E_\nu < E_{\nu'}$. For further comments on the choice of sign in the exponent and on further properties of formula (9.12), especially when taken to higher-order approximations, we refer to [35].

9.2.2. Unbound states A potential with unbound states can be considered as a limiting case of a bound-state potential. In that way, formula (9.12) can be adapted to apply to matrix elements between unbound states pertaining to the radial Schrödinger equation [36].

We assume that we are concerned with a one-turning-point scattering problem, i.e., that the expression (7.2) for Q^2, as well as the expression (7.3) for Q^2_{mod}, has a single zero on the real axis. We define the quantal matrix element as

$$\langle l,k | f(r) | l'k' \rangle = \int_0^\infty u_{l,k}(r)\, f(r)\, u_{l',k'}(r)\, dr , \qquad (9.13)$$

where $u_{l,k}(r)$ is the solution of (7.1) that is zero at the origin, positive immediately to the right of the origin and has an amplitude equal to unity for large positive values of r. Then we have

$$u_{l,k}(r) \sim \sin(kr - \tfrac{1}{2}l\pi + \delta_l) , \qquad r \to +\infty \qquad (9.14)$$

where δ_l is the asymptotic scattering phase shift. We now assume the effective potential for our problem to be the limit of a single-well, bound-state potential, which corresponds to the real zeros r_0 and r_1 of $Q^2_{mod}(l,k;r)$. In the limiting procedure r_1 tends to ∞. Since the expression (9.14), for sufficiently large values of r_1, is valid in essentially the whole interval $0 < r < r_1$, the normalization factor for the solution with the asymptotic behaviour (9.14) is $\sqrt{2/r_1}$. Analogously, if we let r_1' be the outermost zero of $Q^2_{mod}(l',k';r)$, the normalization factor for the pertinent solution is $\sqrt{2/r_1'}$. Thus we can write the right-hand member of (9.13) as the limit, when r_1 and $r_1' \to \infty$, of

$$\tfrac{1}{2}(r_1 r_1')^{\frac{1}{2}} \int_0^\infty \left((2/r_1)^{\frac{1}{2}} u_{l,k}(r)\right) f(r) \left((2/r_1')^{\frac{1}{2}} u_{l',k'}(r)\right) dr \ , \qquad (9.15)$$

where the integral now represents a matrix element of $f(r)$ between normalized bound states, which can be obtained according to (9.12). Noting that, approximately, we have

$$\int_\Gamma \frac{dr}{Q_{mod}(l,k;r)} = \frac{2r_1}{k} \ , \quad \int_{\Gamma'} \frac{dr}{Q_{mod}(l',k';r)} = \frac{2r_1'}{k'} \ , \qquad (9.16)$$

with a relative error which tends to zero as r_1 and r_1' , respectively, tend to zero, we arrive at the following phase-integral formula for the matrix element (9.13),

$$\langle l,k|f(r)|l',k'\rangle =$$
$$= \tfrac{1}{4}(kk')^{\frac{1}{2}} \int_\Gamma f(r) \frac{\exp\{i[w_{l,k}(r) - w_{l',k'}(r)]\}}{Q^{\frac{1}{2}}_{mod}(l,k;r)\, Q^{\frac{1}{2}}_{mod}(l',k';r)}\, dr \ , \qquad (9.17)$$

which is valid when the wave functions are normalized according to (9.14). The sign chosen in the exponential function in the integrand of (9.17) corresponds to the situation that $E_{lk} > E_{l'k'}$. The contour Γ in (9.17) is an infinite non-closed loop starting at $+\infty - i0$, proceeding along the real axis, encircling in the negative sense the two generalized classical turning points r_0 and r_0' (see Fig. 7), proceeding again along the real axis, and ending at $+\infty + i0$. In the first-order approximation, which we are considering for simplicity, the contour integral can be replaced by the real part of an integral along the real axis from r_0 to $+\infty$ ($r_0 < r_0'$, see Fig. 7).

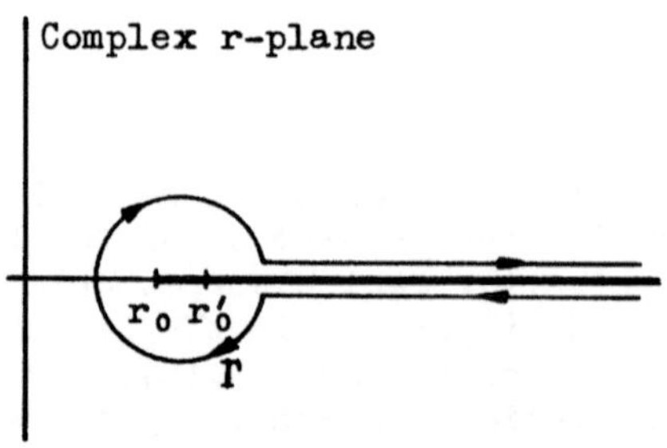

Fig. 7. The integration contour Γ . The generalized classical turning points are r_0 and r_0' .

The first-order formula (9.17) can be obtained by inserting the first-order JWKB wave functions into the integral defining the quantal matrix element, ignoring the singularities at the classical turning points, and then discarding on the basis of simple qualitative arguments an integral with a rapidly oscillating integrand in the classically allowed region and an integral with an exponentially small integrand in the classically forbidden region; see [37]. From such a derivation one would expect that the use of Langer's uniform asymptotic wave functions, which are valid also near the turning points and hence make possible the inclusion of the contributions previously discarded, would improve the result. This has been found, however, not to be the case [38] (see also [39]). Our derivation of (9.17), which is based on the use the phase-integral quantization condition rather than on the use of approximate wave functions, explains this circumstance. Since, loosely speaking, this quantization condition is known to be much more accurate than the corresponding wave functions, it is not surprising that the attempts to improve this first-order formula, in the way sketched above, have failed.

We illustrate in Table V the accuracy of (9.17) and its generalization to higher orders when applied to the Lennard-Jones potential. The figures given are expected to be significant.

Table V. Diagonal and off-diagonal matrix elements for a Lennard-Jones potential. The parameter values and the energy are the same as in [36]. The azimuthal quantum number $l = 50$.

Order of approx.	$\langle r^{-8}\rangle \times 10^5$ in units of $Å^{-7}$	$\langle l\|r^{-2}\|l+1\rangle$ in units of $Å^{-1}$
1	0.6031 4	0.1206 41
3	0.6031 5725 9	0.1206 4318 7165
5	0.6031 5726 1719	0.1206 4318 7174
Exact	0.6031 5	

10. HIGHER-ORDER APPROXIMATIONS

The method for handling connection problems, on which the treatment in the previous sections is based [13], was originally developed for handling the first-order JWKB-approximation and modifications of it. The method can, however, with only minor changes and certain precautions, be applied to any arbitrary order of a type of "symmetric" phase-integral approximations [40], which are simpler and more convenient than the higher-order JWKB-approximations. For instance, the latter approximations, in contrast to the "symmetric" type of phase-integral approximations, do not display the correct relationship between amplitude and phase of the wave function in a classically allowed region, and they do not satisfy the requirement that the Wronskian of two linearly independent solutions shall be constant. The above-mentioned "symmetric" phase-integral approximations and their relationship to the odd-order JWKB-approximations may briefly be described as follows.

We turn again our attention to the derivation of the JWKB-expansion, the first few terms of which were considered in Sec. 2. We first formally introduce a small parameter λ (which is finally to be put equal to unity) into the differential equation (2.1) by replacing $Q^2(z)$ by $Q^2(z)/\lambda^2$, and in the resulting differential equation we then insert

$$\psi = \exp\left\{i \int^{z} dz \frac{1}{\lambda} \sum_{\nu=0}^{\infty} y_{\nu}(z) \lambda^{\nu}\right\}, \tag{10.1}$$

obtaining, quite straightforwardly, $y_0 = \pm Q(z)$ and a recurrence formula from which the functions $y_1, y_2, y_3, \ldots$ can be determined successively. It turns out that the terms y_ν of even order ν appear with the double sign $\pm$, whereas no sign ambiguity occurs for the terms y_ν of odd order ν. Thus one obtains two formal solutions of the differential equation corresponding to the plus and minus signs, respectively. It also turns out that every function y_ν with odd index ν can be written as the derivative with respect to z of an expression containing only functions y_ν with even indices ν. By truncating this series at different terms one obtains JWKB-approximations of different orders. By the order of a JWKB-approximation we mean the value of the index ν for the last term y_ν retained in the JWKB-series. The higher-order JWKB-approximations are rather complicated. For instance, the fifth-order approximation reads

$$\psi \approx \exp\left\{i \int^{z} dz \frac{1}{\lambda} (y_0+y_1\lambda+y_2\lambda^2+y_3\lambda^3+y_4\lambda^4+y_5\lambda^5)\right\} =$$

$$= \text{const} \frac{\exp\left\{i \int^{z} dz \frac{1}{\lambda} (y_0+y_2\lambda^2+y_4\lambda^4)\right\}}{\sqrt{\frac{1}{\lambda} y_0 \exp\left\{\frac{y_2\lambda^2}{y_0} + \left[\frac{y_4\lambda^4}{y_0} - \frac{1}{2}\left(\frac{y_2\lambda^2}{y_0}\right)^2\right]\right\}}}\ . \tag{10.2}$$

where $y_0 = \pm Q(z)$, and the explicit expressions for y_2 and y_4, obtained from the previously mentioned recurrence formula, are to be inserted.

In [40] it was shown how one can obtain related but simpler and in several respects much more convenient phase-integral approximations, for which the connection problems (cf. [41]) can be solved by the method developed in [13]. With the aid of the recurrence formula mentioned above it was proved that, in fact,

$$\exp\left\{ i \int^{z} dz \, \frac{1}{\lambda} \sum_{n=0}^{\infty} y_{2n+1} \, \lambda^{2n+1} \right\} = \frac{\text{const}}{\sqrt{\sum_{n=0}^{\infty} y_{2n} \lambda^{2n}}} \, . \tag{10.3}$$

Hence, except for a constant factor, the formal expression (10.1) for the wave function can be written

$$\psi = \frac{\exp\left\{ i \int^{z} dz \, \frac{1}{\lambda} \sum_{n=0}^{\infty} y_{2n} \, \lambda^{2n} \right\}}{\sqrt{\frac{1}{\lambda} \sum_{n=0}^{\infty} y_{2n} \, \lambda^{2n}}} \, . \tag{10.4}$$

Thus, by a transformation of the usual infinite JWKB-expansion one can achieve the result that the sum of the odd-order terms can be eliminated from the expression for the formal solution. This transformation is exact when the whole series are retained. If the series in (10.4) are truncated after the term $y_{2N} \lambda^{2N}$, one obtains a new kind of approximation:

$$\psi \approx q^{-\frac{1}{2}}(z) \exp\left\{ \int^{z} q(z) \, dz \right\} \tag{10.5a}$$

with

$$q(z) = \frac{1}{\lambda} \sum_{n=0}^{N} y_{2n}(z) \, \lambda^{2n} \, . \tag{10.5b}$$

For $2N + 1 \geqslant 3$ these approximations differ from the JWKB-approximations of the corresponding order. (Compare for instance (10.2) with (10.4) for $2n = 4$.) The distinctive feature of these higher-order approximations is the symmetry between the integrand of the exponentiated integral and the factor in front of the exponential, which does not exist for the higher-order JWKB-approximations. The new "symmetric" approximations (cf. the terminology used by McHugh on p. 280 in [2]) thus have a considerably simpler analytic form than the corresponding higher-order JWKB-approximations, and, what is most important, for the new approximations the connection problems can be solved efficiently by the method described earlier, whereas for the higher-order JWKB-approximations there exists no corresponding satisfactory and sufficiently general method. The further advantages of using the higher-order phase-integral approximations (10.5a,b) instead of the higher-order JWKB-approximations have been clarified in an investigation by Ö. Dammert and P.O. Fröman (to be published).

We have already met with certain difficulties which often appear at the origin when the JWKB-approximations or the above described phase-integral approximations are applied to the radial Schrödinger equation. It was early noticed by Kramers that the first-order JWKB-approximation could be made useful for radial problems by the now well-known replacement of $l(l+1)$ by $(l+\frac{1}{2})^2$. Higher-order approximations, however, cannot be modified consistently by a mere replacement of $l(l+1)$ by $(l+\frac{1}{2})^2$. The problem of modifying consistently the higher-order phase-integral approximations (10.5a,b) has been solved, and a simple modification procedure has been devised (cf. [42] and pp. 126-131 in [43]) which can be used quite generally (not only in radial problems) for making the phase-integral approximations useful at exceptional points where they would otherwise fail. The modified phase-integral approximations of any order are still of the form (10.5a) but with a modified expression for $q(z)$ in which there appears an unspecified function $Q_{mod}(z)$, which is to be chosen conveniently. When one chooses $Q_{mod}(z)$ to be equal to the function $Q(z)$ appearing in the differential equation, one obtains the unmodified phase-integral approximations, which are thus obtained as a particular case of the more general modified phase-integral approximations. The connection problems can be handled by the same method [13], whether one uses modified or unmodified phase-integral approximations of arbitrary order.

According to what has been said above, the unmodified "symmetric" higher-order phase-integral approximations can be obtained by starting from the conventional JWKB-expansion and eliminating the sum of the odd-order terms with the aid of the transformation mentioned above [40]. Alternatively one can, however, start from the assumption of a "symmetric" expansion and then derive a recurrence formula, which yields directly the functions Y_{2n} occurring in the expression for $q(z)$ pertaining to the unmodified or modified phase-integral approximations [42].

The formulas derived in the preceeding sections can be generalized to apply to the arbitrary-order phase-integral approximations in the following way.

The function $Q_{mod}(z)$ is replaced by the function $q(z)$, which is given by (10.5b), generalized to allow for the possibility of a consistent modification. For the (2N+1)th-order phase--integral approximation the expression for $q(z)$ is

$$q(z) = Q_{mod}(z) \sum_{n=0}^{N} Y_{2n} , \qquad (10.6)$$

where $Q_{mod}(z)$ is to be chosen conveniently. The first few functions Y_{2n} are

$$Y_0 = 1 , \qquad (10.7a)$$
$$Y_2 = \tfrac{1}{2}\varepsilon_0 , \qquad (10.7b)$$

$$Y_4 = -\tfrac{1}{8}\varepsilon_0^2 - \tfrac{1}{8}\varepsilon_2 \,, \tag{10.7c}$$

with

$$\varepsilon_0 = \frac{Q^2(z) - Q^2_{mod}(z)}{Q^2_{mod}(z)} + Q^{-\frac{3}{2}}_{mod}(z) \frac{d^2}{dz^2} Q^{-\frac{1}{2}}_{mod}(z) \,, \tag{10.8}$$

and

$$\varepsilon_n = d^n \varepsilon_0 / d\zeta^n \,, \qquad n = 1, 2, 3, \ldots \,, \tag{10.9}$$

where

$$\zeta = \int^z Q_{mod}(z)\, dz \,. \tag{10.10}$$

The properties of the higher-order "symmetric" phase-integral approximations $(2N+1 > 1)$ are described in [40] and [41], and the consistent way of modifying them is treated in [42] and on pp. 126-131 in [43]. It should be noted that the modification, changes the expressions for ε_0 and ζ according to (10.8) and (10.10), while the expressions for Y_{2n} in terms of ε_0, ε_1, ε_2, ... remain unchanged. Since integrals over higher-order expressions for $q(z)$ do not converge at transition points, such as zeros of $Q^2_{mod}(z)$, integrals which in the first-order formulas can be allowed to proceed wholly along the real axis, will, if transition points are involved, be replaced by appropriate contour integrals [40]. In formulas where $q(z)$ occurs in the denominators of integrands, which is the case in formulas for expectation values and matrix elements, those contour integrals must enclose also the zeros of $q(z)$ which are located in the immediate neighbourhood of a zero of $Q^2_{mod}(z)$ [41].

In the way described, it is a straightforward matter to rewrite a formula, derived in the first-order approximation, such that it applies to the arbitrary-order phase-integral approximations in question.

11. EXERCISES

1. a) Prove the relations (4.2a,b).
   b) Show that, for the first-order JWKB-approximation, three anti-Stokes' lines and three Stokes' lines emerge from a simple zero of $Q^2_{mod}(z)$.
   c) Show that on a curve $z = z(t)$, where t is a real parameter, the quantity $|\exp\{iw(z)\}|$ assumes its extreme values when

   $$\mathrm{Im}\{Q_{mod}(z)\, dz/dt\} = 0 \,.$$

   Use this relation to find the location of the extreme values of $|\exp\{iw(z)\}|$ on a circle enclosing a simple zero of $Q^2_{mod}(z)$, when $Q^2_{mod}(z) = z$.

2. Prove the connection formula (5.19) by analogous arguments as used for obtaining (5.16).
(This proof is considerably more intricate than the proof of (5.16).)

3. Derive the expression $T \approx e^{-2K}$ for the transmission coefficient for a thick barrier by tracing an outgoing wave backwards through the barrier with the aid of the appropriate connection formula at each one of the two, well separated, classical turning points.

4. Prove, without calculating the eigenvalues, that the first-order Bohr-Sommerfeld quantization condition (6.2) is exact for the hydrogen atom, if one chooses

$$Q^2_{mod}(r) = Q^2(r) - \frac{1}{4r^2} \; .$$

Hint: Find the asymptotic solution on a large circle in the complex plane and utilize the fact that the solution is single-valued.

5. Derive the quantization condition for a symmetric double-oscillator by using the appropriate connection formula for a single turning point and the connection formula for a barrier.

6. Show that the expression (7.13) for the half-width Γ of a resonance tends to a finite value when $K \to 0$.

7. Show, without explicit calculation, that the first-order phase-integral formula (9.10), with a convenient choice of Q^2_{mod} , yields the following expectation values exactly:
a) $\langle x^0 \rangle$, $\langle x^1 \rangle$ and $\langle x^2 \rangle$ for the harmonic oscillator
b) $\langle r^0 \rangle$, $\langle r^{-1} \rangle$ and $\langle r^{-2} \rangle$ for the hydrogen atom.
Hint: Use the fact that the semi-classical quantization condition is exact for the harmonic oscillator and the hydrogen atom (cf. problem 4).

8. Generalize the first-order formula (9.10) for expectation values to the third-order phase-integral approximation by replacing, according to (10.6) and (10.7a,b), Q_{mod} by $Q_{mod}(1+\frac{1}{2}\varepsilon_0)$, where ε_0 is given by (10.8), and replacing the integral along the real axis by an integral along a closed loop in the complex plane.
Show, by explicit calculation, that the following expectation values are obtained exactly from the third-order phase-integral formula, if Q^2_{mod} is chosen conveniently:
a) $\langle x^4 \rangle$ and $\langle x^6 \rangle$ for the harmonic oscillator,
b) $\langle r^1 \rangle$ and $\langle r^2 \rangle$ for the hydrogen atom.

9. Calculate the probability density at the origin, $\psi_n^2(0)$, for an s-state with the principal quantum number n of the hydrogen atom
   a) by using the exact formula

$$\psi_n^2(0) = \frac{m}{2\pi\hbar^2} \langle dV(r)/dr \rangle$$

   and the first-order phase-integral formula (9.10) for expectation values
   b) by using the normalized JWKB wave function.
   (A numerical comparison of cases a) and b) with the exact result can be found in [34].)

10. Show, by analogous heuristic arguments as used in the derivation of (9.12), that the matrix element of the differential operator $g(z)\, d/dz$ is obtained from a generalization of (9.12), which amounts to the replacement of $f(z)$ in the left-hand member by $g(z)\, d/dz$ and in the right-hand member by

$$\left(-\, iQ_{mod}(z) - \frac{1}{2Q_{mod}(z)} \frac{dQ_{mod}(z)}{dz}\right)_{E=E_{\nu'}} g(z) ,$$

   if the + sign is chosen in the exponent of the integrand, and if $E_\nu \gtrsim E_{\nu'}$. Cf. [35].

ACKNOWLEDGEMENTS

I would like to thank Miss Ebba Johansson for her skilful typing of the manuscript and Dr Finn Karlsson for preparing the figures and the tables, and for providing me with the results, quoted in Table I. My thanks are also due to Professor P.O. Fröman for valuable comments on the manuscript.

REFERENCES

1. Heading, J., An Introduction to Phase Integral Methods. Methuen's Monographs on Physical Subjects, London and New York, 1962.
2. McHugh, J.A.M., Arch. Hist. Exact Sci. 7, 277 (1971).
3. Berry, M.V., Mount, K.E., Rep. Progr. Phys. 35, 315 (1972).
4. Goldstein, H., Classical Mechanics. Addison-Wesley, Cambridge, Mass., 1950.
5. Gans. R.C.. Ann. Phys. (Leipzig) 47, 709 (1915).
6. Jeffreys, H., Proc. London Math. Soc. (2) 23, 428 (1925).
7. Kramers, H.A., Z. f. Physik 39, 828 (1926).
8. Brillouin, L., J. Phys. Radium 7, 353 (1926).

9. Wentzel, G., Z. Physik 38, 518 (1926).
10. Langer, R., Trans. Am. Math. Soc. 33, 23 (1931).
11. Zwaan, A., Intensitäten im Ca-Funkenspektrum. Academisch Proefschrift. Joh. Enschedé en Zonen, Haarlem, 1929.
12. Kemble, E.C., Phys. Rev. 48, 549 (1935).
13. Fröman, N. and Fröman, P.O., JWKB-Approximation, Contributions to the Theory. North-Holland, Amsterdam, 1965.
14. Fröman, N., Arkiv för Fysik 31, 381 (1966).
15. Ford, K.W., Hill, D.L., Wakano, M. and Wheeler, J.A., Ann. Phys. (N.Y.) 7, 239 (1959).
16. Connor, J.N.L., Molec. Phys. 15, 621 (1968).
17. Connor, J.N.L., Molec. Phys. 25, 1469 (1973).
18. Crothers, D.S.F., J. Phys. B: Atom. Molec. Phys. 9, L513 (1976).
19. Dickinson, A.S., Molec. Phys. 18, 441 (1970).
20. Miller, W.H., J. Chem. Phys. 48, 1651 (1968).
21. Fröman, N. and Fröman, P.O., Nucl. Phys. A147, 606 (1970).
22. Fröman, N., Fröman, P.O., Myhrman, U. and Paulsson, R., Ann. Phys. (N.Y.) 74, 314 (1972).
23. Lundborg, B., Math. Proc. Camb. Phil. Soc. 81, 463 (1977).
24. Drukarev, G., Fröman, N. and Fröman, P.O., J. Phys. A: Math. Gen. 12, 171 (1979).
25. Karlsson, F., Phys. Rev. D11, 2120 (1975).
26. Sommerfeld, A., Atombau und Spektrallinien. Wellenmechanischer Ergänzungsband (Vieweg, Braunschweig, 1929) pp. 160-161.
27. Fröman, N. and Fröman, P.O., J. Math. Phys. 18, 96 (1977).
28. Rosenzweig, C. and Krieger, J.B., J. Math. Phys. 9, 849 (1968).
29. Chan, S.I. and Stelman, D.J., J. Chem. Phys. 39, 545 (1963).
30. Furry, W.H., Phys. Rev. 71, 366 (1947).
31. Yngve, S., J. Math. Phys. 13, 324 (1972).
32. Fröman, N., Phys. Lett. 48A, 137 (1974).
33. Fröman, P.O., Ann. Phys. (N.Y.) 88, 621 (1974).
34. Fröman, N., J. Math. Phys. 19, 1141 (1978).
35. Fröman, N. and Fröman, P.O., J. Math. Phys. 18, 903 (1977).
36. Fröman, N., Fröman, P.O. and Karlsson, F., Molec. Phys. 38, 749 (1979).
37. Pack, R.T. and Dahler, J.S., J. Chem. Phys. 50, 2397 (1969).
38. Smith, W.D. and Pack, R.T., J. Chem. Phys. 52, 1381 (1970).
39. Dickinson, A.S. and Shizgal, B., Molec. Phys. 30, 1221 (1975).
40. Fröman, N., Ark. Fys. 32, 541 (1966).
41. Fröman, N., Ann. Phys. (N.Y.) 61, 451 (1970).
42. Fröman, N. and Fröman, P.O., Ann. Phys. (N.Y.) 83, 103 (1974).
43. Fröman, N. and Fröman, P.O., Nuovo Cimento 20B, 121 (1974).

SEMICLASSICAL THEORY OF ELASTIC SCATTERING

J.N.L. Connor

Department of Chemistry
University of Manchester
Manchester M13 9PL
U.K.

These lectures are concerned with the semiclassical theory of elastic scattering. The following topics are discussed: the quantum and classical theory of elastic scattering, partial wave series, semiclassical phase shifts, the semiclassical evaluation of the scattering amplitude using the Poisson sum formula, uniform approximations and their connection with Catastrophe Theory, orbiting collisions, resonances, semiclassical calculations of Regge pole positions and residues, and the surface wave interpretation of elastic scattering.

1. INTRODUCTION

The theory of the elastic scattering of atoms has been one of the most successful applications of semiclassical techniques. The semiclassical theory gives an accurate description and an excellent physical understanding of the various effects that are present in the scattering.

A large amount of research has been carried out on elastic scattering and the semiclassical theory is very fully developed. Various reviews of elastic scattering have been written from time to time, for example, ref. (1-17), but no comprehensive review is available of all the research that has been done.

The purpose of these lectures is to provide an introduction to the semiclassical theory of elastic scattering. No prior knowledge of the subject is assumed. As well as covering some standard topics, the lectures will also deal with several important topics that do not feature prominently in existing reviews.

M. S. Child (ed.), Semiclassical Methods in Molecular Scattering and Spectroscopy, 45–107.

We shall also discuss some recent research that has not been reviewed before.

In general, the lectures will stress the physical significance of the theory as well as deriving useful working formulae in a form suitable for applications. The lectures will also emphasise features of the theory that are important for more general semiclassical theories of inelastic and reactive collisions. These more general theories are reviewed in ref. (13,17-26).

There are a number of important topics that will *not* be treated in these lectures. These include: a rigorous mathematical analysis of the semiclassical formulae complete with error bounds (see ref. (27,28)), the inclusion of higher order terms in the semiclassical formalism (see for example ref. (29)) and a detailed analysis of the class of potentials for which the theory is valid. However, it will be important for us to distinguish between regular and singular potentials.

Section 2 considers the quantum theory of elastic scattering, the definition of differential and total cross sections and the partial wave series representation of the scattering amplitude. Section 3 discusses the corresponding classical theory. As these topics are standard, only a brief discussion is given.

In section 4, the semiclassical formula for the phase shift is derived for the case of a single turning point. This is probably the most important semiclassical formula derived in these lectures. The Langer correction and Born approximation to the phase shift are also discussed.

Section 5 considers the semiclassical evaluation of the scattering amplitude. Topics discussed include: the Poisson sum representation of the scattering amplitude, the stationary phase and saddle point methods for the asymptotic evaluation of integrals, transitional approximations and the uniform semiclassical evaluation of integrals with two coalescing saddle points. The theory of cusped rainbows is briefly discussed and contact is made with Catastrophe Theory.

In section 6, the semiclassical theory of orbiting collisions is considered. The semiclassical phase shift for the case of three turning points is derived. It is also shown how the phase shift can be reduced to a Breit-Wigner resonance form. The effect of resonances on the total cross section is also discussed.

Finally, section 7 discusses the Regge pole representation of the scattering amplitude. It is shown how semiclassical methods can be used for the accurate calculation of Regge pole positions and their associated residues. The Regge approach allows a description of the scattering in terms of "surface waves", which propagate around the core of the potential. This approach is particularly useful for the description of the large angle elastic scattering of chemical reactive systems, where a new class of interference effects is predicted.

2. QUANTUM THEORY OF ELASTIC SCATTERING

This section considers the boundary conditions for the Schrödinger equation and shows how observable quantities, such as the differential cross section, can be related to the scattering amplitude. The partial wave series representation for the scattering amplitude is obtained; this is the basic result on which the rest of the chapter depends. As the results obtained in this section are standard, only brief derivations will be presented. More details can be found in ref. (1,5,13,30-33) for example.

2.1. Schrödinger Equation and Boundary Conditions

The scattering calculations are always carried out in the *centre of mass system*. The Schrödinger equation is

$$\left[-\frac{\hbar^2}{2\mu}\nabla^2+V(r)\right]\Psi(\underset{\sim}{r})=E\Psi(\underset{\sim}{r}) \qquad \underset{\sim}{r}=(r,\theta,\phi) \tag{1}$$

where μ is the reduced mass of the system, E is the collision energy and V(r) is the interaction potential which is assumed to depend only on r, i.e. we are working with a spherically symmetric potential. It is also assumed that V(r) falls off more rapidly than r^{-1} to exclude potentials with coulombic tails.

We seek a particular solution $\Psi^+(\underset{\sim}{r})$ of equation (1) which is regular at the origin and which for $r\to\infty$ has the asymptotic form

$$\Psi^+(\underset{\sim}{r}) \underset{r\to\infty}{\sim} e^{ikz} + f(\theta)\frac{e^{ikr}}{r} \tag{2}$$

where

$$k=(2\mu E)^{\frac{1}{2}}/\hbar \tag{3}$$

is the wavenumber. That such a solution of equation (1) exists and is unique can be shown using the theory of integral equations see ref. (34,35). The first term in equation (2) represents a *plane wave* of momentum $\hbar\underset{\sim}{k}$ moving in the positive z direction. The second term represents an *outgoing spherical wave*. The coefficient f(θ) is the *scattering amplitude*. f(θ) also depends on

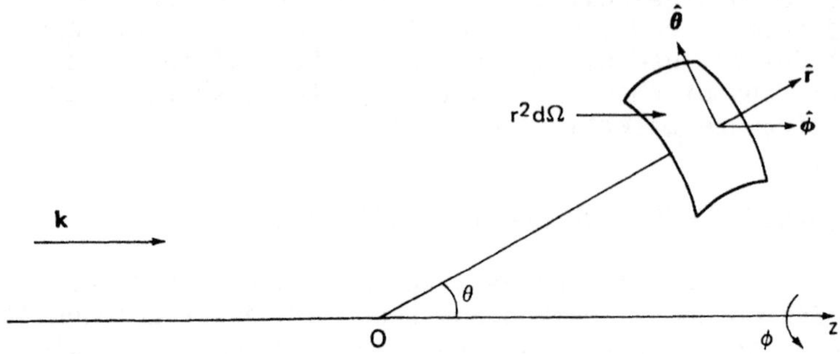

Figure 1. Spherical polar coordinate system used for the quantum theory of elastic scattering. O indicates the origin of the coordinate system. $(\hat{r},\hat{\theta},\hat{\phi})$ are unit vectors. The vector $\underset{\sim}{k}$ indicates a plane wave propagating in the positive z direction.

E(or k) but this will not usually be indicated explicitly. θ is the scattering angle and is the angle between the z axis and the direction of the scattered particle as shown in figure 1. Because $V(r)$ is assumed to be spherically symmetric, $f(\theta)$ is independent of ϕ.

2.2. Differential and Total Cross Sections

The *differential cross section* $I(\theta)$ is defined as

$$I(\theta)d\Omega = \frac{\text{number of particles scattered into } d\Omega \text{ per unit time}}{\text{incident particle flux}} \tag{4}$$

where $d\Omega$ is an element of solid angle as shown in figure 1. Since we are dealing with a spherically symmetric potential we can integrate over ϕ to get

$$d\Omega = 2\pi \sin\theta d\theta \tag{5}$$

In order to obtain the numerator and denominator of equation (4), we calculate the *probability current density* using

$$j(\underset{\sim}{r}) = \frac{\hbar}{2\mu i}[\Psi^*(\underset{\sim}{r})\nabla\Psi(\underset{\sim}{r}) - \Psi(\underset{\sim}{r})\nabla\Psi^*(\underset{\sim}{r})] \tag{6}$$

Applying equation (6) with $\nabla=\partial/\partial z$ to the first term in $\Psi^+(\underset{\sim}{r})$, we find the incident flux is $\hbar k/\mu = v$, the velocity. For the numerator, we calculate the outgoing radial flux through the surface element $r^2 d\Omega$. With $\nabla=\partial/\partial r$ applied to the second term in $\Psi^+(\underset{\sim}{r})$, we obtain

$$v|f(\theta)|^2/r^2$$

where terms in higher powers of r^{-1} have been neglected.

The definition (4) then gives for the differential cross section

$$I(\theta) = |f(\theta)|^2 \tag{7}$$

The *total cross section* $\sigma(E)$ is obtained by integrating $I(E,\theta)$ over all possible scattering angles. Thus for a spherically symmetric potential we have

$$\sigma(E) = 2\pi\int_0^\pi I(E,\theta)\sin\theta d\theta \tag{8}$$

Equations (7) and (8) are of basic importance. They show how the experimental observables $I(\theta)$ and $\sigma(E)$ are related to the asymptotic form of the wavefunction $\Psi^+(\underset{\sim}{r})$ via the scattering amplitude $f(\theta)$.

2.3. Partial Wave Representation of f(θ)

In section 2.2, it was shown that $I(\theta)$ or $\sigma(E)$ can be calculated once $f(\theta)$ is known. In this section, the *partial wave series representation* of $f(\theta)$ will be derived.

Consider first the case when $V(\underset{\sim}{r})\equiv 0$ in the Schrödinger equation (1). The solution is a plane wave moving in the positive z direction, namely exp(ikz). In polar coordinates, exp(ikz) can be written (see ref. (36))

$$\begin{aligned}\exp(ikz) &= \exp(ikr\cos\theta)\\ &= \sum_{\ell=0}^{\infty} i^\ell(2\ell+1)j_\ell(kr)P_\ell(\cos\theta)\end{aligned} \tag{9}$$

This is the partial wave series representation for exp(ikz). In equation (9), $j_\ell(kr)$ is the spherical Bessel function of order ℓ and $P_\ell(\cos\theta)$ is the Legendre polynomial of degree ℓ.

For large values of r, we can use the asymptotic result (see ref. (36))

$$j_\ell(kr) \underset{r\to\infty}{\sim} (kr)^{-1}\sin(kr-\tfrac{1}{2}\ell\pi) \tag{10}$$

so that equation (9) becomes

$$\exp(ikz) \underset{r\to\infty}{\sim} (2ikr)^{-1} \sum_{\ell=0}^{\infty} i^{\ell}(2\ell+1)P_{\ell}(\cos\theta)$$

$$\times [\exp(ikr-i\tfrac{1}{2}\ell\pi)-\exp(-ikr+i\tfrac{1}{2}\ell\pi)] \qquad (11)$$

which represents an infinite superposition of incoming and outgoing spherical waves.

We now exploit the fact that $V(r)$ is spherically symmetric to expand $\Psi^{+}(\underset{\sim}{r})$ in the form

$$\Psi^{+}(\underset{\sim}{r}) = (kr)^{-1}\sum_{\ell=0}^{\infty} a_{\ell}\, i^{\ell}(2\ell+1)\psi_{\ell}(r)P_{\ell}(\cos\theta) \qquad (12)$$

where a_{ℓ} is an expansion coefficient and the factor $i^{\ell}(2\ell+1)/kr$ has been included explicitly because the equations that follow will simplify cf. equations (9) and (11).

On substituting the expansion (12) into the Schrödinger equation (1), an equation for the radial wavefunction $\psi_{\ell}(r)$ is obtained

$$\frac{d^2\psi_{\ell}(r)}{dr^2} + [k^2-\frac{2\mu}{\hbar^2}V(r)-\frac{\ell(\ell+1)}{r^2}]\psi_{\ell}(r) = 0 \qquad (13)$$

The *boundary conditions* for $\psi_{\ell}(r)$ are

$$\psi_{\ell}(0) = 0 \qquad (14)$$

and

$$\psi_{\ell}(r) \underset{r\to\infty}{\sim} \sin(kr-\tfrac{1}{2}\ell\pi+\delta_{\ell}) \qquad (15)$$

where δ_{ℓ} is a real quantity.

The boundary condition (14) ensures that $\psi_{\ell}(r)$ is finite everywhere i.e. we are dealing with the *regular solution*. In equation (15), $\delta_{\ell}(k)$ is the *phase shift*. The term $-\frac{1}{2}\ell\pi$ is present in (15) so that $\delta_{\ell}=0$ when $V(r)=0$, see equation (10). Note that the regular solution of the radial Schrödinger equation when $V(r)\equiv 0$ is just $krj_{\ell}(kr)$. Also in equation (15), we have fixed the normalization of the sine wave as unity.

We require the partial wave expansion (12) for $\Psi^{+}(\underset{\sim}{r})$ to satisfy the boundary condition (2) when $r\to\infty$. Substituting equation (15) into equation (12) gives

$$\Psi^+(\underset{\sim}{r}) \underset{r\to\infty}{\sim} (2ikr)^{-1}\sum_{\ell=0}^{\infty} a_\ell i^\ell(2\ell+1)P_\ell(\cos\theta)$$

$$\times\ [\exp(ikr-i\tfrac{1}{2}\ell\pi+i\delta_\ell)-\exp(-ikr+i\tfrac{1}{2}\ell\pi-i\delta_\ell)] \qquad (16)$$

We next choose a_ℓ so that (16) and (2) are equivalent. For the incoming spherical wave exp(-ikr)/r we must have

$$a_\ell = \exp(i\delta_\ell) \qquad (17)$$

if (16) is to be equivalent to (2) and (11). Having fixed a_ℓ, the partial wave expansion for $f(\theta)$ is obtained from

$$f(\theta)\exp(ikr)/r \underset{r\to\infty}{\sim} \Psi^+(\underset{\sim}{r})-\exp(ikz)$$

Using the expansions for $\Psi^+(\underset{\sim}{r})$ and exp(ikz) already given, it is found that

$$f(\theta) = (2ik)^{-1}\sum_{\ell=0}^{\infty}(2\ell+1)[\exp(2i\delta_\ell)-1]P_\ell(\cos\theta) \qquad (18)$$

The quantity

$$S_\ell = \exp(2i\delta_\ell) \qquad (19)$$

is the elastic S matrix element.

Equation (18) is the partial wave series representation for $f(\theta)$. It can be used to obtain accurate values for $f(\theta)$ provided (a) accurate phase shifts δ_ℓ are available, and (b) a sufficiently large number of partial waves can be included to converge the sum to (say) three or four significant figures. We shall see that semiclassical techniques can be used in connection with (a) and (b).

The total cross section $\sigma(E)$ is obtained from $f(\theta)$ by equation (8). Substituting the partial wave series (18) into equation (8) gives rise to a double sum. However with the help of the orthogonality relation

$$\int_0^\pi P_\ell(\cos\theta)P_{\ell'}(\cos\theta)\sin\theta d\theta = \frac{2}{2\ell+1}\delta_{\ell\ell'} \qquad (20)$$

where $\delta_{\ell\ell'}$ is the Kronecker delta, the double sum collapses to a single sum and we find

$$\sigma(E) = \frac{4\pi}{k^2}\sum_{\ell=0}^{\infty}(2\ell+1)\sin^2\delta_\ell \qquad (21)$$

Equation (21) is the partial wave series representation for the total cross section.

Finally, it should be noted that the above derivations can be generalized to include complex valued optical potentials of the type

$$V(r) = V_R(r) - iV_I(r) \tag{22}$$

where

$$V_I(r) \geq 0 \tag{23}$$

The imaginary part of V(r) absorbs part of the incident flux. Optical potentials are used to study the effect that other inelastic or reactive channels have on the elastic scattering.

The partial wave series (18) remains valid for optical potentials except that the phase shift is now a complex quantity

$$\delta_\ell = \eta_\ell + i\xi_\ell \tag{24}$$

with

$$\xi_\ell \geq 0 \tag{25}$$

3. CLASSICAL THEORY OF ELASTIC SCATTERING

3.1. Differential and Total Cross Sections

The differential cross section $I(\theta)$ can be calculated from equation (4). The number of particles scattered into $d\Omega$ per unit time is $FI(\theta)d\Omega$ where F is the incident flux. The particles entering $d\Omega=2\pi\sin\theta d\theta$ originally passed through the ring $2\pi bdb$, where b is the *impact parameter* and $0\leq b<\infty$ (see figure 2). Thus initially the number of particles passing through $2\pi bdb$ per unit time is $F2\pi bdb$. Hence

$$FI(\theta)d\Omega = F2\pi bdb \tag{26}$$

or

$$I(\theta) = b/\left(\sin\theta \frac{d\theta}{db}\right) \tag{27}$$

This formula needs to be modified in three ways to deal with scattering for atom-atom interactions of the Lennard-Jones type:

(a) $d\theta/db$ may be negative. For example, for repulsive potentials, θ decreases as b increases. Since $I(\theta)$ is by convention a positive quantity, we replace $d\theta/db$ by $|d\theta/db|$ to deal with this situation.

(b) The particle may wind around the scattering centre several times before making its exit. In this situation, it is convenient to distinguish between the experimentally measurable scattering angle θ where $0\leqslant\theta\leqslant\pi$ and the *deflection function* Θ where $-\infty<\Theta\leqslant\pi$. We replace $|d\theta/db|$ in equation (27) by $|d\Theta/db|$.

(c) Particles with different impact parameters may all scatter into the same angle θ. We must then sum over all the contributing impact parameters.

We finally obtain

$$I(\theta) = \sum_i b_i/\left(\sin\theta\left|\frac{d\Theta}{db_i}\right|\right) \tag{28}$$

Thus if we know how $\Theta(b)$ depends on b, we can calculate $I(\theta)$ by equation (28). Note the singular behaviour of $I(\theta)$ when $\sin\theta=0$ (a glory) or $d\Theta/db_i=0$ (a rainbow). The avoidance of these infinities in a semiclassical analysis is discussed in section 5.

The total cross section $\sigma(E)$ is again given by equation (8). Unless the potential is of finite range, $\sigma(E)$ is infinite. Physically, this corresponds to the fact that a potential with an infinite tail like g/r^n always gives rise to some scattering, no matter how large b is. A semiclassical analysis again rounds off this infinity.

3.2. The Deflection Function

In this section we will use conservation of energy and angular momentum to obtain the dependence of Θ on b.

Since the angular momentum $\underset{\sim}{L}$ is conserved, the trajectories lie in a plane. We can then use polar coordinates (r,ϕ) as illustrated in figure 2. We have

$$L = \mu vb = \mu r^2\dot{\phi} \tag{29}$$

and

$$\begin{aligned} E &= \tfrac{1}{2}\mu v^2 \\ &= \tfrac{1}{2}\mu(\dot{r}^2+r^2\dot{\phi}^2)+V(r) \end{aligned} \tag{30}$$

or using equation (29)

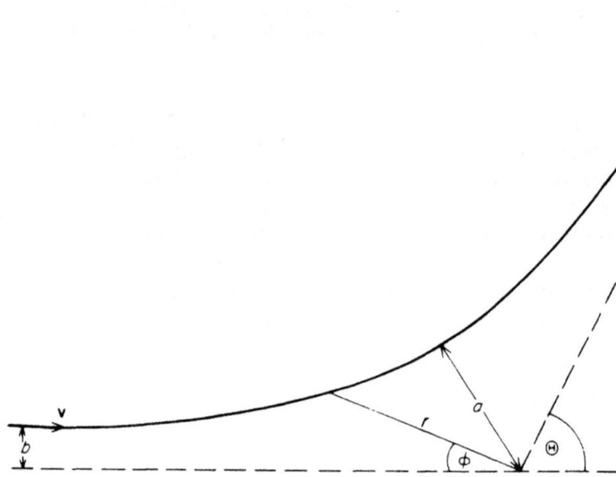

Figure 2. Plane polar coordinate system used for the classical theory of elastic scattering. A classical particle with velocity $\underset{\sim}{v}$ and impact parameter b is deflected through angle Θ. The classical turning point is at r=a.

$$E = \tfrac{1}{2}\mu\dot{r}^2 + \frac{L^2}{2\mu r^2} + V(r) \tag{31}$$

where the dots indicate differentiation with respect to time.

From equations (29)-(31), we can write

$$\frac{d\phi}{dr} = \frac{\dot{\phi}}{\dot{r}} \tag{32}$$

$$= \pm L/\{r^2[2\mu E - L^2/r^2 - 2\mu V(r)]^{\frac{1}{2}}\} \tag{33}$$

The $\pm$ sign in equation (33) corresponds to the sign of the radial velocity in (32). On the inward part of the trajectory, $\dot{r}$ is negative, and on the outward part, $\dot{r}$ is positive.

Equation (33) is next integrated from $\phi=0$, $r=\infty$ to $\phi=\phi_a$, r=a where a is the distance of closest approach (see figure 2) and is the outermost root of

$$2\mu[E-V(r)]-L^2/r^2 = 0 \tag{34}$$

From equation (33), we obtain

$$\phi_a = -L\int_{\infty}^{a} r^{-2}\{2\mu[E-V(r)]-L^2/r^2\}^{-\frac{1}{2}}dr \quad (35)$$

The geometry of figure 2 shows that

$$\Theta = \pi-2\phi_a \quad (36)$$

and from equation (35) we obtain

$$\Theta(L) = \pi-2L\int_{a}^{\infty} r^{-2}\{2\mu[E-V(r)] -L^2/r^2\}^{-\frac{1}{2}}dr \quad (37)$$

Equation (37) shows explicitly the dependence of $\Theta(L)$ on L (or b, since $L=\mu vb$). The integral on the right hand side of (37) can only be evaluated explicitly for a few simple potentials (see ref. (37) for example). However, for potentials of the Lennard-Jones type, it is straightforward to integrate equation (37) by Gauss-Mehler quadrature (see ref. (38)). The derivatives required in equation (28) can then be obtained by numerical differentiation. Alternatively, an analytic expression for $d\Theta/dL$ can be derived in terms of an integral which can be evaluated by a quadrature formula (see ref. (39,40)).

Finally we note that equation (37) can be inverted to find V(r) provided V(r) falls off monotonically with r. This method, which is due to Firsov (41), is explained in ref. (13,15,42).

4. THE SEMICLASSICAL PHASE SHIFT

It was shown in Section 2 that the problem of calculating $f(\theta)$, and hence $I(\theta)$ can be reduced to that of calculating phase shifts δ_ℓ and then summing over ℓ in the partial wave series. In this section, we will consider the semiclassical approximation for δ_ℓ.

We have to solve the radial Schrödinger equation

$$\frac{d^2\psi_\ell(r)}{dr^2} + [k^2-\frac{2\mu}{\hbar^2}V_\ell(r)]\psi_\ell(r) = 0 \quad (38)$$

with boundary conditions

$$\psi_\ell(0) = 0 \quad (39)$$

$$\psi_\ell(r) \underset{r\to\infty}{\sim} \sin(kr-\tfrac{1}{2}\ell\pi+\delta_\ell) \quad (40)$$

where we have introduced the *effective potential*

$$V_\ell(r) = V(r) + \frac{\hbar^2 \ell(\ell+1)}{2\mu r^2} \tag{41}$$

4.1. Turning Points

Before applying semiclassical methods, it is first necessary to examine the turning point structure of equation (38). That is, we must examine the real and complex roots of

$$k^2 - \frac{2\mu}{\hbar^2} V_\ell(r) = 0 \tag{42}$$

or equivalently

$$E - V_\ell(r) = 0 \tag{43}$$

Assuming that $V(r)$ is a potential of the Lennard-Jones (12,6) form, then $V_\ell(r)$ is shown in figure 3 for various ℓ values. It can be seen that for small ℓ values, $V_\ell(r)$ has a barrier. Consequently for low values of E, there are three real turning points. As E increases, two of the turning points move closer together until they coalesce at the top of the barrier. These two turning points then move into the complex r plane, where they become complex conjugates of each other. In general, there are also many other complex roots of equation (43) in the complex r plane.

In the remainder of this section, we shall suppose that equation (43) has only one real turning point and that the other complex turning points are far removed from the real r axis, so their contribution to the phase shift is negligible. The case of three turning points is discussed again in sections 6 and 7.

In the neighbourhood of the turning point a_ℓ, the effective potential can be Taylor expanded in the form

$$V_\ell(r) = V_\ell(a_\ell) + V_\ell'(a_\ell)(r-a_\ell) + \ldots \tag{44}$$

with

$$V_\ell'(a_\ell) \neq 0$$

That is, we can locally approximate $V_\ell(r)$ by a linear potential. This suggests we use the *comparison equation method* (see ref.(9,43)) to obtain the semiclassical solution of equation (38). In this method, $\psi_\ell(r)$ is mapped in a uniform (although local) one-to-one manner onto the solution of the Schrödinger equation for a linear potential, namely the regular Airy function equation

$$\frac{d^2 Ai(x)}{dx^2} = x Ai(x) \tag{45}$$

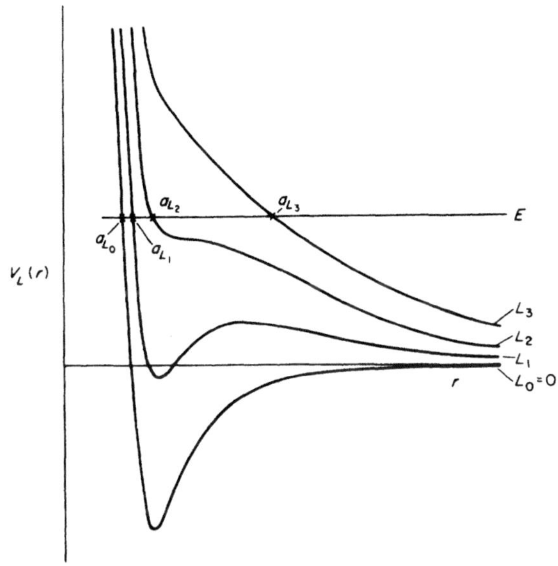

Figure 3. The effective potential $V_L(r)$ for four values of L. The corresponding classical turning points are a_{L_0}, a_{L_1}, a_{L_2}, and a_{L_3} at the collision energy E.

From the asymptotic form of the solution of the Airy function, we obtain the well known *linear connection formula*. However, before applying this formula to the solution of the radial equation, we must first discuss the *Langer transformation*.

4.2. Langer Transformation

If the linear connection formula is applied to the radial Schrödinger equation (38), a problem arises for regular potentials, that is, those that satisfy

$$\lim_{r\to 0} r^2 V(r) = 0 \tag{46}$$

The difficulty is related to the fact that the usual linear connection formula is obtained from a Schrödinger equation defined on $(-\infty,+\infty)$ whereas the radial equation is defined on $[0,+\infty)$.

To see the nature of the problem, consider the following estimate for the validity of the semiclassical approximation

$$\left|\frac{d\lambda\!\!\!^{-}_\ell(r)}{dr}\right| \ll 1 \tag{47}$$

where $\lambda\!\!\!^{-}_\ell(r) = 1/k_\ell(r)$, and

$$k_\ell(r) = [k^2 - \frac{2\mu}{\hbar^2}V_\ell(r)]^{\frac{1}{2}} \tag{48}$$

Substituting equation (48) into (47) gives

$$\left| \frac{2\mu}{\hbar^2}\frac{dV_\ell(r)}{dr} \Big/ [k^2 - \frac{2\mu}{\hbar^2}V_\ell(r)]^{3/2} \right| \ll 1 \tag{49}$$

Consider now the case when $r \to 0$, so that $V_\ell(r)$ dominates k^2 in the denominator of (49). For r close to zero, we can write

$$V_\ell(r) = g/r^s \tag{50}$$

where $s<2$ for a regular potential and $\ell=0$, and $s>2$ for a singular potential, that is one that satisfies

$$\lim_{r\to 0} r^2 V(r) = \infty \tag{51}$$

The case $s=2$ represents a transition potential between the regular and singular classes.

Substituting equation (50) into the estimate (49) gives for the left hand side

$$\left[\frac{g2\mu}{\hbar^2}\right]^{-\frac{1}{2}} s r^{s/2-1} \tag{52}$$

This quantity tends to zero if $s>2$, is small for $s=2$ if g is large, but tends to infinity for $0<s<2$. As a result, the condition for the validity of the semiclassical approximation fails for regular potentials near the origin.

Langer (44) showed how to deal with this problem by first transforming the radial equation from $[0,+\infty)$ to $(-\infty,+\infty)$ before applying the semiclassical approximation.

Making the change of variables

$$r = e^x \tag{53}$$

and

$$\psi_\ell(x) = \exp(x/2)\phi_\ell(x)$$

the radial Schrödinger equation becomes

$$\frac{d^2\phi_\ell(x)}{dx^2} + \kappa_\ell^2(x)\phi_\ell(x) = 0 \tag{54}$$

where

$$\kappa_\ell^2(x) = e^{2x}[k^2-\frac{2\mu}{\hbar^2}V(e^x)]-(\ell+\tfrac{1}{2})^2 \tag{55}$$

A semiclassical phase integral for equation (54) and (55) is then of the form

$$\int^x \kappa_\ell(x)dx \tag{56}$$

or changing back to the original variable r, we find

$$\int^r [k^2-\frac{2\mu}{\hbar^2}V(r)-\frac{(\ell+\frac{1}{2})^2}{r^2}]^{\frac{1}{2}}dr \tag{57}$$

The net effect of the change of variables is thus to introduce a term $(\ell+\frac{1}{2})^2$ instead of $\ell(\ell+1)$ in the effective potential:

$$V_\ell(r) = V(r)+\frac{\hbar^2(\ell+\frac{1}{2})^2}{2\mu r^2} \tag{58}$$

From now on we will define $V_\ell(r)$ by equation (58) rather than the earlier definition of equation (41). The definition of $k_\ell(r)$ thus becomes

$$k_\ell(r) = [k^2-\frac{2\mu}{\hbar^2}V(r)-\frac{(\ell+\frac{1}{2})^2}{r^2}]^{\frac{1}{2}} \tag{59}$$

Further discussion on the Langer transformation can be found in ref. (9,27,45).

4.3. Semiclassical Phase Shift

We now return to the problem of calculating the semiclassical phase shift. The boundary condition (39) at the origin requires an exponentially decreasing wavefunction in the region $r \ll a_\ell$:

$$\psi_\ell(r) \sim \tfrac{1}{2}[k/|k_\ell(r)|]^{\frac{1}{2}}\exp[-\int_r^{a_\ell}|k_\ell(r')|dr'] \tag{60}$$

where $k_\ell(r)$ is given by equation (59). The normalization factor $k^{\frac{1}{2}}/2$ has been chosen to simplify the following equations.

The use of the linear connection formula shows that in the region $r \gg a_\ell$ the wavefunction (60) connects with

$$\psi_\ell(r) \sim [k/k_\ell(r)]^{\frac{1}{2}}\sin[\int_{a_\ell}^{r} k_\ell(r')dr' + \tfrac{1}{4}\pi] \tag{61}$$

Comparing equations (40) and (61) then gives for the phase shift

$$\delta_\ell = \lim_{r\to\infty} [\int_{a_\ell}^{r} k_\ell(r')dr' - kr] + \tfrac{1}{2}(\ell+\tfrac{1}{2})\pi \tag{62}$$

which can also be written in the alternative form

$$\delta_\ell = \int_{a_\ell}^{\infty} k_\ell(r)dr - \int_{(\ell+\frac{1}{2})/k}^{\infty} [k^2-(\ell+\tfrac{1}{2})^2/r^2]^{\frac{1}{2}}dr \tag{63}$$

Equations (62) and (63) are the semiclassical approximation for the phase shift. Note that δ_ℓ is discontinuous everywhere since it is only defined for $\ell=0,1,2,\ldots$ However, in the radial equation, ℓ occurs merely as a parameter and the preceding theory holds for ℓ continuous as well as discrete values. In this case we can write

$$L = (\ell+\tfrac{1}{2})\hbar \tag{64}$$

where L is the classical angular momentum introduced in Section 3. The impact parameter b is then related to ℓ by

$$b = (\ell+\tfrac{1}{2})/k \tag{65}$$

In addition, on differentiating equation (62), it is found that

$$\frac{2d\delta(\ell)}{d\ell} = \Theta(\ell) \tag{66}$$

where $\Theta(\ell)$ is the classical deflection function.

For singular potentials, it was noted in section 4.2 that it is not necessary to make the Langer transformation. However, it is the usual practice to do so, in order that relations (64)-(66) are again valid.

4.4. Numerical Evaluation

In order for the semiclassical phase shift to be useful in practice, we must be able to evaluate δ_ℓ efficiently. One

		Quantum		Semiclassical	
Re ℓ	Im ℓ	Re $\delta(\ell)$	Im $\delta(\ell)$	Re $\delta(\ell)$	Im $\delta(\ell)$
80	0π	-26.191	0.000	-26.186	0.000
80	1π	-26.146	2.822	-26.141	2.822
80	2π	-26.012	5.644	-26.006	5.644
80	3π	-25.787	8.468	-25.782	8.469
80	4π	-25.474	11.296	-25.469	11.296
80	5π	-25.071	14.127	-25.066	14.127
80	6π	-24.579	16.963	-24.574	16.963

Table 1. Comparison of exact quantum and semiclassical results for $\delta(\ell)$ for a Lennard-Jones (12,6) potential with reduced parameters of K=5.0 and A=141.43 (taken from ref. (52)).

	Quantum		Semiclassical	
ℓ	Re $\delta(\ell)$	Im $\delta(\ell)$	Re $\delta(\ell)$	Im $\delta(\ell)$
0	-15.104	0.873	-15.081	0.873
1	-13.592	0.872	-13.569	0.872
2	-12.140	0.871	-12.117	0.871
3	-10.748	0.869	-10.725	0.869
4	- 9.417	0.868	- 9.393	0.867
5	- 8.145	0.864	- 8.122	0.863
6	- 6.935	0.860	- 6.911	0.860
7	- 5.787	0.855	- 5.763	0.855

Table 2. Comparison of exact quantum and semiclassical results for $\delta(\ell)$ for a complex optical potential. The real part of the optical potential is a Lennard-Jones (12,6) potential and the imaginary part is of the form $-iW(\sigma/r)^{10}$. The reduced parameters are K=2.592, A=18.0 and $2\mu W\sigma^2/\hbar^2$=125.0 (taken from ref. (53)).

straightforward method for doing this uses gaussian quadrature. However, more efficient methods are available using Clenshaw-Curtis or Gauss-Mehler quadrature. These methods exploit the square root singularity in the integrand in equations (62) and (63). They are described in ref. (46-51). Another method solves Hamilton's equations instead of using a quadrature (see ref. (52)). This approach is frequently used to evaluate action integrals in more general theories of inelastic and reactive collisions.

4.5. Results

It was mentioned in section 4.3 that the semiclassical ex-

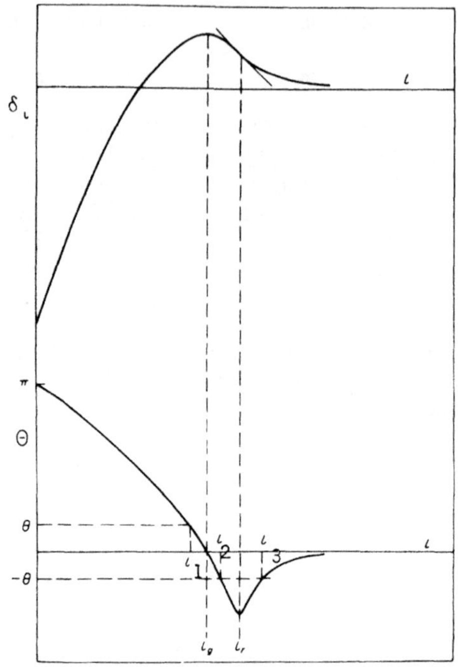

Figure 4. Plot of the semiclassical phase shift δ_ℓ and classical deflection function Θ_ℓ for a Lennard-Jones potential. The three ℓ values that contribute to θ, namely ℓ_1, ℓ_2, and ℓ_3 are shown. The glory ℓ_g and rainbow ℓ_r values of ℓ are also shown.

pression (63) for the phase shift is valid not only for ℓ=0,1,2,.. but also for ℓ continuous. In fact, the theory described above can be extended to complex values of ℓ. For this case, it is found the semiclassical approximation for $\delta(\ell)$ is again given by equations (62) and (63), provided ℓ is not too far from the real axis. In a similar way, the theory can also be extended to complex optical potentials of the type (22). In these problems, the turning point a_ℓ now lies off the real axis in the complex plane.

Tables 1 and 2 compare some exact and semiclassical results for $\delta(\ell)$ (taken from ref. (52) and (53)). It can be seen the semiclassical $\delta(\ell)$ are very accurate. In general, it has been found for real ℓ and for scattering off Lennard-Jones type potentials that the semiclassical $\delta(\ell)$ are accurate to within about 0.1%. This is sufficient for all but the most exacting applications.

Figure 4 shows $\delta(\ell)$ for a Lennard-Jones potential together with $\Theta(\ell)$. The graphs illustrate the semiclassical equivalence relation (66). The region where $\delta(\ell)<0$, corresponds to scattering by the repulsive core of the potential whereas the region where $\delta(\ell)>0$, corresponds to scattering by the attractive part of the potential. Figure 4 will play an important role in the semi-classical analysis of section 5.

4.6. Born Approximation

For large values of ℓ, the contribution of V(r) to the semi-classical phase shift is small, since the centrifugal term dominates in the effective potential $V_\ell(r)$. In this situation, the two integrals in equation (63) are both large, whereas their difference, which is $\delta(\ell)$, is small as can be seen in figure 4.

For this case, a perturbation treatment of equation (63) is appropriate. It is equivalent to the Born approximation. In equation (59) for $k_\ell(r)$, assume that V(r) is small and expand by the binomial theorem. We find

$$k_\ell(r) \simeq [k^2-\frac{(\ell+\frac{1}{2})^2}{r^2}]^{\frac{1}{2}} - \frac{\mu V(r)}{\hbar^2}/[k^2-\frac{(\ell+\frac{1}{2})^2}{r^2}]^{\frac{1}{2}} \quad (67)$$

If we replace a_ℓ by $(\ell+\frac{1}{2})/k$ in equation (63), we find that $\delta(\ell)$ simplifies to

$$\delta(\ell) = -\frac{\mu}{\hbar^2}\int_{(\ell+\frac{1}{2})/k}^{\infty} V(r)[k^2-\frac{(\ell+\frac{1}{2})^2}{r^2}]^{-\frac{1}{2}}dr \quad (68)$$

which is the Born approximation for $\delta(\ell)$.

For certain simple forms of V(r), the integral in equation (68) can be evaluated explicitly. Thus for

$$V(r) = -c_s/r^s \quad (69)$$

it is found that

$$\delta(\ell) = \frac{\mu c_s}{\hbar^2}\frac{k^{s-2}}{(\ell+\frac{1}{2})^{s-1}}\frac{\pi^{\frac{1}{2}}}{2}\frac{\Gamma(\frac{1}{2}s-\frac{1}{2})}{\Gamma(\frac{1}{2}s)} \quad (70)$$

Again for an exponential potential of the form

$$V(r) = Ae^{-\alpha r}$$

$\delta(\ell)$ simplifies to

$$\delta(\ell) = -\frac{\mu A}{\hbar^2 k^2}(\ell+\tfrac{1}{2})K_1((\ell+\tfrac{1}{2})\alpha/k) \qquad (71)$$

where $K_1(x)$ is a modified Bessel function (see ref. (36)).

In practice, it is straightforward to find the range of ℓ values where the Born approximation can be used by comparing $\delta(\ell)$ calculated from equation (63) with $\delta(\ell)$ calculated from equation (68). It is usually found that the Born approximation is valid for $|\delta(\ell)| \lesssim 0.5$ rad.

5. SEMICLASSICAL EVALUATION OF $f(\theta)$

In this section, it will be shown how semiclassical techniques can be used for the accurate evaluation of $f(\theta)$.

5.1. Summation of the Partial Wave Series

It was shown in section 4, that for the case of a single turning point, the semiclassical approximation for δ_ℓ is a reliable one. Thus an accurate way to evaluate $f(\theta)$ is to sum the partial wave series (18) using semiclassical δ_ℓ.

An estimate of the number of terms required can be obtained as follows. If the potential is assigned a radius R, then about

$$\ell+\tfrac{1}{2} \approx kR \qquad (72)$$

partial waves are expected to be strongly affected by the potential. Since most interatomic potentials have a tail, more terms than the estimate (72) will actually be required to converge the partial wave sum. Under typical scattering conditions for atom-atom collisions, several hundred to several thousand partial waves may be required.

For some systems, it is not possible to obtain converged values of $f(\theta)$ by summing the partial waves series because of its very slow convergence. This can occur, for example, in the elastic scattering of chemically reactive systems. This case is discussed in more detail in section 7.

Figures 5-8 show differential cross sections $I(\theta)$ calculated using semiclassical phase shifts. The most notable features are:

(a) the large peak in $I(\theta)$ at small angles

(b) the smooth constant behaviour of $I(\theta)$ at large angles

(c) the presence of oscillations in $I(\theta)$.

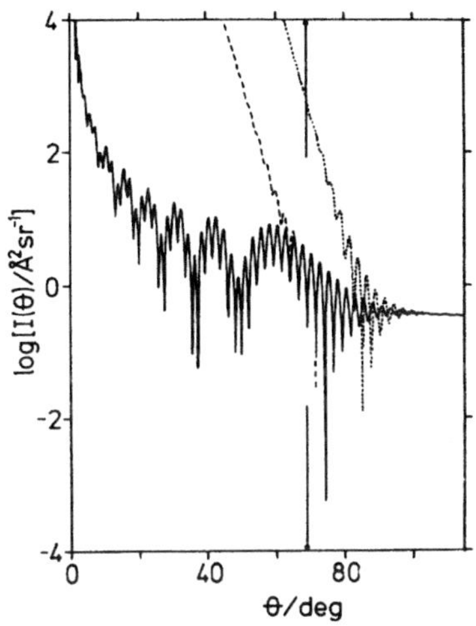

Figure 5. Plot of log I(θ) against θ for a Lennard-Jones (6,4) potential with reduced parameters of K=1.149 and A=63.64. Solid line gives the partial wave results using semiclassical phase shifts. Dashed line is the Regge theory with several poles in the pole sum. Dotted line is the Regge theory with one pole retained in the pole sum. The arrows indicate the classical rainbow angle. (Taken from ref. (54).)

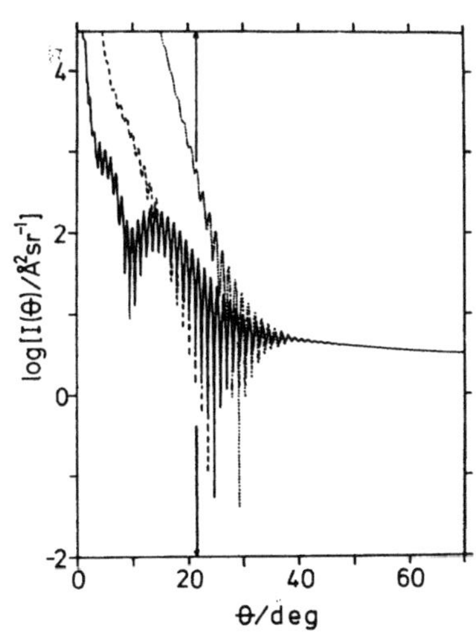

Figure 6. Same as figure 5 except for a Lennard-Jones (18,6) potential with reduced parameters of K=6.153 and A=154.5 (taken from ref. (54)).

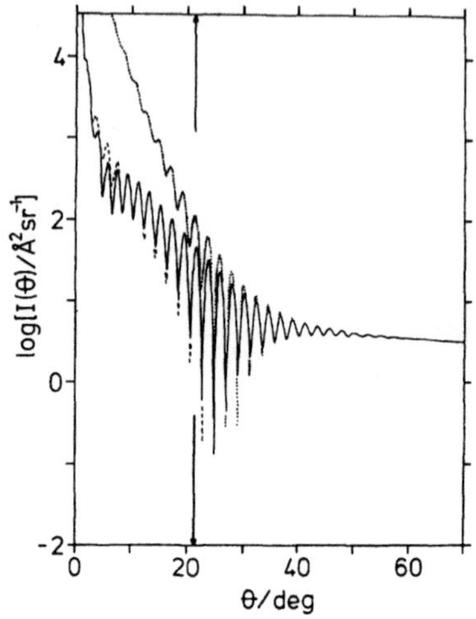

Figure 7. Same as figure 5 except for a Lennard-Jones (18,6) potential with reduced parameters of K=6.153 and A=82.38 (taken from ref. (54)).

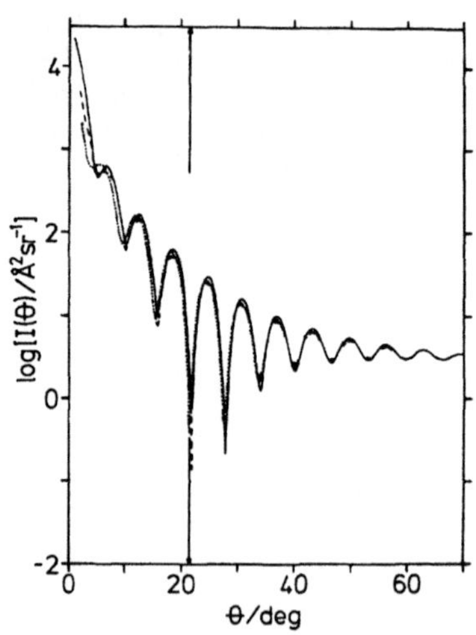

Figure 8. Same as figure 5 except for a Lennard-Jones (18,6) potential with reduced parameters of K=6.153 and A=29.13 (taken from ref. (54)).

Figure 5 shows two kinds of oscillations in $I(\theta)$. There are low frequency oscillations comprising a broad *rainbow* peak near the classical rainbow angle θ_r with *supernumerary rainbows* at smaller angles. Superimposed on the oscillations are high frequency oscillations which damp out beyond θ_r. Figures 6-8 show that as we proceed towards more quantum like systems, the width of the rainbow oscillation becomes broader and the high frequency oscillations become more widely spaced, until in figure 8 they dominate the scattering pattern. The oscillations in figure 8 are then sometimes referred to as "diffraction oscillations."

Note that although the partial wave series allows accurate values of $f(\theta)$ to be calculated, because of the large number of terms that must be summed, it does not provide a physical understanding of the various effects observed in the scattering. We shall now derive a new (exact) representation for $f(\theta)$ such that keeping only a few terms is a good approximation. A physical interpretation can then be attached to each term. For the present problem, it is the *Poisson sum representation* of $f(\theta)$ that we require.

5.2. Poisson Sum Representation

The Poisson sum formula is (see ref. (55,56))

$$\sum_{\ell=0}^{\infty} f(\ell+\tfrac{1}{2}) = \sum_{m=-\infty}^{\infty} (-1)^m \int_0^{\infty} f(\lambda)\exp(i2m\pi\lambda)d\lambda \tag{73}$$

where

$$\lambda = \ell+\tfrac{1}{2} \tag{74}$$

Applied to the partial wave series (18), we obtain

$$f(\theta) = (ik)^{-1} \sum_{m=-\infty}^{\infty} (-1)^m \int_0^{\infty} d\lambda\lambda[S(\lambda)-1]P_{\lambda-\frac{1}{2}}(\cos\theta)\exp(i2m\pi\lambda) \tag{75}$$

Notice that on the right hand side of equation (75) $S(\ell)$ and $P_\ell(\cos\theta)$ have been interpolated from $\ell=0,1,2,\ldots$ to continuous ℓ values. $P_\ell(\cos\theta)$ is now a Legendre *function* rather than a Legendre *polynomial*. Equation (75) is an exact representation of $f(\theta)$. We now proceed to introduce semiclassical approximations.

5.3. Semiclassical Approximations

We will assume we are dealing with a single turning point scattering problem of the kind discussed in section 4. We now introduce the following approximations into the Poisson sum formula (75):

(a) the phase shift is replaced by its semiclassical approximation (62);

(b) since the integrals in (75) are expected to receive contributions from large values of ℓ, we replace $P_{\lambda-\frac{1}{2}}(\cos\theta)$ by the following asymptotic approximation (see ref. (57))

$$P_{\lambda-\frac{1}{2}}(\cos\theta) \sim \left[\frac{2}{\pi\lambda\sin\theta}\right]^{\frac{1}{2}}\cos(\lambda\theta-\tfrac{1}{4}\pi) \tag{76}$$

which is valid for large λ and θ not close to 0 or π.

Because the following analysis does not apply to values of θ close to 0 or π, we can also drop the term involving unity in the partial wave series. This follows from the relation (see ref. (58))

$$\sum_{\ell=0}^{\infty}(2\ell+1)P_\ell(\cos\theta) = 4\delta(1-\cos\theta) \tag{77}$$

where $\delta(x)$ is the Dirac delta function.

We now obtain for $f(\theta)$ the expression

$$f(\theta) = (ik)^{-1}(2\pi\sin\theta)^{-\frac{1}{2}}\sum_{m=-\infty}^{\infty}(-1)^m[\exp(-i\tfrac{1}{4}\pi)I_m^+ + \exp(i\tfrac{1}{4}\pi)I_m^-] \tag{78}$$

where

$$I_m^{\pm} = \int_0^{\infty} d\lambda\,\lambda^{\frac{1}{2}}\exp\{i[2\delta(\lambda)\pm\lambda\theta+2m\pi\lambda]\} \tag{79}$$

The integrand of the integral (79) is highly oscillatory, which suggests the *stationary phase method* is an appropriate technique for the approximate evaluation of the $I_m^{\pm}$.

5.4. Stationary Phase Integration

The stationary phase method is an approximate method for the asymptotic evaluation of integrals of the form (see ref. (59))

$$J = \int_a^b g(\lambda)\exp[if(\lambda)]\,d\lambda \tag{80}$$

If the integrand is rapidly varying for a range of λ values, the contribution to J from this range is expected to be small. Suppose, however, $f(\lambda)$ has a real stationary point λ_0 defined by

$$\frac{\partial f(\lambda)}{\partial\lambda} = 0 \text{ for } \lambda = \lambda_0 \tag{81}$$

and which lies in (a,b).

Then for values of λ close to λ_0, the phase is slowly varying and the integral is expected to receive its main contribution from these values of λ.

Taylor expanding $f(\lambda)$ around λ_0 gives

$$f(\lambda) = f(\lambda_0)+\tfrac{1}{2}f''(\lambda_0)(\lambda-\lambda_0)^2+ \ldots\ldots \quad (82)$$

where it is assumed that

$$f''(\lambda_0) \neq 0 \quad (83)$$

We next substitute equation (82) with equation (80), replace $g(\lambda)$ by $g(\lambda_0)$ and extend the integration range to $(-\infty,+\infty)$. This gives

$$J \approx g(\lambda_0)\exp[if(\lambda_0)]\int_{-\infty}^{\infty} \exp[i\tfrac{1}{2}f''(\lambda_0)(\lambda-\lambda_0)^2]d\lambda \quad (84)$$

With the help of the result

$$\int_{-\infty}^{\infty} \exp(\pm i\tfrac{1}{2}au^2)du = (2\pi/a)^{\frac{1}{2}}\exp(\pm i\tfrac{1}{4}\pi) \quad a>0 \quad (85)$$

equation (84) becomes

$$J = \left[\frac{2\pi}{|f''(\lambda_0)|}\right]^{\frac{1}{2}}g(\lambda_0)\exp[if(\lambda_0)+i\tfrac{1}{4}\pi\,\mathrm{sign}f''(\lambda_0)] \quad (86)$$

Notice that the success of the method depends on the inequality (83).

If there are several stationary phase points, then there is a contribution of the type (86) from each of them.

It may happen there is no real solution to equation (81), but there is a solution if λ_0 is complex. In this case, the *saddle point method* may be applicable (see ref. (59)). In this method, the contour of integration is deformed to pass through the saddle point along a path of steepest descent, together with a quadratic approximation of $f(\lambda)$. The final result for J can be written

$$J \approx \left[\frac{2\pi i}{f''(\lambda_0)}\right]^{\frac{1}{2}}g(\lambda_0)\exp[if(\lambda_0)] \quad (87)$$

The stationary phase method can be regarded as a special case of the saddle point method in which the contour of integration is deformed to cross the real axis at $\pm 45^{\circ}$ (see ref. (59)).

We will now apply the above approximations to the integrals $I_m^{\pm}$ of section 5.3. The resulting formulae will be called *primitive semiclassical approximations*.

5.5. Primitive Semiclassical Approximations

The stationary phase condition for the integrals $I_m^{\pm}$ is

$$\frac{2d\delta(\lambda)}{d\lambda} = \mp\theta - 2m\pi \qquad m=0,\pm 1,\pm 2,\ldots. \tag{88}$$

or using the semiclassical equivalence relationship (66),

$$\Theta(\lambda) = \mp\theta - 2m\pi \qquad m=0,\pm 1,\pm 2,\ldots. \tag{89}$$

Consider now a deflection function of the type shown in figure 4. The only real values of λ that satisfy equation (89) are for m=0, so that

$$\Theta(\lambda) = \mp\theta \tag{90}$$

Figure 4 shows that for $\theta<\theta_r$, where θ_r is the rainbow angle, there are three real roots of equation (90), $\lambda=\lambda_{i\theta}$, with i=1,2,3 whereas for $\theta>\theta_r$, there is only one real root, namely $\lambda_{1\theta}$. Furthermore, $\lambda_{1\theta}$ contributes to I_0^- and $\lambda_{2\theta}$ and $\lambda_{3\theta}$ contribute to I_0^+.

When $\theta<\theta_r$, we can write

$$f(\theta) = f_1(\theta)+f_2(\theta)+f_3(\theta) \tag{91}$$

where the expressions for $f_i(\theta)$ are obtained by applying the stationary phase result (86) to $I_0^{\pm}$. We find

$$f_1(\theta) = -\frac{i}{k}\left[\frac{\lambda_{1\theta}}{\sin\theta\,|d\Theta(\lambda_{1\theta})/d\lambda|}\right]^{\frac{1}{2}}\exp[2i\delta(\lambda_{1\theta}) - i\lambda_{1\theta}\theta] \tag{92}$$

$$f_2(\theta) = -\frac{1}{k}\left[\frac{\lambda_{2\theta}}{\sin\theta\,|d\Theta(\lambda_{2\theta})/d\lambda|}\right]^{\frac{1}{2}}\exp[2i\delta(\lambda_{2\theta}) + i\lambda_{2\theta}\theta] \tag{93}$$

and

$$f_3(\theta) = -\frac{i}{k}\left[\frac{\lambda_{3\theta}}{\sin\theta\, d\Theta(\lambda_{3\theta})/d\lambda}\right]^{\frac{1}{2}}\exp[2i\delta(\lambda_{3\theta}) + i\lambda_{3\theta}\theta] \tag{94}$$

Note that we can write

$$f_i(\theta) = [I_i(\theta)]^{\frac{1}{2}}\exp[i\alpha_i(\theta)] \qquad i=1,2,3 \tag{95}$$

where $I_i(\theta)$ is the classical differential cross section and $\alpha_i(\theta)$ is a phase. When $I(\theta)$ is calculated from equation (91), oscillations result from interference between the three terms.

When $\theta>\theta_r$, it was noted above that the only real root of equation (90) is $\lambda_{1\theta}$. However, there are also complex roots, in particular, $\lambda_{2\theta}$ and $\lambda_{3\theta}$. If the saddle point result is applied to these complex roots it is found they give an exponentially decreasing contribution to $I(\theta)$.

As $\theta\to\theta_r$, equations (93) and (94) start to break down because $\Theta'(\lambda)\to 0$. In fact for $\theta=\theta_r$, we have $\Theta'(\lambda_r)=0$ and they fail catastrophically because condition (83) has been violated. We will now discuss the *transitional Airy approximation* which is valid for values of θ close to θ_r.

5.6. Transitional Airy Approximation

In order to derive the transitional Airy approximation, we return to the general integral (80), which is now written as

$$J(\theta) = \int_a^b g(\lambda)\exp[if(\theta;\lambda)]d\lambda \tag{96}$$

to indicate explicitly that the exponent depends on the parameter θ. For values of (θ,λ) close to (θ_r,λ_r), we can Taylor expand $f(\theta;\lambda)$ in the form

$$f(\theta;\lambda) = f_r+f_r'(\lambda-\lambda_r)+\tfrac{1}{6}f_r'''(\lambda-\lambda_r)^3 \tag{97}$$

where an abbreviated notation has been used

$$f_r = f(\theta;\lambda_r) \quad f_r' = f'(\theta;\lambda_r)$$

and so on. On replacing $g(\lambda)$ by $g(\lambda_r)$ the integral J becomes

$$J(\theta) \approx g_r\exp(if_r)\int_a^b \exp[if_r'(\lambda-\lambda_r)+i\tfrac{1}{6}f_r'''(\lambda-\lambda_r)^3]\,d\lambda \tag{98}$$

Changing the limits from (a,b) to $(-\infty,+\infty)$ and making the change of variable

$$v = (f_r'''/2)^{1/3}(\lambda-\lambda_r) \tag{99}$$

allows equation (98) to be written in terms of the regular Airy

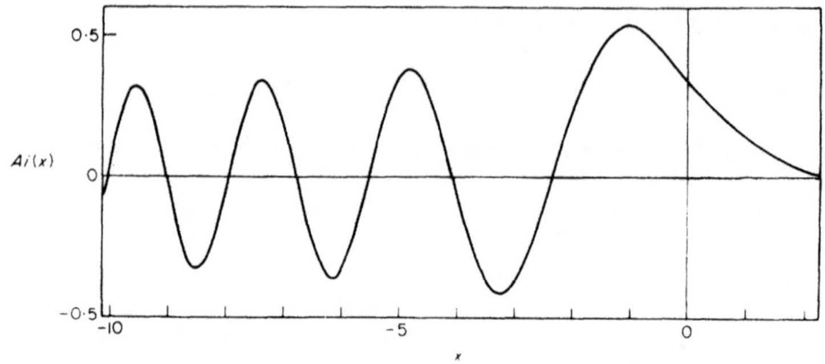

Figure 9. The regular Airy function Ai(x)

function Ai(x). We find

$$J(\theta) = 2\pi g_r(2/f_r''')^{1/3}\exp(if_r)\mathrm{Ai}[f_r'(2/f_r''')^{1/3}] \qquad (100)$$

A graph of the regular Airy function

$$\mathrm{Ai}(x) = (2\pi)^{-1}\int_{-\infty}^{\infty}\exp[i(u^3/3+xu)]du \qquad (101)$$

is shown in figure 9.

Applying equation (100) to I_0^+ gives for the rainbow contribution, $f_r(\theta)$, to the scattering amplitude

$$f_r(\theta) = \frac{(2\pi\lambda_r)^{\frac{1}{2}}}{k(\sin\theta)^{\frac{1}{2}}}\, q_r^{-1/3}\exp\{i[2\delta(\lambda_r)+\lambda_r\theta-\tfrac{3}{4}\pi]\}\mathrm{Ai}[(\theta-\theta_r)q_r^{-1/3}] \qquad (102)$$

where

$$q_r = \tfrac{1}{2}\Theta''(\lambda_r) \qquad (103)$$

Equation (102) requires information only at (θ_r,λ_r) for its evaluation, and was first derived by Ford and Wheeler (60) (note however a phase error of $-\pi/2$ in their formula).

We now have two different approximate formulae for I_0^+. The primitive semiclassical approximations (93) and (94) which are valid for θ not close to θ_r, and the transitional Airy approximation (102) which is valid for θ close to θ_r. Unfortunately, these two approximations do not join together smoothly as θ varies, that is, they are not uniform in θ. We will now derive the *uniform Airy approximation* which is uniformly valid as θ varies.

5.7. Uniform Airy Approximation

In the integral (96), there are two stationary phase or saddle points $\lambda_2(\theta)$ and $\lambda_3(\theta)$, whose position depends on the parameter θ. As θ varies, the stationary phase points move around until for $\theta=\theta_r$, they coalesce. The following method for the uniform asymptotic evaluation of the integral (96) is due to Chester, Friedman and Ursell (61) (see also ref. (62,63)). We follow the development of the theory presented in ref. (64).

The first step is to introduce a new integration variable $y=y(\theta;\lambda)$ and new parameters $\xi(\theta)$ and $A(\theta)$ so that

$$f(\theta;\lambda) = \tfrac{1}{3}y^3-\xi y+A \tag{104}$$

This transformation is a *locally exact uniformly analytic* mapping that is valid in the neighbourhood of the two coalescing stationary phase points. The integral (96) becomes

$$J(\theta) = \exp(iA)\int_{a(y)}^{b(y)} g[\lambda(y)]\frac{d\lambda}{dy}\exp[i(\tfrac{1}{3}y^3-\xi y)]\,dy \tag{105}$$

We require $d\lambda/dy$ to be finite which implies that the transformation should also be *one-to-one* as well as uniformly analytic. This is achieved by requiring the stationary points on either side of (104) to correspond. We have

$$\begin{aligned} \lambda &= \lambda_2 \leftrightarrow y_2 = -\xi^{\frac{1}{2}} \\ \lambda &= \lambda_3 \leftrightarrow y_3 = +\xi^{\frac{1}{2}} \end{aligned} \tag{106}$$

Substituting these values back into equation (104) and taking the sum and difference allows ξ and A to be determined. We find that

$$A = \tfrac{1}{2}(f_2+f_3) \tag{107}$$

and

$$\xi = [3(f_2-f_3)/4]^{2/3} \tag{108}$$

The branch of ξ is chosen so that $\xi>0$ on the bright side of the rainbow where $\theta<\theta_r$ and $\xi<0$ on the dark side where $\theta>\theta_r$.

The second step is to expand the pre-exponential factor in equation (105) in the form (see ref. (61))

$$g\frac{d\lambda}{dy} = p+qy+\sum_{m=1}^{\infty}(p_m+q_m y)(y^2-\xi)^m \qquad (109)$$

The last term in (109) is zero at a stationary point. Since the main contribution to J comes from the stationary points, it is reasonable to suppose the contribution from this term is small. We therefore make the approximation

$$g\frac{d\lambda}{dy} = p+qy \qquad (110)$$

If the integration limits in equation (105) are extended to $(-\infty, +\infty)$, the resulting approximation for J can be written in terms of Ai(x) and its derivative Ai'(x). We find

$$J(\theta) \simeq 2\pi p\exp(iA)Ai(-\xi)-2\pi iq\exp(iA)Ai'(-\xi) \qquad (111)$$

The next step is to evaluate p and q. Inserting the saddle point equation (106) into (110) and taking the sum and difference gives

$$p = \tfrac{1}{2}\left[g_3\left(\frac{d\lambda}{dy}\right)_3+g_2\left(\frac{d\lambda}{dy}\right)_2\right] \qquad (112)$$

and

$$q = \tfrac{1}{2}\xi^{-\frac{1}{2}}\left[g_3\left(\frac{d\lambda}{dy}\right)_3-g_2\left(\frac{d\lambda}{dy}\right)_2\right] \qquad (113)$$

In order to obtain the derivatives in equations (112) and (113) we next differentiate the transformation equation (104) twice. This gives

$$f''\left[\frac{d\lambda}{dy}\right]^2+f'\frac{d^2\lambda}{dy^2} = 2y \qquad (114)$$

Inserting the saddle point conditions (106) into equation (114) gives

$$\left[\frac{d\lambda}{dy}\right]_2 = \left[\frac{-2\xi^{\frac{1}{2}}}{f''_2}\right]^{\frac{1}{2}} \qquad (115)$$

$$\left[\frac{d\lambda}{dy}\right]_3 = \left[\frac{2\xi^{\frac{1}{2}}}{f''_3}\right]^{\frac{1}{2}} \qquad (116)$$

Combining together equations (111)-(116) we finally obtain

$$J(\theta) \approx 2^{\frac{1}{2}}\pi\exp(iA)\{\left[\frac{g_3}{(f_3'')^{\frac{1}{2}}}+\frac{g_2}{(-f_2'')^{\frac{1}{2}}}\right]\xi^{1/4}Ai(-)$$

$$-i\left[\frac{g_3}{(f_3'')^{\frac{1}{2}}} - \frac{g_2}{(-f_2'')^{\frac{1}{2}}}\right]\xi^{-1/4}Ai'(-\xi)\} \qquad (117)$$

where A and ξ are given by equations (107) and (108) respectively. Equation (117) is the uniform asymptotic approximation to the integral (96) for the case of two coalescing saddle points. This remarkable formula contains as limiting cases the three formulae discussed in sections 5.4 and 5.6. Thus when $\xi \gg 0$, equation (117) goes over to the stationary phase result, as can be verified from the asymptotic formulae

$$Ai(-\xi) \sim \pi^{-\frac{1}{2}}\xi^{-1/4}\sin(\tfrac{2}{3}\xi^{3/2}+\tfrac{1}{4}\pi) \qquad (118)$$

$$Ai'(-\xi) \sim -\pi^{-\frac{1}{2}}\xi^{1/4}\cos(\tfrac{2}{3}\xi^{3/2}+\tfrac{1}{4}\pi) \qquad (119)$$

When $\xi \ll 0$, the saddle point result (87) is obtained and when $\xi \simeq 0$, equation (117) reduces to the transitional Airy approximation (100).

Applying equation (117) to the integral I_0^+ gives for the rainbow scattering amplitude

$$f_r(\theta) = \pi^{\frac{1}{2}}\exp[i(A-\tfrac{3}{4}\pi)]\{[I_2(\theta)^{\frac{1}{2}}+I_3(\theta)^{\frac{1}{2}}]\xi^{1/4}Ai(-\xi)$$

$$+i[I_2(\theta)^{\frac{1}{2}}-I_3(\theta)^{\frac{1}{2}}]\xi^{-1/4}Ai'(-\xi)\} \qquad (120)$$

where $I_2(\theta)$ and $I_3(\theta)$ are classical differential cross sections and

$$A = \tfrac{1}{2}[2\delta(\lambda_{2\theta})+2\delta(\lambda_{3\theta})+(\lambda_{2\theta}+\lambda_{3\theta})\theta] \qquad (121)$$

$$\xi = (\tfrac{3}{4}\{2[\delta(\lambda_{2\theta})-\delta(\lambda_{3\theta})]+(\lambda_{2\theta}-\lambda_{3\theta})\theta\})^{2/3} \qquad (122)$$

Equation (120) was first obtained by Berry (65) (note however an error in his formula). The expression for the differential cross section is

$$I_r(\theta) = \pi[I_2(\theta)^{\frac{1}{2}}+I_3(\theta)^{\frac{1}{2}}]^2\xi^{\frac{1}{2}}Ai^2(-\xi)$$

$$+\pi[I_2(\theta)^{\frac{1}{2}}-I_3(\theta)^{\frac{1}{2}}]^2\xi^{-\frac{1}{2}}Ai'^2(-\xi) \qquad (123)$$

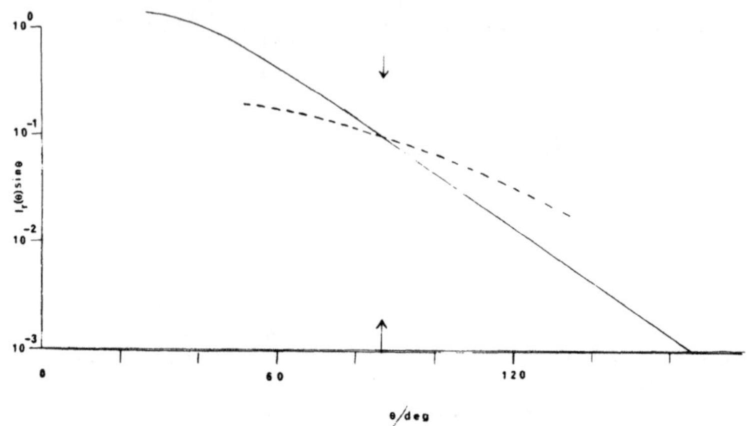

Figure 10. Plot of the rainbow contribution to the differential cross section for a Lennard-Jones (12,6) potential with reduced parameters of K=1.59 and A=14.1. The solid line is the uniform Airy approximation and the dashed line is the transitional Airy approximation. The arrows indicate the position of the classical rainbow angle.

5.8. Results

Figure (10) compares the transitional Airy approximation (102) with the uniform Airy approximation (123). On the dark side of the rainbow, it is necessary to analytically continue equation (120) for $f_r(\theta)$ using complex valued classical trajectories. It can be seen that the transitional approximation only agrees with the uniform approximation over a very small angular range.

The full differential cross section $I(\theta)$ is obtained from

$$f(\theta) = f_1(\theta)+f_r(\theta) \tag{124}$$

where $f_1(\theta)$ is given by equation (92). The rainbow and supernumerary rainbow oscillations in figure 5 arise from $f_r(\theta)$, whereas the high frequency oscillations arise from the interference between $f_1(\theta)$ and $f_r(\theta)$. As we progress from figure 5 to figure 8, the width of the rainbow oscillation becomes very broad (as in figure 10) and only the interference between $f_1(\theta)$ and $f_r(\theta)$ is visible in $I(\theta)$.

As θ moves beyond θ_r, the term $|f_1(\theta)|$ becomes larger than $|f_r(\theta)|$, which explains the damping of the oscillations beyond

θ_r (although in cases like figure 8 the damping is rather slow). This fact can be exploited in practical applications of equation (124). Thus if the transitional Airy approximation (102) is used for $f_r(\theta)$ for angles $\theta \gtrsim \theta_r$ and the uniform Airy approximation (120) for $\theta < \theta_r$, it is found the resulting $I(\theta)$ is practically indistinguishable from the partial wave results of figures 5-8 (see ref. (54,66)), except at small angles where the theory presented above breaks down. Note that for $\theta > \theta_r$, use of the transitional Airy approximation is computationally much easier than the uniform approximation, because the latter approximation requires complex valued trajectories on the dark side of the rainbow.

5.9. Cusped Rainbows

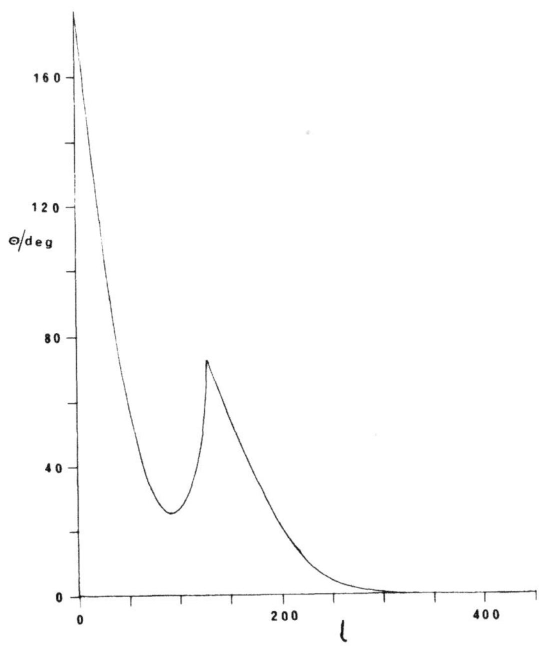

Figure 11. Deflection function curve showing a maximum and a minimum (taken from ref. (69)).

The semiclassical rainbow theory described in sections 5.6 and 5.7 has been for the case where the deflection function possesses a single minimum (or maximum). However, in some problems, such as those involving a curve crossing (see ref. (67,68)), a deflection function possessing both a maximum and a minimum can

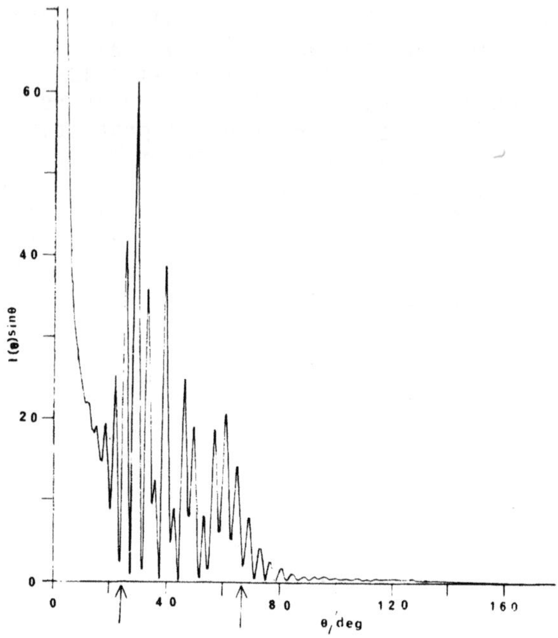

Figure 12. Differential cross section for the curve crossing problem whose deflection function is shown in figure 11 (taken from ref. (69)). The arrows indicate the positions of the classical rainbow angles.

occur. A typical example is shown in figure 11 (69). The differential cross section associated with this deflection function is shown in figure 12 (69).

The uniform Airy analysis of section 5.7 can be applied, for example, around the minimum of the deflection function curve, but it then fails in the neighbourhood of the maximum of $\Theta(\ell)$. Moreover, the maximum and minimum may coalesce to form a point of inflection and again the preceding theory is inapplicable.

A uniform semiclassical analysis for the case of three contributing real or complex trajectories has been developed in ref. (70). It is based on the rigorous asymptotic theory of Ursell (71). For the deflection function shown in figure 11, the function f in the integral (96) now depends on two parameters and possesses three coalescing saddle points.

Instead of the transformation (104) the semiclassical theory makes use of the mapping

$$f(\alpha,\beta;\lambda) = t^4+yt^2+xt+A \tag{125}$$

and the *canonical integral* is no longer the regular Airy function but is *Pearcey's integral* (72)

$$P(x,y) = \int_{-\infty}^{\infty} \exp[i(xt+yt^2+t^4)]dt \tag{126}$$

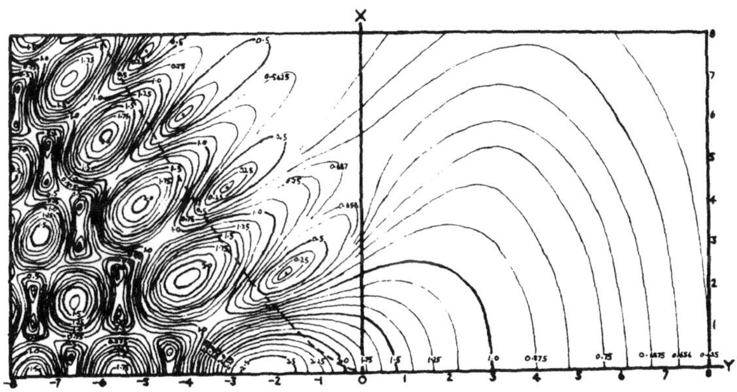

Figure 13. Contour plot of $|P(x,y)|$. Since $P(-x,y)= P(x,y)$ it is only necessary to plot the upper half plane. The caustic where the corresponding classical problem is singular is shown as a dashed line (taken from ref. (70,72)).

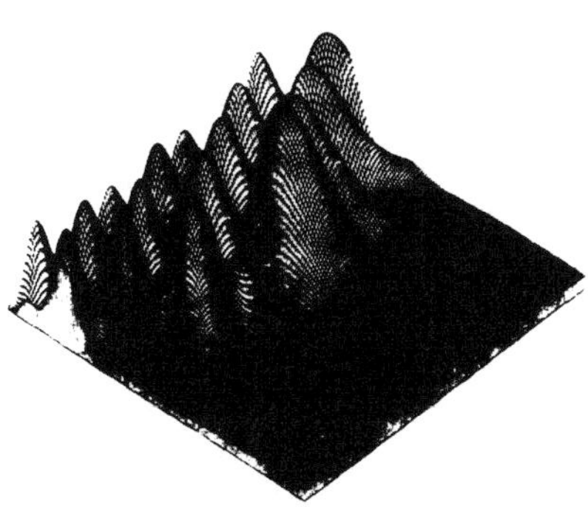

Figure 14. Three dimensional view of $|P(x,y)|$ (taken from ref. (73)).

Plots of $|P(x,y)|$ are shown in figures 13 and 14. The uniform Pearcey theory has been applied to the deflection function shown in figure 11 (69). The results for $I(\theta)$ are practically indistinguishable from those obtained from the partial wave series (69).

In the corresponding classical problem, $I(\theta)$ is singular on a cusped curve rather than just at a point. Finally, we note that cusped rainbows are easy to observe as pointed out by Larmor (74) in 1891: "The cusped caustic bands are easily seen when a distant street lamp is viewed through a spectacle lens with minute raindrops deposited on it."

5.10. Catastrophes

The uniform Airy semiclassical analysis of section 5.7 and the uniform Pearcey semiclassical analysis of section 5.9 can also be discussed from the point of view of Catastrophe Theory. The exact local transformations (104) and (125), which played a key role in the analysis, are examples of the fold and cusp unfoldings of elementary Catastrophe Theory (75).

Catastrophe Theory is a branch of differential topology concerned with the singularities of smooth mappings. It can be used to classify the stable unfoldings of a function that depends on four or fewer parameters (75). The first five unfoldings of elementary Catastrophe Theory are given in Table 3.

These unfoldings define the canonical integrals which, together with their derivatives, give rise to the uniform semiclassical approximations. In addition, they define the singularities or caustics which occur in the corresponding classical theory. Thus, Catastrophe Theory can be used to build a general semiclassical theory in a systematic way. It should be noted, however, that all the canonical forms in Table 3, with the exception of the elliptic umbilic were actually studied in a semiclassical context without the help of Catastrophe Theory (see ref. (71,76-79)). Further information on semiclassical mechanics and Catastrophe Theory can be found in ref. (80-82).

An important case not covered by Catastrophe Theory concerns glory scattering. In figure 4, this occurs when ℓ_1 and ℓ_2 coalesce. The classical differential cross section (28) is then infinite because $\sin\theta=0$. A uniform semiclassical approximation can be derived in which the regular Bessel function of order zero is the canonical integral (see ref. (83,84)). More general Bessel uniform approximations are important in the semiclassical theory of inelastic collisions (85-87).

Name	Unfolding
fold	$x^3+\xi_1 x+A$
cusp	$x^4+\xi_2 x^2+\xi_1 x+A$
swallowtail	$x^5+\xi_3 x^3+\xi_2 x^2+\xi_1 x+A$
hyperbolic umbilic	$x^3+y^3+\xi_3 xy+\xi_2 y+\xi_1 x+A$
elliptic umbilic	$x^3-3xy^2+\xi_3(x^2+y^2)+\xi_2 y+\xi_1 x+A$

Table 3. The first five unfoldings of elementary Catastrophe Theory.

6. ORBITING COLLISIONS AND RESONANCES

The semiclassical theory presented in sections 3-5 has always assumed that the semiclassical phase shift (62) is valid i.e. we are dealing with the single turning point case in figure 3. In this section, we shall derive the phase shift when there are three turning points and discuss the effect of resonances on the total cross section.

6.1. Turning Points

The effective potential $V_\ell(r)$ in the three turning point case is shown in figure 15. When E lies below the top of the barrier but above the bottom of the well, there are three real turning points. As E increases, the turning points c_ℓ and e_ℓ coalesce at the top of the barrier. They then move into the complex r plane where they become complex conjugates of each other. In the classical theory, as E approaches the top of the barrier, the particles orbit around each other for longer periods of time; i.e. we have $\Theta \to -\infty$.

6.2. Phase Shift

There have been a number of derivations of the semiclassical phase shift for three turning points, which are uniformly valid as E crosses the maximum of the effective potential (see ref. (88-94) for example). We shall follow the derivation of ref. (92).

In the region between the turning points a_ℓ and c_ℓ, the semiclassical wavefunction (61) can be written

$$\psi_\ell(r) \sim [k/k_\ell(r)]^{\frac{1}{2}}\cos[\int_{a_\ell}^{r} k_\ell(r')dr' - \tfrac{1}{4}\pi] \qquad (127)$$

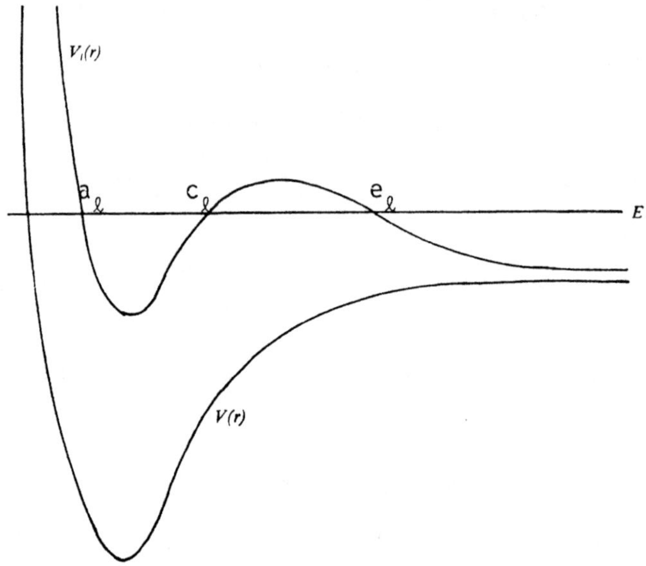

Figure 15. The effective potential energy curve for the three turning point problem.

We next write the phase integral in equation (127) in terms of c_ℓ rather than a_ℓ. We find

$$\psi_\ell(r) \sim \tfrac{1}{2}[k/k_\ell(r)]^{\frac{1}{2}}\{\exp[i\alpha_\ell - i\int_r^{c_\ell} k_\ell(r')dr' - i\tfrac{1}{4}\pi]$$

$$+\exp[-i\alpha_\ell + i\int_r^{c_\ell} k_\ell(r')dr' + i\tfrac{1}{4}\pi]\} \tag{128}$$

where

$$\alpha_\ell = \int_{a_\ell}^{c_\ell} k_\ell(r)dr \tag{129}$$

Next, suppose that in the neighbourhood of d_ℓ, the position of the maximum, $V_\ell(r)$ can be expanded in the form

$$V_\ell(r) = V_\ell(d_\ell) - \tfrac{1}{2}\mu\omega_\ell^{*2}(r-d_\ell)^2+\ldots.. \tag{130}$$

where

$$\omega_\ell^* \neq 0 \tag{131}$$

is the classical angular vibration frequency in the inverted barrier. We can again use the comparison equation method (9,43) to locally map $\psi_\ell(r)$ in a uniform one-to-one manner onto the solution for a parabolic barrier. For the present problem the comparison equation is *Weber's equation* (95)

$$\frac{d^2\psi}{dx^2} + \left(\varepsilon + \frac{x^2}{4}\right)\psi = 0 \tag{132}$$

with solutions in terms of Weber parabolic cylinder functions. From the asymptotic form of the parabolic cylinder function, the following *parabolic connection formula* can be deduced: the semiclassical wavefunction valid for $a_\ell \ll r \ll c_\ell$, namely

$$[k_\ell(r)]^{-\frac{1}{2}}\exp[\pm i\int_r^{c_\ell} k_\ell(r')dr' \pm i\tfrac{1}{4}\pi] \tag{133}$$

connects with the semiclassical wavefunction

$$[k_\ell(r)]^{-\frac{1}{2}}\exp[\pm i\int_{e_\ell}^{r} k_\ell(r')dr' \mp i\tfrac{1}{4}\pi]\exp(-\pi\varepsilon_\ell) \tag{134}$$

$$+[k_\ell(r)]^{-\frac{1}{2}}\exp[\mp i\int_{e_\ell}^{r} k_\ell(r')dr' \pm i\tfrac{1}{4}\pi]\exp[\pm i\phi(\varepsilon_\ell)][1+\exp(-2\pi\varepsilon_\ell)]^{\frac{1}{2}}$$

which is valid for $r \gg e_\ell$. In (134),

$$\phi(\varepsilon_\ell) = \varepsilon_\ell + \arg\Gamma(\tfrac{1}{2}+i\varepsilon_\ell) - \varepsilon_\ell \ln|\varepsilon_\ell| \tag{135}$$

$$= -\phi(-\varepsilon_\ell)$$

and

$$\pi\varepsilon_\ell = -i\int_{c_\ell}^{e_\ell} k_\ell(r)dr \tag{136}$$

The branch of $k_\ell(r)$ in equation (136) is chosen so that $\varepsilon>0$ for $E>V_\ell(d_\ell)$ and $\varepsilon<0$ for $E<V_\ell(d_\ell)$. If the barrier is exactly parabolic, the integral in equation (136) can be evaluated explicitly to give

$$\varepsilon_\ell = \frac{E-V_\ell(d_\ell)}{\hbar\omega_\ell^*} \tag{137}$$

The function $\phi(\varepsilon_\ell)$ acts as a phase correction factor. Table 4 gives numerical values of $\phi(\varepsilon_\ell)$. For larger values of ε_ℓ, the asymptotic result

ε	$\phi(\varepsilon)$	ε	$\phi(\varepsilon)$
0.00	0.000	0.70	0.068
0.05	0.102	0.80	0.058
0.10	0.137	0.90	0.051
0.15	0.149	1.00	0.045
0.20	0.150	1.50	0.029
0.25	0.144	2.00	0.021
0.30	0.135	2.50	0.017
0.40	0.115	3.00	0.014
0.50	0.096	4.00	0.011
0.60	0.080	5.00	0.008

Table 4. Values of the phase correction factor $\phi(\varepsilon)$.

$$\phi(\varepsilon_\ell) \sim \frac{1}{24\varepsilon_\ell} + \frac{7}{2880\varepsilon_\ell^3} + \ldots\ldots \tag{138}$$

can be used. Note that ϕ has a maximum of $\simeq 0.15$ at $\varepsilon_\ell \simeq 0.2$, and that the slope of $\phi(\varepsilon_\ell)$ is infinite at $\varepsilon_\ell = 0$.

The parabolic connection formulae given above are for the case when a_ℓ, c_ℓ and e_ℓ are real. However they are also valid when E is above the barrier maximum. The turning points c_ℓ and e_ℓ in the action integral (136) are then complex conjugates. A convenient phase reference point in (129), (133) and (134) is where the Stokes' line from c_ℓ to e_ℓ crosses the real axis.

Applying the parabolic connection formulae to the wavefunction (128) gives for the semiclassical wavefunction valid for $r \gg e_\ell$

$$\psi_\ell(r) \sim \tfrac{1}{2}[k/k_\ell(r)]^{\frac{1}{2}}(\{\exp[i(\alpha_\ell - \phi_\ell)][1+\exp(-2\pi\varepsilon_\ell)]^{\frac{1}{2}} + \exp[-i(\alpha_\ell + \pi\varepsilon_\ell)]\}\exp[i\int_{e_\ell}^{r} k_\ell(r')dr' - i\tfrac{1}{4}\pi] + \text{c.c.}) \tag{139}$$

where c.c. denotes complex conjugate.

Writing equation (139) in the form of a sine wave and comparing with equation (40) allows δ_ℓ to be determined. We find

$$\delta_\ell(E) = \delta_\ell^{(0)}(E) + \delta_\ell^{(r)}(E) \tag{140}$$

where

$$\delta_\ell^{(0)}(E) = \lim_{r\to\infty} [\int_{e_\ell}^{r} k_\ell(r')dr' - kr] + \tfrac{1}{2}(\ell+\tfrac{1}{2})\pi - \tfrac{1}{2}\phi(\varepsilon_\ell) \tag{141}$$

and

$$\delta_\ell^{(r)}(E) = \tan^{-1}\left(\left[\frac{[1+\exp(2\pi\varepsilon_\ell)]^{\frac{1}{2}}-1}{[1+\exp(2\pi\varepsilon_\ell)]^{\frac{1}{2}}+1}\right]\tan[\alpha_\ell - \tfrac{1}{2}\phi(\varepsilon_\ell)]\right) \tag{142}$$

Equations (140)-(142) are the semiclassical phase shift for the three turning point problem.

6.3. Limiting Cases

We will now consider two limiting cases of the three turning point phase shift (140).

When $\varepsilon_\ell \gg 0$, so that E is well above the barrier maximum, the term (142) yields a contribution α_ℓ to the phase shift and equation (140) goes over to the single turning point result (62).

On the other hand when $\varepsilon_\ell \ll 0$, so that E is well below the barrier maximum, the contribution from the term (142) is very small, except for energies close to

$$\alpha_\ell(E_{n\ell}) - \tfrac{1}{2}\phi[\varepsilon_\ell(E_{n\ell})] = (n+\tfrac{1}{2})\pi \quad n=0,1,2,\ldots \tag{143}$$

when $\delta_\ell^{(r)}(E)$ rapidly increases by π. We then have a resonance. We will now show how equations (140)-(142) can be reduced to a Breit-Wigner resonance form.

6.4. Resonances

Consider the case when E is well below the top of the barrier. For energies close to a resonance energy we can write

$$E = E_{n\ell} + \delta E \tag{144}$$

where $E_{n\ell}$ satisfies equation (143). In the phase shift (142), we can expand $\alpha_\ell(E)$ to first order obtaining

$$\alpha_\ell(E) = \alpha_\ell(E_{n\ell}) + \frac{\pi(\delta E)}{\hbar\omega_{n\ell}} \tag{145}$$

where $\omega_{n\ell}$ is classical angular frequency of vibration in the potential well

$$\omega_{n\ell} = \pi/(\mu\int_{a_\ell}^{c_\ell}\{2\mu[E_{n\ell}-V_\ell(r)]\}^{-\frac{1}{2}}dr) \tag{146}$$

Using equations (143) and (144) and neglecting the energy dependence of ϕ, the tan term in equation (142) can be approximated as

$$\tan[\alpha_\ell-\tfrac{1}{2}\phi(\varepsilon_\ell)] \approx \frac{\hbar\omega_{n\ell}}{\pi(\delta E)} \tag{147}$$

We can then write the phase shift in the Breit-Wigner resonance form

$$\delta_\ell(E) = \delta_\ell^{(0)}(E) + \tan^{-1}\left[\frac{\Gamma_{n\ell}}{2(E_{n\ell}-E)}\right] \tag{148}$$

where the width $\Gamma_{n\ell}$ is

$$\Gamma_{n\ell} = \frac{2\hbar\omega_{n\ell}}{\pi}\;\frac{[1+\exp(2\pi\varepsilon_{n\ell})]^{\frac{1}{2}}-1}{[1+\exp(2\pi\varepsilon_{n\ell})]^{\frac{1}{2}}+1} \tag{149}$$

Making use of the approximation

$$\ell n\ x \approx 2(x-1)(x+1)^{-1} \tag{150}$$

equation (149) can also be written

$$\Gamma_{n\ell} = \frac{\hbar\omega_{n\ell}}{2\pi}\ \ell n[1+\exp(2\pi\varepsilon_{n\ell})] \tag{151}$$

It is important to note that the approximations (149) and (151) are not valid close to the top of the barrier because we have neglected the energy dependence of ϕ. When $\varepsilon_{n\ell} \ll 0$, equation (151) simplifies to

$$\Gamma_{n\ell} = \frac{\hbar\omega_{n\ell}}{2\pi}\exp(-2\pi|\varepsilon_{n\ell}|) \tag{152}$$

a result first obtained by Gamow.

Numerical applications of the above formulae can be found in ref. (23,90,93,96-101). The discrepancies reported in ref. (101)

arise in part because equations (149)-(152) are not valid close to top of the barrier (23) or because semiclassical approximations are used for a different set of boundary conditions from those of equations (39) and (40).

The theory presented above has been for the case where c_ℓ and e_ℓ can coalesce and move into the complex r plane. However, figure 15 shows that another possibility is for a_ℓ and c_ℓ to coalesce. The uniform semiclassical theory for this case is described in ref. (92).

6.5. Total Cross Section

The total cross section is given in terms of the phase shifts by equation (21). For a phase shift of the Breit-Wigner form (148), we have

$$\sin^2\delta_\ell = \frac{(E_{n\ell}-E)^2}{(E_{n\ell}-E)^2+(\Gamma_{n\ell}/2)^2}\sin^2\delta_\ell^{(0)}$$

$$+\frac{(\Gamma_{n\ell}/2)^2}{(E_{n\ell}-E)^2+(\Gamma_{n\ell}/2)^2}\cos^2\delta_\ell^{(0)}$$

$$+\frac{(\Gamma_{n\ell}/2)(E_{n\ell}-E)}{(E_{n\ell}-E)^2+(\Gamma_{n\ell}/2)^2}\sin 2\delta_\ell^{(0)} \qquad (153)$$

Equation (153) shows explicitly that a resonance occurs at $E_{n\ell}$ and that the shape of the resonance depends on the value of the background phase $\delta_\ell^{(0)}$. When $\delta_\ell^{(0)}=n\pi$ a large positive spike is present, whereas for $\delta_\ell^{(0)} = (n+\frac{1}{2})\pi$ there is a large negative spike. For other values of $\delta_\ell^{(0)}$ there are both positive and negative contributions. The experimental detection of these orbiting resonances in the total cross section is reported in ref. (102) for example. The contribution of resonances to the differential cross section is discussed in ref. (103,104).

7. REGGE POLE REPRESENTATION OF f(θ)

In the previous sections, the partial wave series representation and Poisson sum representation of f(θ) have been derived and discussed. In this section, a third exact representation of f(θ) will be introduced, namely the *Regge pole representation*. It will be shown that semiclassical techniques are a useful way of exploiting the properties of the Regge pole representation.

7.1. Watson Transformation

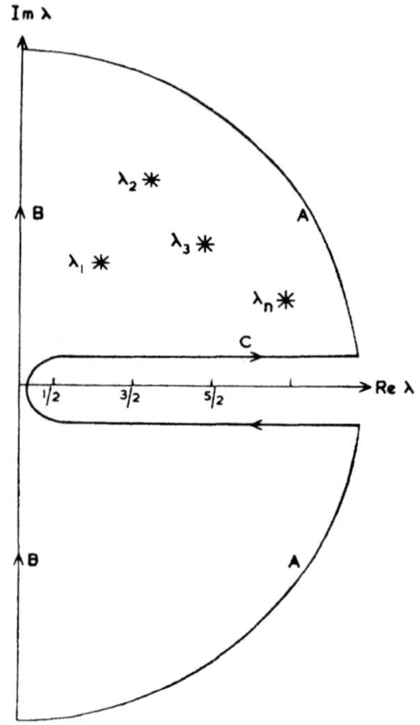

Figure 16. Contours in the complex λ plane for the Watson transformation.

The Regge pole representation makes use of the Watson transformation to convert an infinite series into a contour integral (see ref. (56)). The Watson transformation exploits the fact that $\pi/\cos(\pi\lambda)$ has simple poles at $\lambda = \ell+\frac{1}{2} = \frac{1}{2},\frac{3}{2},\ldots$ with residues of $(-1)^{\ell+1}$. It then follows that

$$\sum_{\ell=0}^{\infty}(-1)^{\ell}f(\ell+\tfrac{1}{2}) = \frac{1}{2i}\int_{C}d\lambda f(\lambda)/\cos(\pi\lambda) \tag{154}$$

where the contour C is shown in figure 16.

Applying the formula (154) to the partial wave series (18) and assuming that $S(\ell)$ is an analytic function of ℓ we obtain

$$f(\theta) = -\frac{1}{2k}\int_C \frac{\lambda[S(\lambda)-1]P_{\lambda-\frac{1}{2}}(-\cos\theta)}{\cos(\pi\lambda)}d\lambda \tag{155}$$

The partial wave series (18) is regained from equation (155) by application of the residue theorem together with the relation

$$(-1)^{\ell}P_{\ell}(\cos\theta) = P_{\ell}(-\cos\theta) \qquad \ell=0,1,2,\dots. \tag{156}$$

In equation (155), $P_{\lambda-\frac{1}{2}}(-\cos\theta)$ is a Legendre function.

7.2. Regge Pole Representation.

The next step is to deform the contour C away from the real axis. Assuming the integrand of equation (155) is well behaved for $|\lambda|\to\infty$, the contour C can be deformed into the contour B of figure 16 which lies along (or in some cases, close to) the imaginary λ axis. In the process of doing this, contributions from the poles of $S(\lambda)$ in the right hand λ plane are picked up. For singular potentials of the Lennard-Jones type, it is known there are an infinite number of poles in the first quadrant, but the fourth quadrant is free of poles (see ref. (105)).

The scattering amplitude can now be written

$$f(\theta) = f_B(\theta) + f_P(\theta) \tag{157}$$

where

$$f_B(\theta) = -\frac{1}{2k}\int_B \frac{\lambda[S(\lambda)-1]P_{\lambda-\frac{1}{2}}(-\cos\theta)}{\cos(\pi\lambda)}d\lambda \tag{158}$$

and

$$f_P(\theta) = -\frac{\pi i}{k}\sum_n \frac{\lambda_n r_n}{\cos(\pi\lambda_n)} P_{\lambda_n-\frac{1}{2}}(-\cos\theta) \tag{159}$$

In equation (159), λ_n is the position of the nth pole and r_n is the corresponding residue. $f_B(\theta)$ is the "background" integral and $f_P(\theta)$ is the pole sum. Equations (157)-(159) are the Regge pole representation of the scattering amplitude.

7.3. Physical Interpretation

Each term in the pole sum can be given a physical interpretation involving *surface waves* (see ref. (35)). This follows from the asymptotic approximation (76) for $P_{\lambda-\frac{1}{2}}(-\cos\theta)$, namely

$$P_{\lambda_n-\frac{1}{2}}(-\cos\theta) \sim (2\pi\lambda_n\sin\theta)^{-\frac{1}{2}}\{\exp[-\lambda_i(\pi-\theta)]\exp[i\lambda_r(\pi-\theta)-i\tfrac{1}{4}\pi]$$

$$+ \exp[\lambda_i(\pi-\theta)]\exp[-i\lambda_r(\pi-\theta)+i\tfrac{1}{4}\pi]\} \quad (160)$$

where we have written

$$\lambda_n = \lambda_r + i\lambda_i \quad (161)$$

The approximation (160) is valid for $|\lambda_n|$ large and θ not close to 0 or π. The two terms in (160) represent surface waves traveling in opposite directions. They decay exponentially as they propagate and their *angular life* is $1/\text{Im}\lambda_n$. When $\text{Im}\lambda_n \ll 1$, the surface waves undergo many revolutions before they decay, corresponding to an orbiting type of collision. On the other hand, when $\text{Im}\lambda_n \gg 1$ the surface waves decay rapidly. A schematic illustration of the surface waves propagating around the core of a potential is shown in figure 17.

What physical interpretation can be attached to the background integral? The analysis presented in Section 5 suggests that the surface waves are in some sense associated with the stationary phase trajectories $\lambda_{2\theta}$ and $\lambda_{3\theta}$. The background integral would then be associated with the direct scattering from the repulsive core of the potential, which is represented by the stationary phase trajectory $\lambda_{1\theta}$. A rigorous analysis shows that this physical argument is correct (see ref. (106)). We can therefore make the approximation

$$f_B(\theta) \approx f_1(\theta) \quad (162)$$

The directly reflected trajectory is also shown in figure 17.

In order to apply equations (157)-(159), we must calculate λ_n, r_n and the Legendre function as well as approximate the background integral. The semiclassical calculation of λ_n and r_n is described in the next section. An extensive discussion of methods for calculating Legendre functions of complex degree can be found in ref. (107).

7.4. Semiclassical Calculation of Regge Pole Positions and Residues

The positions ℓ_n of the Regge poles are determined by the radial Schrödinger equation (38) with boundary conditions

$$\psi(\ell_n;0) = 0 \quad (163)$$

and

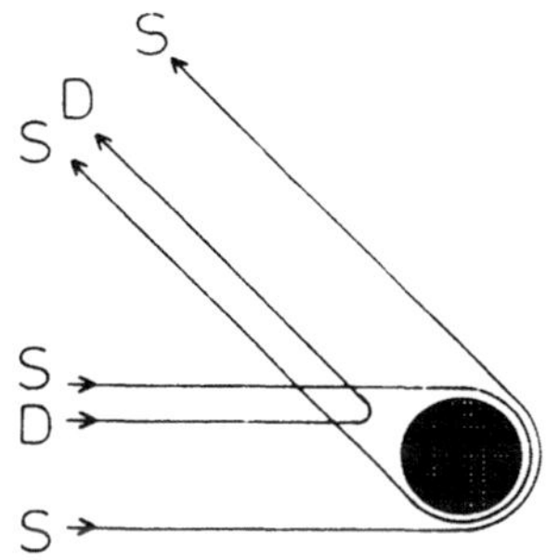

Figure 17. Schematic representation of surface wave (S) and direct (D) elastic scattering. The hatched circle represents the repulsive core of the potential or the region of chemical reaction (absorption).

$$\psi(\ell_n;r) \underset{r\to\infty}{\sim} \exp(ikr) \tag{164}$$

Equation (164) represents an *outgoing wave only* boundary condition. For values of ℓ close to ℓ_n, the S matrix has the form

$$S(\ell) = r_n/(\ell-\ell_n) \tag{165}$$

which determines the residue r_n.

In order to obtain a semiclassical solution of the radial Schrödinger equation with boundary conditions (163) and (164), we again start by examining the turning point distribution. Figures 18a and 18c show the turning points in the complex r plane for the orbiting case of section 6, whilst figures 18b and 18d show them for the outgoing wave only boundary condition (164). In the latter case, all three turning points are complex, but there are still three regions I, II and III where the semiclassical wavefunction is valid.

In region II, the semiclassical solution (128) is valid. In order to connect uniformly across c_ℓ and e_ℓ into region III, the following parabolic connection formulae are required which are analytic continuations of equations (133)-(136) (see ref. (108, 109)).

In region II, the semiclassical wavefunction

$$[k_\ell(r)]^{-\frac{1}{2}}\exp[\pm i\int_r^{c_\ell} k_\ell(r')dr' \pm i\tfrac{1}{4}\pi] \tag{166}$$

connects with the following semiclassical wavefunction in region III

$$[k_\ell(r)]^{-\frac{1}{2}}\exp[\pm i\int_{e_\ell}^{r} k_\ell(r')\,dr' \mp i\tfrac{1}{4}\pi]\exp(-\pi\varepsilon_\ell)$$
$$+[k_\ell(r)]^{-\frac{1}{2}}A^{\pm}(\varepsilon_\ell)\exp[\mp i\int_{e_\ell}^{r} k_\ell(r')dr' \pm i\tfrac{1}{4}\pi] \tag{167}$$

where

$$A^{\pm}(\varepsilon_\ell) = \frac{(2\pi)^{\frac{1}{2}}}{\Gamma(\frac{1}{2}\mp i\varepsilon_\ell)}\exp(-\tfrac{1}{2}\pi\varepsilon_\ell)\exp\{\pm i[\varepsilon_\ell-\varepsilon_\ell \ln(-\varepsilon_\ell)]\} \tag{168}$$

and ε_ℓ is again given by equation (136). When ε_ℓ is real, so that the identity

$$|\Gamma(\tfrac{1}{2}+i\varepsilon_\ell)| = [\pi/\cosh(\pi\varepsilon_\ell)]^{\frac{1}{2}} \qquad \varepsilon_\ell \text{ real} \tag{169}$$

can be employed, we regain the connection formulae (133)-(135) used in section 6.

Applying the connection formulae (166) and (167) to the wavefunction (128), we find the semiclassical wavefunction valid in region III is

$$\tfrac{1}{2}[k/k_\ell(r)]^{\frac{1}{2}}\{[\exp(i\alpha)A^{-}(\varepsilon)+\exp(-i\alpha-\pi\varepsilon)]\exp[i\int_{e_\ell}^{r} k_\ell(r')dr'-i\tfrac{1}{4}\pi]$$
$$+[\exp(-i\alpha)A^{+}(\varepsilon)+\exp(i\alpha-\pi\varepsilon)]\exp[-i\int_{e_\ell}^{r} k_\ell(r')dr'+i\tfrac{1}{4}\pi]\} \tag{170}$$

where α is again given by equation (129).

Applying the outgoing wave boundary condition (164) to the wavefunction (170) gives

$$\exp(-i\alpha)A^{+}(\varepsilon)+\exp(i\alpha-\pi\varepsilon) = 0 \tag{171}$$

which can be rearranged to

$$\Phi(\ell_n) = (n+\tfrac{1}{2})\pi \qquad n=0,1,2,\ldots. \tag{172}$$

where

$$\Phi(\ell_n) = \alpha-\tfrac{1}{2}[\varepsilon-\varepsilon\ln(-\varepsilon)] + i\tfrac{1}{2}\ln\left[\frac{(2\pi)^{\frac{1}{2}}\exp(\frac{1}{2}\pi\varepsilon)}{\Gamma(\frac{1}{2}-i\varepsilon)}\right] \tag{173}$$

and

$$\alpha = \alpha(\ell_n) \qquad\qquad \varepsilon = \varepsilon(\ell_n) \tag{174}$$

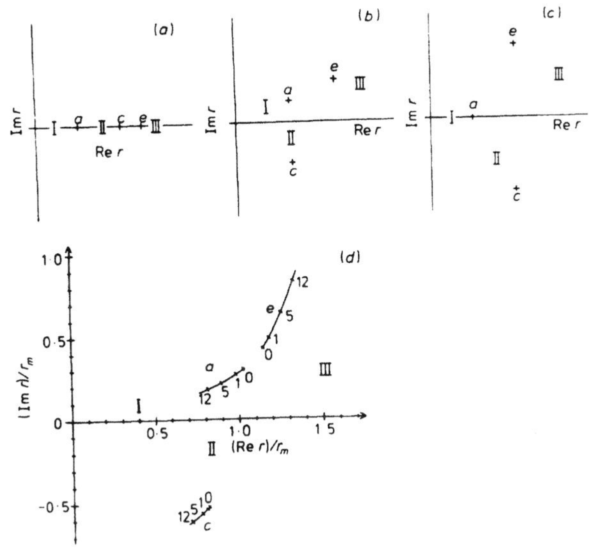

Figure 18. Turning point distributions in the complex r plane. (a) and (c) are for the boundary condition (40), whereas (b) and (d) are for the outgoing wave only boundary condition (164). (d) indicates the turning points for the Regge poles n=0,1,5 and 12. The Roman numerals I, II and III indicate the regions in which the semiclassical approximation to the wavefunction is valid (taken from ref. (52)).

Equation (172) is a generalized Bohr-Sommerfeld quantization formula for the Regge pole positions.

The residues can also be obtained from equations (165) and (170). It is found that (see ref. (52))

$$r_n = i\left[\frac{2d\Phi(\ell_n)}{d\ell}\right]^{-1}\exp\{2[\pi\varepsilon(\ell_n)-i\alpha(\ell_n)+i\delta(\ell_n)]\} \qquad (175)$$

where

$$\delta(\ell) = \lim_{r\to\infty}\left[\int_{e_\ell}^{r} k_\ell(r')dr'-kr\right]+\tfrac{1}{2}(\ell+\tfrac{1}{2})\pi \qquad (176)$$

7.5. Limiting Cases

Some limiting cases of equations (172) and (175) are interesting. When the turning points c_ℓ and e_ℓ in figure 18 are well separated from one another, Stirling's approximation can be applied to the complex gamma function in equation (173):

$$\ell n\Gamma(\tfrac{1}{2}+z) \sim \tfrac{1}{2}\ell n2\pi+z\ell nz-z \qquad (177)$$

The quantization formula (172) simplifies to

$$\gamma(\ell_n) = (n+\tfrac{1}{2})\pi \qquad n=0,1,2,.... \qquad (178)$$

where

$$\gamma(\ell) = \int_{a_\ell}^{e_\ell} k_\ell(r)dr \qquad (179)$$

and the formula for the residue becomes (see ref. (52))

$$r_n = -i\left[\frac{2d\gamma(\ell_n)}{d\ell}\right]^{-1}\exp[2i\delta(\ell_n)] \qquad (180)$$

Equation (178) is a simple Bohr-Sommerfeld quantization formula for the Regge pole positions.

Tables 4 and 5 compare semiclassical ℓ_n and r_n from equations (172) and (175) with exact quantum results (110,111, see also ref. (112,113)). The semiclassical ℓ_n are in excellent agreement with the quantum ℓ_n. The semiclassical $|r_n|$ are in error by about 6% for n=0, but their accuracy rapidly improves as n increases. Very similar results are obtained using equations (178) and (180) instead of equations (172) and (175).

The formalism developed above has been for complex ℓ and real E but essentially the same equations apply to the case of real ℓ and complex E. The outgoing wave only boundary condition (164) then gives rise to complex energy eigenvalues, which can

	Quantum		Semiclassical	
n	Re ℓ_n	Im ℓ_n	Re ℓ_n	Im ℓ_n
0	97.519	12.398	97.520	12.398
1	96.726	14.025	96.727	14.025
2	95.948	15.687	95.948	15.687
3	95.187	17.384	95.188	17.384
4	94.447	19.114	94.448	19.114
5	93.730	20.877	93.731	20.877

Table 4. Comparison of exact quantum and semiclassical Regge pole positions for a Lennard-Jones (6,4) potential with reduced parameters of K=1.149 and A=63.64 (taken from ref. (52)).

	Quantum		Semiclassical	
n	$\|r_n\|$	arg r_n	$\|r_n\|$	arg r_n
0	0.144 (9)	1.546π	0.136 (9)	1.578π
1	0.287(10)	0.762π	0.283(10)	0.794π
2	0.263(11)	-0.016π	0.264(11)	0.017π
3	0.151(12)	1.214π	0.151(12)	1.247π
4	0.592(12)	0.458π	0.598(12)	0.490π
5	0.174(13)	1.715π	0.176(13)	1.746π

Table 5. Same as table 4 except for the residues. The numbers in parentheses indicate the power of 10 by which the entry must be multiplied (taken from ref. (52)).

be written

$$E = E_{n\ell} - i\tfrac{1}{2}\Gamma_{n\ell} \tag{181}$$

When the width $\Gamma_{n\ell}$ is small, the quantization formula (173) can be simplified. Let us write it in the form

$$\alpha(E) = (n+\tfrac{1}{2})\pi + \tfrac{1}{2}[\varepsilon - \varepsilon\ell n(-\varepsilon)] - i\tfrac{1}{2}\ell n\left[\frac{(2\pi)^{\frac{1}{2}}\exp(\frac{1}{2}\pi\varepsilon)}{\Gamma(\frac{1}{2}-i\varepsilon)}\right] \tag{182}$$

On the right hand side of equation (182), replace E by $E_{n\ell}$, so that $\varepsilon(E_{n\ell})$ is a real quantity. The identity (169) can then be used to simplify (182). We find

$$\alpha(E) = (n+\tfrac{1}{2})\pi + \tfrac{1}{2}\phi(\varepsilon_{n\ell}) - i\tfrac{1}{4}\ell n[1+\exp(2\pi\varepsilon_{n\ell})] \tag{183}$$

Expanding $\alpha(E)$ to first order gives (compare with equation (145))

$$\alpha(E) = \alpha(E_{n\ell}) - \frac{\pi\Gamma_{n\ell}}{2\hbar\omega_{n\ell}} \tag{184}$$

where $\omega_{n\ell}$ is given by equation (146). On comparing real and imaginary parts of equations (183) and (184) we obtain equations (143) and (151) of section 6. Note that the approximations (183) and (184) are only valid for $E_{n\ell}$ well below the barrier maximum because we have neglected the energy dependence of ϕ.

7.6. Elastic Scattering from Lennard-Jones Potentials

Figures 5-8 show differential cross sections calculated from the Regge representation (157). The ℓ_n and r_n are calculated by

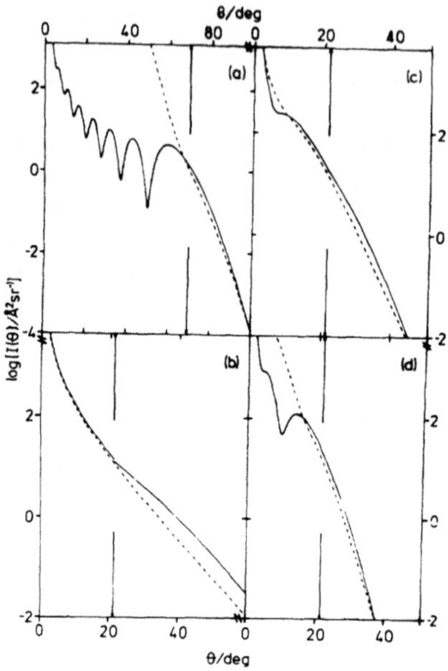

Figure 19. Plot of I(θ) against θ for the semiclassical Airy approximation of equations (102) and (120) (solid line ——), and Regge theory with several poles retained in the pole sum (dashed line ---). (a) Lennard-Jones (6,4) potential of figure 5, (b) Lennard-Jones (18,6) potential of figure 8, (c) Lennard-Jones (18,6) potential of figure 7, (d) Lennard-Jones (18,6) potential of figure 6 (taken from ref. (54)).

the semiclassical techniques described in section 7.4 and the approximation (162) is used for the background integral.

When widely spaced oscillations are present in I(θ), the Regge pole closest to the real ℓ axis is sufficient to characterize the scattering (see figure 8). As rainbow maxima become more prominent, the angular range for the validity of the one pole approximation decreases, until only the oscillations for $\theta > \theta_r$ are reproduced. However, the range of validity can be improved by including more pole terms, which successfully reproduces the oscillations down to $\theta \simeq \theta_r$.

The Regge pole formalism does not reproduce the supernumerary rainbows though. This appears to be related to the fact that the $|r_n|$ are very large when supernumerary rainbows are present (see table 5). Since considerable cancellation of terms is expected in the pole sum (159), this implies that the $|r_n|$ must be known to more significant figures than we are able to accurately compute. This point is also apparent in figure 19 which compares the pole sum with the uniform Airy approximation for $\theta<\theta_r$ and the transitional Airy approximation for $\theta\gtrsim\theta_r$.

In figures 5-8 and 19, the Im ℓ_n are large, even for the leading n=0 pole. Consequently only a single surface wave in figure 17 contributes to the scattering. The surface wave that goes the long way round the core of the potential is damped out and does not contribute.

7.7. Elastic Scattering in Chemically Reactive Systems

When reactive channels are open, the phase shift becomes a complex quantity (see equations (24) and (25)). The elastic $S(\ell)$ matrix element can then be written

$$S(\ell) = \exp[2i\delta(\ell)] \tag{185}$$

$$= \exp[2i\eta(\ell)]\exp[-2\xi(\ell)] \tag{186}$$

By choosing suitable parametrized forms for $\eta(\ell)$ and $\xi(\ell)$, the effect of chemical reaction on the elastic scattering can be studied.

In the calculations described below, $\eta(\ell)$ is taken to be the phase shift for scattering from a Lennard-Jones (12,6) potential and for $\exp[-2\xi(\ell)]$ we use (see ref. (114))

$$\exp[-2\xi(\ell)] = 1 - \frac{P_0}{1+\exp[(\ell-L)/d]} \tag{187}$$

Equation (187) has the form of a smooth step function in which d is the "diffuseness" parameter, L controls the "cut off" value of ℓ and P_0 determines the amount of absorption at low partial waves.

Figure 20 shows $I(\theta)$ calculated from the partial wave series (18) with parameters that approximate the elastic scattering of K by HBr. Up to 15000 partial waves are required to get converged values of $I(\theta)$. It is immediately apparent there is a rich interference structure in the large angle scattering. Similar interference effects occur for other parametrizations of the S matrix that are of the smooth step function type (see ref. (115)).

The parametrization (187) has an infinite number of poles located at

$$\ell_n = L \pm in\pi d \qquad n=1,3,5,\ldots \tag{188}$$

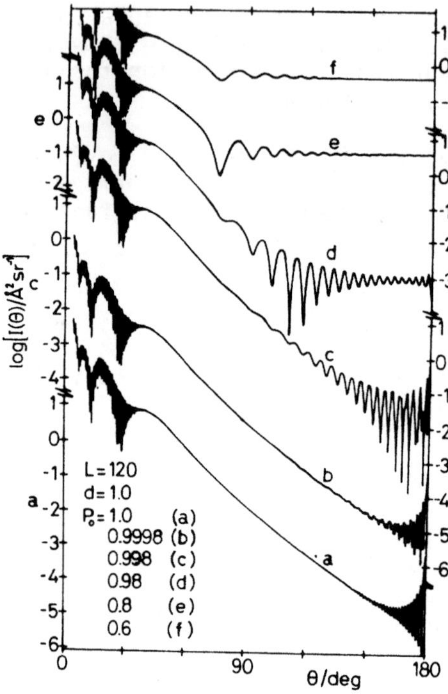

Figure 20. Interference effects in the large angle elastic scattering of chemically reactive systems (taken from ref. (114)).

with residues of

$$r_n = P_0 d \exp[2i\eta(\ell_n)] \qquad n=1,3,5,.... \tag{189}$$

However to account for the interference effects in figure 20 only the two poles closest to the real ℓ axis are required. A physical interpretation of the large angle oscillations can be given in terms of the surface wave and directly reflected trajectories in figure 17. When P_0=1, the contribution from the direct term is negligible and the interference effects arise from the surface waves propagating around the region of chemical reaction. As P_0 decreases, the direct term becomes more important and the interference effects move to smaller angles. Figure 21 compares the surface wave and direct contributions to $I(\theta)$ for the six cases in figure 20. It is remarkable that only three terms are required to explain quantitatively the interference effects when the Regge pole representation is employed, compared with up to 15000 terms for the partial wave series. Further discussion of large angle

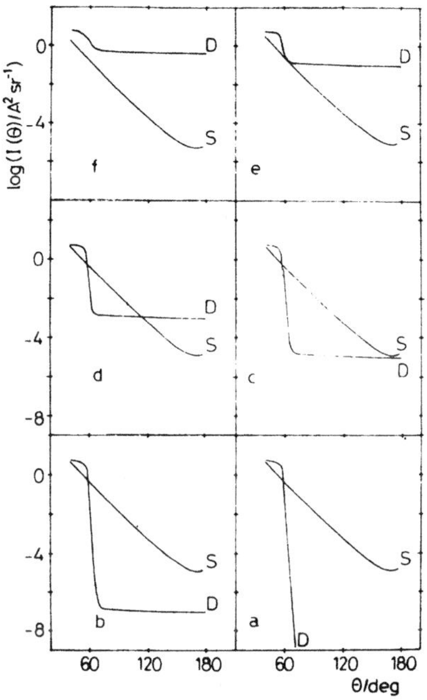

Figure 21. Contribution of surface wave (S) and direct (D) elastic scattering to I(θ) for the six cases of figure 20. (a) P_0=1.0 (b) P_0=0.9998 (c) P_0=0.998 (d) P_0=0.98 (e) P_0=0.8 (f) P_0=0.6, Oscillations associated with the surface wave scattering at backward angles have been averaged over for clarity of presentation (taken from ref. (114)).

interference effects in the elastic scattering of chemically reactive systems can be found in ref. (114-117).

ACKNOWLEDGEMENTS

I am grateful to Mrs E. Morris for her rapid and accurate typing of the manuscript. I also thank the Science Research Council for its research support.

REFERENCES

1. Mott, N.F., and Massey, H.S.W.: 1965, *The Theory of Atomic Collisions*, 3rd edition (Clarendon Press, Oxford) Chap. 5.
2. Pauly, H, and Toennies, J.P.: 1965, Adv.Atom.Molec.Phys. 1, pp. 195-344.

3. Bernstein, R.B.: 1966, Adv.Chem.Phys. 10, pp. 75-134.
4. Bernstein, R.B., and Muckerman, J.T.: 1967, Adv.Chem.Phys. 12, pp. 389-486.
5. McDowell, M.R.C., and Coleman, J.P.: 1970, *Introduction to the Theory of Ion-Atom Collisions* (North-Holland, Amsterdam) Chap. 1 and 2.
6. Beck, D.: 1970, Proc.Int.School Phys. Enrico Fermi, 44, pp.15-49.
7. Massey, H.S.W., Burhop, E.H.S., and Gilbody, H.B.: 1971, *Electronic and Ionic Impact Phenomena* 2nd edition, Vol. 3 *Slow Collisions of Heavy Particles* by H.S.W. Massey (Clarendon Press, Oxford) Chap. 16.
8. Massey, Sir H.: 1971, Contemp.Phys. 12, pp. 537-558.
9. Berry, M.V., and Mount, K.E.: 1972, Rep.Prog.Phys. 35, pp.315-397.
10. Weise, H.P.: 1973, Ber.Bunsenges.Physik.Chem. 77, pp.578-593.
11. Buck, U.: 1974, Rev.Mod.Phys. 46, pp. 369-389.
12. Toennies, J.P.: 1974, *Physical Chemistry - An Advanced Treatise,* Vol VIA, *Kinetics of Gas Reactions,* edited by W. Jost (Academic, New York) Chap. 5.
13. Child, M.S.: 1974, *Molecular Collision Theory* (Academic, New York).
14. Ding, A.: 1975, *Interaction between Ions and Molecules*, edited by P. Ausloos (Plenum, New York) pp. 33-73.
15. Buck, U.: 1975, Adv.Chem.Phys. 30, pp. 313-388.
16. Pauly, H.: 1975, *Physical Chemistry - An Advanced Treatise*, Vol. VIB, *Kinetics of Gas Reactions,* edited by W. Jost (Academic, New York) Chap. 8.
17. Child, M.S.: 1978, Adv.Atom.Molec.Phys. 14, pp. 225-280.
18. Miller, W.H. : 1971, Accounts Chem.Res. 4, pp. 161-167.
19. Marcus, R.A.: 1973, Faraday Disc.Chem.Soc. 55, pp. 34-44.
20. Miller, W.H.: 1973, *The Physics of Electronic and Atomic Collisions, Invited Lectures and Progress Reports,* 8th ICPEAC, edited by B.C. Čobič and M.V. Kurepa (Institute of Physics, Beograd) pp. 503-528.
21. Miller, W.H.: 1974, Adv.Chem.Phys. 25, pp.69-177.
22. Miller, W.H.: 1975, Adv.Chem.Phys. 30, pp.77-136.
23. Connor, J.N.L.: 1976, Chem.Soc.Rev. 5, pp. 125-148.
24. Child, M.S.: 1976, *Dynamics of Molecular Collisions* Part B, edited by W.H. Miller (Plenum, New York) Chap. 4.
25. Miller, W.H.: 1976, *The New World of Quantum Chemistry*, Proc. 2nd Intern.Cong. of Quantum Chemistry, New Orleans, U.S.A., 19-24 April 1976, edited by B. Pullman and R. Parr (Reidel, Dordrecht) pp. 87-102.
26. Connor, J.N.L.: 1979, Comput.Phys.Comm. 17, pp. 117-144.
27. Fröman, N., and Fröman, P.O.: 1965, *JWKB Approximation: Contributions to the Theory* (North Holland, Amsterdam).
28. Olver, F.W.J.: 1974, *Asymptotics and Special Functions* (Academic, New York).
29. Fröman, N., and Mrazek, W.: 1977, J.Phys.A: Math.Gen. 10, pp. 1287-1295.

30. Bransden, B.H.: 1970, *Atomic Collision Theory* (Benjamin, New York) Chap. 1.
31. Hasted, J.B.: 1972, *Physics of Atomic Collisions* 2nd edition (Butterworth, London) Chap. 2.
32. Farina, J.E.G.: 1973, *Quantum Theory of Scattering Processes* (Pergamon, Oxford) Chap. 1.
33. Joachain, C.J.: 1975, *Quantum Collision Theory* (North-Holland, Amsterdam) Part II.
34. Newton, R.G.: 1964, *The Complex j plane* (Benjamin, New York).
35. De Alfaro, V., and Regge, T.: 1965, *Potential Scattering* (North-Holland, Amsterdam).
36. Abramowitz, M., and Stegun, I.A.: 1964, *Handbook of Mathematical Functions* (Dover, New York) Chap. 10.
37. Goldstein, H.: 1950, *Classical Mechanics* (Addison-Wesley, Reading, Mass.) Chap. 3.
38. Smith, F.J.: 1964, Physica, 30, pp. 497-504.
39. Ioup, G.E., and Thomas, B.S.: 1969, J.Chem.Phys. 50, pp. 5009-5015.
40. Boyle, J.F.: 1977, Physica C, 92, pp. 281-289.
41. Firsov, O.B.: 1953, Zh.Eksp.Teor.Fiz. 24, pp. 279-283.
42. Landau, L.D., and Lifshitz, E.M.: 1969, *Mechanics* 2nd edition (Pergamon, Oxford) Chap. IV.
43. Miller, S.C., and Good, R.H.: 1953, Phys.Rev. 91, pp. 174-179.
44. Langer, R.E.: 1937, Phys.Rev. 51, pp. 669-676.
45. Frank, W.M., Land, D.J., and Spector, R.M.: 1971, Rev.Mod. Phys. 43, pp. 36-98.
46. Kennedy, M., and Smith, F.J.: 1967, Molec.Phys. 13, pp.443-448.
47. O'Hara, H., and Smith, F.J.: 1968, Comput.J. 11, pp. 213-219.
48. O'Hara, H., and Smith, F.J.: 1970, J.Comput.Phys. 5, pp. 328-344.
49. O'Hara, H., and Smith, F.J.: 1971, Comput.Phys.Comm. 2, pp. 47-54.
50. Pack, R.T.: 1974, J.Chem.Phys. 60, pp. 633-639.
51. Cohen, J.S.: 1978, J.Chem.Phys. 68, pp.1841-1843.
52. Connor, J.N.L., Jakubetz, W., and Sukumar, C.V.: 1976, J.Phys.B: Atom.Molec.Phys. 9, pp. 1783-1799.
53. Connor, J.N.L., and Mackay, D.C., to be published.
54. Connor, J.N.L., and Jakubetz, W.: 1978, Molec.Phys. 35, pp. 949-964.
55. Morse, P.M., and Feshbach, H.: 1953, *Methods of Theoretical Physics* (McGraw-Hill, New York) p. 467.
56. Boas, R.P., and Stutz, C.: 1971, Am.J.Phys. 39, pp. 745-753.
57. Reference 36, p. 336.
58. Landau, L.D., and Lifshitz, E.M.: 1965, *Quantum Mechanics* second edition (Pergamon, Oxford) p. 474.
59. Carrier, G.F., Krook, M., and Pearson, C.E.: 1966, *Functions of a Complex Variable: Theory and Technique* (McGraw-Hill, New York) Chap. 6.
60. Ford, K.W., and Wheeler, J.A.: 1959, Ann.Phys. 7, pp. 259-322.

61. Chester, C., Friedman, B., and Ursell, F.: 1957, Proc.Camb. Phil.Soc. 53, pp. 599-611.
62. Friedman, B., 1959, J.Soc.Ind.Appl.Math.: 7, pp. 280-289.
63. Ursell, F.: 1965, Proc.Camb.Phil.Soc. 61, pp. 113-128.
64. Connor, J.N.L., and Marcus, R.A.: 1971, J.Chem.Phys. 55, pp. 5636-5643.
65. Berry, M.V.: 1966, Proc.Phys.Soc. 89, pp. 479-490.
66. Mullen, J.M., and Thomas, B.S.: 1973, J.Chem.Phys. 58, pp. 5216-5221.
67. Delos, J.B.: 1974, Phys.Rev.A 9, pp. 1626-1634.
68. Saxon, R.B., and Olson, R.E.: 1975, Phys.Rev.A 12, pp. 830-834.
69. Connor, J.N.L., and Farrelly, D.: to be published.
70. Connor, J.N.L.: 1973, Molec.Phys.26, pp. 1217-1231.
71. Ursell, F.: 1972, Proc.Camb.Phil.Soc. 72, pp. 49-65.
72. Pearcey, T.: 1946, Phil.Mag. 37, pp. 311-317.
73. Trinkhaus, H., and Drepper, F.: 1977, J.Phys.A: Math.Gen. Phys. 10, pp. L11-L16.
74. Larmor, J.: 1891, Proc.Camb.Phil.Soc. 7, pp. 131-137.
75. Thom, R.: 1975, *Structural Stability and Morphogenesis* (Benjamin, Reading, Mass).
76. Connor, J.N.L.: 1973, Faraday Dis.Chem.Soc. 55, pp. 51-58.
77. Connor, J.N.L.: 1973, Molec.Phys. 26, pp. 1371-1377.
78. Connor, J.N.L.: 1974, Molec.Phys. 27, pp. 853-866.
79. Ursell, F.: 1979, "Integrals with a large parameter. A double integral in two complex variables with four nearly coincident saddle points", preprint.
80. Connor, J.N.L.: 1976, Molec.Phys. 31, pp. 33-55.
81. Berry, M.V.: 1976, Adv.Phys. 25, pp. 1-26.
82. Poston, T., and Stewart, I.: 1978, *Catastrophe Theory and its Applications* (Pitman, London) Chap. 12.
83. Berry, M.V.: 1969, J. Phys. B: Atom.Molec.Phys. 2, pp. 381-392.
84. Mount, K.E., 1973, J.Phys.B.: Atom.Molec.Phys. 6, pp. 1397-1411.
85. Stine, J., and Marcus, R.A.: 1973, J.Chem.Phys. 59, pp. 5145-5150.
86. Connor, J.N.L., and Mayne, H.R.: 1979, Molec.Phys. 37, pp. 1-13.
87. Connor, J.N.L., and Mayne, H.R.: 1979, Molec.Phys. 37, pp. 15-25.
88. Ford, K.W., Hill, D.L. Wakano, M., and Wheeler, J.C.: 1959, Ann.Phys. 7, pp. 239-258.
89. Dubrovskii, G.V.: 1964, Opt.Spekt. 17, pp. 771-775. (English translation: 1964, Opt.Spect. 17, pp. 416-418.)
90. Herm, R.R.: 1967, J.Chem.Phys. 47, pp. 4290-4296.
91. Miller, W.H.: 1968, J.Chem.Phys. 48, pp. 1651-1658.
92. Connor, J.N.L.: 1968, Molec.Phys. 15, pp. 621-631.
93. Dickinson, A.S.: 1970, Molec.Phys. 18, pp. 441-449.
94. Drukarev, G., Fröman, N., and Fröman, P.O.: 1979, J.Phys.A: Math.Gen.Phys. 12, pp. 171-186.

95. Miller, J.C.P.: 1955, *Tables of Weber Parabolic Cylinder Functions* (HMSO: London).
96. Le Roy, R.J., and Bernstein, R.B.: 1971, J.Chem.Phys. 54, pp. 5114-5126.
97. Bain, R.A.,and Bardsley, J.N.: 1971, J.Chem.Phys. 55, pp. 4535-4538.
98. Child, M.S.: 1974, *Molecular Spectroscopy - A Specialist Periodical Report* edited by R.F. Barrow, D.A. Long and D.J. Millen (The Chemical Society, London) Vol.2, Chap. 7.
99. Price, R.I.: 1977, Molec.Phys. 33, pp. 559-571.
100. Davis, J.P., and Thorson, W.R.: 1978, Can.J.Phys. 56, pp. 996-1020.
101. Le Roy, R.J., and Liu, W.K.: 1978, J.Chem.Phys. 69, pp. 3622-3631.
102. Toennies, J.P., Welz, W., and Wolf. G.: 1976, J.Chem.Phys. 64, pp. 5305-5307.
103. Sheen, S.H., Skofronick, J.G., and Mueller, C.R.: 1975, Int.J.Quant.Chem. 9, pp. 817-824.
104. Düren, R., Hoppe, H.O., and Tischer, H.: 1979, Chem.Phys. Lett. 64, pp. 357-359.
105. Dombey, N., and Jones, R.H.: 1968, J.Math.Phys. 9, pp. 986-995.
106. Connor, J.N.L., and Mackay, D.C.: 1978, Chem.Phys.Lett. 59, pp.163-167.
107. Connor, J.N.L., and Mackay, D.C.: 1979, Molec.Phys. 37. pp. 1703-1712.
108. Connor, J.N.L.: 1973, Molec.Phys. 25, pp. 1469-1473.
109. Crothers, D.S.F.: 1976, J. Phys. B.: Atom.Molec.Phys. 9, pp. L513-L517.
110. Sukumar, C.V., and Bardsley, J.N.: 1975, J.Phys.B:Atom. Molec.Phys. 8, pp. 568-576.
111. Sukumar, C.V., Lin, S.L., and Bardsley, J.N.: 1975, J.Phys. B: Atom.Molec.Phys. 8, pp. 577-587.
112. Delos, J.B., and Carlson, C.E.: 1975, Phys.Rev.A, 11, pp. 210-220.
113. Bosanac, S.: 1978, J.Math.Phys. 19, pp. 789-797.
114. Connor, J.N.L., and Jakubetz, W.: 1977, Molec.Phys. 33, pp. 1619-1640.
115. Connor, J.N.L., and Mackay, D.C.: 1979, Chem. Phys. 40, pp. 11-18.
116. Connor, J.N.L.: 1975, Molec.Phys. 29, pp. 745-759.
117. Connor, J.N.L. and Jakubetz, W.: 1975, Chem.Phys.Lett. 36, pp. 29-33.

PROBLEMS

Problem 1

Use the partial wave series representation (18) to verify the *optical theorem*

$$\sigma = (4\pi/k)\,\mathrm{Im}f(\theta=0)$$

Problem 2

Use the partial wave series representation (18) to verify the *unitarity relation*

$$(4\pi/k)\,\mathrm{Im}f(\theta) = \int f^*(\theta')f(\psi)d\Omega'$$

where ψ is the angle between the directions $(\theta,0)$ and (θ',ϕ'). The following relations will be required

$$P_\ell(\cos\psi) = \frac{4\pi}{2\ell+1}\sum_{m=-\ell}^{\ell} Y^*_{\ell m}(\theta',\phi')Y_{\ell m}(\theta,\phi)$$

$$P_\ell(\cos\theta) = \left(\frac{4\pi}{2\ell+1}\right)^{\frac{1}{2}} Y_{\ell 0}(\theta)$$

and the orthonormality property of the spherical harmonics.

Problem 3

For scattering by a complex optical potential (22) show that the total elastic cross section is given by

$$\sigma = (\pi/k^2)\sum_{\ell=0}^{\infty}(2\ell+1)|1-S_\ell|^2$$

Problem 4

Instead of Taylor expanding the function $f(\lambda)$ in equation (82), show that the stationary phase result (86) can also be obtained by making an exact local transformation of the form

$$f(\lambda) = \pm u^2 + A$$

Problem 5

Verify that the uniform Airy approximation (117) reduces to the transitional Airy approximation (100) when λ_2 and λ_3 are close to λ_r.

Problem 6

Verify that the uniform Airy approximation (117) goes over to the stationary phase result when λ_2 and λ_3 are well separated from one another.

Problem 7

Show that another exact representation for $f(\theta)$ is

$$f(\theta) = \frac{1}{2k}\int_C \left[\frac{S(\lambda)P_{\lambda-\frac{1}{2}}(\cos\theta)\exp[-\pi i(\lambda+\frac{1}{2})]}{\cos(\pi\lambda)} + \frac{P_{\lambda-\frac{1}{2}}(-\cos\theta)}{\cos(\pi\lambda)}\right]\lambda d\lambda$$

where C is defined in figure 16.

Problem 8

Verify that the uniform approximations (172) and (175) for ℓ_n and r_n reduce to (178) and (180) when the turning points c_ℓ and e_ℓ are well separated from one another.

Problem 9

In two dimensions, using plane polar coordinates (r,ϕ), the asymptotic form of the wavefunction is

$$\Psi(r,\phi)\underset{r\to\infty}{\sim} e^{ikr\cos\phi}+f(\phi)\frac{e^{ikr}}{r^{\frac{1}{2}}}$$

Show that the differential cross length and total cross length are given by

$$I(\phi) = |f(\phi)|^2$$

$$\sigma(E) = \int_0^{2\pi} I(E,\phi)d\phi$$

Problem 10

In two dimensions, the radial Schrödinger equation is

$$\frac{d^2R_m(r)}{dr^2} + \{\frac{2\mu}{\hbar^2}[E-V(r)] - \frac{(m^2-\frac{1}{4})}{r^2}\}R_m(r) = 0$$

with

$$m = 0, \pm1, \pm2, \ldots.$$

The boundary conditions for $R_m(r)$ are

$$R_m(0) = 0$$

$$R_m(r) \underset{r\to\infty}{\sim} \cos[kr-\tfrac{1}{2}\pi(m+\tfrac{1}{2})+\delta_m]$$

where δ_m is the phase shift.

Using the two dimensional partial wave expansion of a plane wave

$$\exp(ikr\cos\phi) = \sum_{m=-\infty}^{+\infty} \exp[i(m\phi+\tfrac{1}{2}m\pi)]J_m(kr)$$

$$\underset{r\to\infty}{\sim} \left(\frac{2}{\pi kr}\right)^{\frac{1}{2}} \sum_{m=-\infty}^{\infty} \exp[i(m\phi+\tfrac{1}{2}m\pi)]\cos[kr-\tfrac{1}{2}\pi(m+\tfrac{1}{2})]$$

show that

$$\Psi(r,\phi) = \left(\frac{2}{\pi kr}\right)^{\frac{1}{2}} \sum_{m=-\infty}^{\infty} \exp[i(m\phi+\tfrac{1}{2}m\pi+\delta_m)]R_m(r)$$

$$f(\phi) = (2\pi ik)^{-\frac{1}{2}} \sum_{m=-\infty}^{\infty} \exp(im\phi)[\exp(2i\delta_m)-1]$$

$$\sigma(E) = (4/k) \sum_{m=-\infty}^{\infty} \sin^2\delta_m$$

Problem 11

By applying the Langer transformation to the radial Schrödinger equation for two dimensions, show that the semi-classical phase shift for a single turning point a_m is given by

$$\delta_m = \lim_{r\to\infty}\left\{\int_{a_m}^{r} \left[\frac{2\mu}{\hbar^2}[E-V(r')]-\frac{m^2}{r'^2}\right]^{\frac{1}{2}}dr'-kr\right\}+\tfrac{1}{2}m\pi$$

Problem 12

With the help of the formulae

$$\sum_{m=-J}^{+J} e^{im\phi} = \sin[(J+\tfrac{1}{2})\phi]/\sin(\tfrac{1}{2}\phi) \qquad \phi \neq 0$$

and

$$\lim_{J\to\infty} \sin[(J+\tfrac{1}{2})\phi]/\sin(\tfrac{1}{2}\phi) = 0 \quad \phi \neq 0$$

show that $f(\phi)$ for scattering in two dimensions can be written

$$f(\phi) = (2\pi ik)^{-\frac{1}{2}} \sum_{m=0}^{\infty}{}' \exp(2i\delta_m)[\exp(im\phi)+\exp(-im\phi)]$$

where the prime on the summation sign indicates a factor of $\frac{1}{2}$ when m=0.

By making approximations similar to those in section 5.3, show that $f(\phi)$ can be reduced to the classical result

$$I(\phi) = \sum_i \left|\frac{db_i}{d\Theta}\right|$$

where Θ is the classical deflection function.

APPLICATIONS OF BOHR QUANTIZATION IN DIATOMIC MOLECULE SPECTROSCOPY

Robert J. Le Roy

Guelph-Waterloo Centre for Graduate Work in Chemistry,
University of Waterloo, Waterloo, Ontario, Canada N2L 3G1

Use of the first- and higher-order JWKB quantization conditions for determining the diatomic molecule level **energies** and rotational constants implied by any given potential energy function is described. The widely used first-order RKR inversion procedure for determining the potential energy function implied by a given set of experimental vibrational energies and rotational constants is then derived and a procedure for including the effect of higher-order terms is outlined. Finally, simple analytic expressions for the limiting near-dissociation behaviour of vibrational energies and rotational constants are derived and their use for determining accurate molecular dissociation energies and long-range potential energy constants is described.

1. INTRODUCTION

Elsewhere in this volume, Fröman (1) has discussed the basis of the JWKB approximation and derived the first- and higher-order JWKB quantization conditions for the allowed eigenvalues of a one-dimensional potential. The present chapter will describe certain applications of these expressions in diatomic molecule spectroscopy, paying special attention to the problem of determining the molecular dissociation energy and potential energy curve implied by a given set of observed vibration-rotation level energies. As in any inversion problem, it is essential to first understand the properties of the equation one is trying to invert, so we will begin with an examination of the JWKB quantization condition and of the way in which the spectroscopic properties of a given system may be calculated from a knowledge of its potential energy function. This ability to calculate properties of an assumed-known potential

M. S. Child (ed.), Semiclassical Methods in Molecular Scattering and Spectroscopy, 109–126.

energy function is also important in its own right, as it often greatly facilitates the analysis of experimental data.

2. SPECTROSCOPIC PROPERTIES OF A GIVEN POTENTIAL ENERGY FUNCTION

The two-term (or third-order) JWKB quantization condition for a particle of mass μ with energy E bound by the one-dimensional potential energy function U(r) may be written as (1):

$$v+\tfrac{1}{2} = \left(\frac{2\mu}{\hbar^2}\right)^{\frac{1}{2}} \oint_{\Gamma} dr\,[E-U(r)]^{\frac{1}{2}} + \frac{1}{96\pi}\left(\frac{\hbar^2}{2\mu}\right)^{\frac{1}{2}} \oint_{\Gamma} dr\; U''(r)/[E-U(r)]^{3/2} \qquad (1)$$

where primes denote differentiation with respect to the distance coordinate r and the contour Γ encloses the segments of the real line (see Fig. 1) on which E>U(r). While eq. (1) may be applied much more generally, the present chapter will only consider cases such as that shown in Fig. 1 where there exists only one "classically accessible" interval $[r_1, r_2]$ enclosed by the contour Γ.

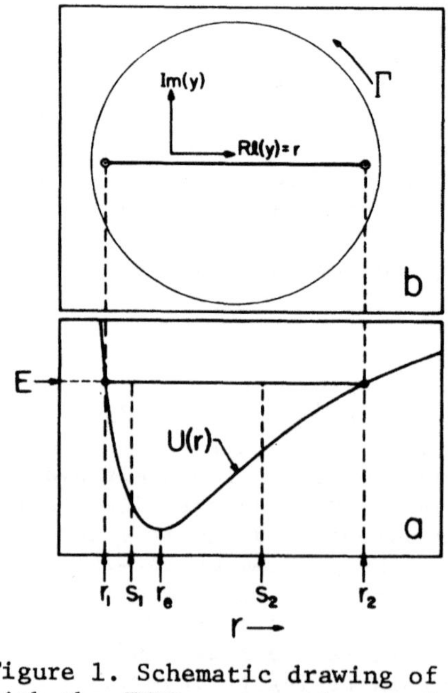

Figure 1. Schematic drawing of quantities associated with the JWKB contour integrals of eq. (1) (Reproduced by permission of the American Institute of Physics)

To obtain the one-term quantization condition, the second integral is deleted from the right hand side of eq. (1), while the higher-order eigenvalue equations are obtained by adding appropriate higher-order contour integrals to this equation. However, with a few notable exceptions (2), it is the single-term or first-order version of the JWKB quantization condition which has been used in most practical applications to diatomic molecule spectroscopy. Much of the discussion in this chapter is therefore based on this simple case. However, since the greater accuracy provided by the higher-order expressions is becoming increasingly important in modern work, the theory is formulated in so as to make extension to the higher-order approximations quite straightforward.

Within any chosen order of the JWKB approximation, the allowed eigenvalues of the potential U(r) are defined as the values of E for which the sum of terms on the right hand side of eq. (1) yields an exact integer for the vibrational index v (=0,1,2,...etc.). For a diatomic molecule, $\mu=m_1m_2/(m_1+m_2)$ is the reduced mass associated with the stretching and rotation of the molecular axis, while the total potential energy function

$$U(r) = V(r) + J(J+1)\hbar^2/2\mu r^2 \qquad (2)$$

is the sum of the electronic potential energy V(r) plus the centrifugal potential term due to the rotational motion characterized by total angular momentum quantum number J. Thus, values of v calculated from eq. (1) in general depend on the two independent variables E and J.

Before proceeding further, it is convenient to introduce the notation

$$I^{m,n}_{k,\ell} = I^{m,n}_{k,\ell}(E,J) = \oint_{\Gamma} dr[\partial^m U(r)/\partial r^m]^n/r^{\ell}[E-U(r)]^{k+\frac{1}{2}} \qquad (3)$$

for integrals of the type appearing in eq. (1), and to note that for any well-behaved function U(r) such integrals may be readily evaluated to arbitrary accuracy using either of two simple numerical quadrature procedures (3). The first of these procedures (3,4) makes use of the fact that any $I^{m,n}_{k,\ell}$ integral may be written as a numerical constant times the k-th partial derivative with respect to E (at constant J) of $I^{m,n}_{0,\ell}$. The integrand of the latter is integrable on the real line across the whole of the interval bounded by the classical turning points r_1 and r_2. Subdividing this interval $[r_1,r_2]$ at points s_1 and s_2 defined such that $r_1 < s_1 < r_e < s_2 < r_2$ (see Fig. 1) and performing integration by parts on the subintervals $[r_1,s_1]$ and $[s_2,r_2]$ prior to each of the k stages of differentiation with respect to E allows these $I_{k,\ell}^{m,n}$ integrals to be expressed as a linear combination of simple integrals across the subintervals $[r_1,s_1]$, $[s_1,s_2]$ and $[s_2,r_2]$, plus a few edge effect terms involving values of the integrand at s_1 and s_2. The second integration method (3,5) is even simpler to use in that it involves a simple Gaussian-like quadrature formula for the integral as an appropriately weighted sum of values of the integrand evaluated at certain specified quadrature mesh points. The development of these methods is a matter of some significance, since while the $k\leqslant 0$ integrals arising in many applications of the single-term quantization condition may be readily evaluated using a variety of techniques, difficulties associated with evaluation of such integrals for $k\geqslant 1$ had inhibited general use of the higher-order JWKB approximation.

Defining $\beta=(\hbar^2/2\mu)^{\frac{1}{2}}$, allows the two-term quantization condition of eq. (1) to be rewritten as

$$v+\tfrac{1}{2} \;=\; (1/2\pi\beta)\,I_{-1,0}^{0,0} \;+\; (\beta/96\pi)\,I_{1,0}^{2,1} \qquad (4)$$

In applications of this expression, it is important to remember that the natural independent variables associated with the $I_{k,\ell}^{m,n}$ integrals are E and $[J(J+1)]$. The partial derivatives of $I_{k,\ell}^{m,n}$ with respect to these variables are readily expressed in terms of other integrals of this form:

$$\left(\frac{\partial}{\partial E}I_{k,\ell}^{m,n}\right)_J \;=\; -(k+\tfrac{1}{2})\,I_{k+1,\ell}^{m,n} \qquad (5)$$

$$\left(\frac{\partial}{\partial\,[J(J+1)]}I_{k,\ell}^{m,n}\right)_E \;=\; \beta^2\left((k+\tfrac{1}{2})\,I_{k+1,\ell+2}^{m,n} \;+\; (-1)^m n(m+1)!\;I_{k,\ell+m+2}^{m,n-1}\right) \qquad (6)$$

However, in the calculation of many spectroscopic properties, it is more appropriate to hold v constant when taking partial derivatives with respect to $[J(J+1)]$, so we note that

$$\left(\frac{\partial}{\partial\,[J(J+1)]}I_{k,\ell}^{m,n}\right)_v \;=\; \left(\frac{\partial}{\partial\,[J(J+1)]}I_{k,\ell}^{m,n}\right)_E \;+\; \left(\frac{\partial E}{\partial\,[J(J+1)]}\right)_v\left(\frac{\partial}{\partial E}I_{k,\ell}^{m,n}\right)_J \qquad (7)$$

For any given value of J, the vibration-rotation eigenvalues E(v,J) are readily determined by evaluating the integrals on right hand side of eq. (3) at a sequence of trial energies until their sum converges on the desired value of $v+\tfrac{1}{2}$. The convergence of this search is greatly accelerated if at each trial energy the partial derivative of v with respect to energy

$$1/\omega_v \equiv \left(\frac{\partial(v+\frac{1}{2})}{\partial E}\right)_J = (1/4\pi\beta)\{I^{0,0}_{0,0} - (\beta^2/16)I^{2,1}_{2,0}\} \tag{8}$$

is calculated and used to estimate the required correction to E. Since both $(v+\frac{1}{2})$ and $1/\omega_v$ are smooth monotonically increasing functions of E, this convergence should be very rapid.

The traditional way of expressing the dependence of diatomic molecule level energies on the rotational quantum number J is via the expression

$$E(v,J) = G(v) + B_v[J(J+1)] - D_v[J(J+1)]^2 + H_v[J(J+1)]^3 + \ldots \tag{9}$$

The rotational constant B_v and the centrifugal distortion constants D_v, H_v,...etc. are therefore defined as derivatives of the energy with respect to [J(J+1)] at constant v, all evaluated at J=0. Application of eqs. (4)-(6) and (9) then yields:

$$B_v = \left(\frac{\partial E}{\partial[J(J+1)]}\right)_v = -\left(\frac{\partial(v+\frac{1}{2})}{\partial[J(J+1)]}\right)_E \Bigg/ \left(\frac{\partial(v+\frac{1}{2})}{\partial E}\right)_J$$

$$= (\beta/4\pi)\omega_v\{I^{0,0}_{0,2} - (\beta^2/16)[I^{2,1}_{2,2} + 4I^{0,0}_{1,4}]\} \tag{10}$$

Expressions for the centrifugal distortion constants are then obtained by repeated partial differentiation of eq. (10) with respect to [J(J+1)] at constant v (using eqs. (5-7)). In particular, within the single-term quantization condition, the leading centrifugal distortion constant is (4):

$$D_v = -(\beta^4/4)[I^{0,0}_{1,4}\big/I^{0,0}_{0,0} - 2I^{0,0}_{0,2}I^{0,0}_{1,2}\big/(I^{0,0}_{0,0})^2 + I^{0,0}_{1,0}(I^{0,0}_{0,2})^2\big/(I^{0,0}_{0,0})^3] \tag{11}$$

and analogous expressions for H_v, L_v,...etc. are readily obtained (4) by repeated differentiation of eq. (9).

The ability to calculate reliable values of the centrifugal distortion constants associated with any given potential energy function is often essential for optimizing the analysis of experimental data. In practise, it is often difficult to determine reliable values of these constants from direct fits of experimental data to eq. (9). This is particularly true for highly excited vibrational levels for which relatively few rotational sublevels are often observed, but for which the distortion constants have the largest magnitude. This in turn introduces uncertainties into the values of G(v) and B_v obtained from these fits and into the potential energy curve generated from them (see section 3 below). However, in the absence of curve crossings or other perturbations, centrifugal distortion constants calculated from eq. (11) and its analogs using any reasonably good trial potential will usually be quite reliable. Holding D_v, H_v, ...etc. fixed at these calculated values in a subsequent fit of experimental data to eq. (9) should yield improved G(v) and B_v values, and thence an improved potential energy curve. The latter can then be used to calculated new centrifugal distortion constants and the procedure iterated to convergence. While a number of other quantum mechanical and semiclassical methods of calculating centrifugal distortion constants have been proposed (6)-(12), the present approach is unique in that it is equally stable and reliable for vibrational levels lying near dissociation and for those lying near the potential minimum, since the $I_{k,\ell}^{m,n}$ integrals may be computed equally readily in either region (3).

3. THE RKR INVERSION PROCEDURE

The first general inversion procedure for determining a diatomic molecule potential energy curve from a set of experimental vibration-rotation level energies is that due to Dunham (2). He showed that if the potential energy expansion

$$V(r) = a_0\xi^2[1 + a_1\xi + a_2\xi^2 + ...] \quad , \tag{12}$$

where $\xi \equiv (r-r_e)/r_e$, is substituted into the two-term JWKB quantization condition, the level energies are given by

$$E(v,J) = \sum_{\ell,m} Y_{\ell m} (v+\tfrac{1}{2})^{\ell} [J(J+1)]^m \tag{13}$$

where the "Dunham coefficients" $Y_{\ell m}$ are explicitly-known (2) functions of the potential energy parameters a_0, a_1, a_2,...etc. Values of $Y_{\ell m}$ obtained from a fit of eq. (13) to experimental data may then be used to determine these potential parameters.

One limitation of the Dunham procedure is the fact that the potential expansion of eq. (12) diverges at large r, rather than approach an asymptote. To remove this difficulty, Simons et al. (13) and Thakkar (14) replaced Dunham's variable ξ by the expansion parameters $(r-r_e)/r$ and $[1-(r_e/r)^P]$, respectively, and used the two-term quantization condition to determine the relationships between their new potential expansion coefficients and the $Y_{\ell m}$ constants of eq. (13). However, all three of these Dunham-type methods are rather complicated to apply when more than a few terms are included in the potential expansion. Moreover, while the second type of potential (13) may in principle be constrained to have the correct theoretically-known long-range behaviour, introduction of the necessary constraints makes the resulting formulation sufficiently complex to discourage its use. Thus, potentials obtained from this type of analysis usually do not have the correct long-range functionality, and hence they cannot be extrapolated reliably beyond the range of the data used in the analysis.

The Rydberg-Klein-Rees or RKR method (15-20) is a semi-classical inversion procedure which is much easier to apply to cases in which the data are very extensive, and it is not restricted by the assumption of any particular analytic form for either the potential energy function or the vibrational energy level equation. It is based on a formal inversion of the one-term quantization condition of eqs. (1) or (4) to yield expressions for the values of the classical turning points associated with given values of the vibrational energy (or of the vibrational index v).

Following Vanderslice et al. (18), the present derivation of the RKR procedure begins by identifying the variable A as the area on Fig. 1a enclosed by the potential energy curve and the total energy,

$$A(E,J) = \int_{r_1}^{r_2} dr[E-U(r)] \quad , \tag{14}$$

where the turning points r_1 and r_2 are functions of E and J. Partial differentiation of eq. (14) then yields:

$$r_2 - r_1 = (\partial A/\partial E)_J \tag{15}$$

and

$$1/r_1 - 1/r_2 = -(1/\beta^2)(\partial A/\partial\,[J(J+1)])_E \tag{16}$$

However, the definition of an Eulerian integral of the first kind allows eq. (14) to be expressed as

$$A(E,J) = (1/\pi) \int_{U_0}^{E} dE'(E-E')^{\frac{1}{2}}\, I_{0,0}^{0,0}(E',J) \tag{17}$$

where $U_0=U_0(J)$ is the energy at the potential minimum. Substituting eq. (17) into (15) and (16) then yields

$$r_2 - r_1 = (1/2\pi) \int_{U_0}^{E} dE'(E-E')^{-\frac{1}{2}} I^{0,0}_{0,0}(E',J) \qquad (18)$$

and

$$\frac{1}{r_1} - \frac{1}{r_2} = -(1/2\pi) \int_{U_0}^{E} dE'(E-E')^{\frac{1}{2}} I^{0,0}_{1,2}(E',J)$$

$$+ (1/\pi\beta^2)(E-U_0)^{\frac{1}{2}} I^{0,0}_{0,0}(U_0,J) \frac{dU_0}{d[J(J+1)]} \qquad (19)$$

Making use of eq. (5), eq. (19) may be integrated by parts and transformed into

$$\frac{1}{r_1} - \frac{1}{r_2} = \frac{1}{2\pi} \int_{U_0}^{E} dE'(E-E')^{-\frac{1}{2}} I^{0,0}_{0,2}(E',J) + (E-U_0)^{\frac{1}{2}}\Delta(U_0,J) \qquad (20)$$

where

$$\Delta(U_0,J) = (1/\pi)\left(-I^{0,0}_{0,2}(U_0,J) + (1/\beta^2)I^{0,0}_{0,0}(U_0,J)\frac{dU_0}{d[J(J+1)]}\right) \qquad (21)$$

According to the one-term quantization condition, $v_0=-\frac{1}{2}$ and $\Delta(U_0,J)\equiv 0$; while this is not exact, the correction term $(E-U_0)^{\frac{1}{2}}\Delta(U_0,J)$ is always fairly small.

Substituting for $I^{0,0}_{0,0}$ from eq. (8) and noting that $(1/\omega_{v'})dE'=dv'$, eq. (18) becomes

$$r_2 - r_1 = 2\beta \int_{v_0}^{v} dv'[E(v)-E(v')]^{-\frac{1}{2}}\{1 + \frac{\beta\omega_{v'}}{64\pi}I^{2,1}_{2,0}(E',J)\} \qquad (22)$$

where v_0 is the non-integer vibrational index associated with the potential minimum and $E'=E(v',J)$. Similarly, substitution of $I^{0,0}_{0,2}$ from eq. (10) into eq. (20) yields:

$$\frac{1}{r_1} - \frac{1}{r_2} = \frac{2}{\beta} \int_{v_0}^{v} dv'B_{v'}[E(v)-E(v')]^{-\frac{1}{2}}\{1 + \frac{\beta^3\,\omega_{v'}}{64\pi B_{v'}}\left(I^{2,1}_{2,2}(E',J)\right.$$

$$\left. + 4I^{0,0}_{1,4}(E',J)\right)\} + (E-U_0)^{\frac{1}{2}}\Delta(U_0,J) \qquad (23)$$

These are the generalized RKR equations which include effects due to the second (i.e., third-order) term in the JWKB quantization condition.

Within the first-order JWKB approximation, the integrand factors in parentheses {...} appearing in eqs. (22) and (23) collapse to 1 and $\Delta(U_0,J)=0$. Evaluation of the resulting integrals therefore merely requires a knowledge of the experimental vibrational energies E(v) and rotational constants B_v. While the integrands of these integrals are singular at the upper limit v'=v, the integrals are finite and may be readily evaluated using a variety of techniques. It is interesting to note that while the basic formulation of the "one-term" RKR procedure was developed by 1932, the method saw very little use for almost 30 years, at least partly because of the apparent difficulty of evaluating these improper integrals accurately. However, it is now clear (19,20) that with an appropriately chosen numerical integration technique, these integrals present no problem. In particular, making the change of variables $v'=v_0+(v-v_0)x$ reduces the integrals of eqs. (22) and (23) to the form

$$\int_0^1 dx\, F(x)\,[1-x]^{-\frac{1}{2}} \qquad (24)$$

where the function F(x) is well-behaved and finite across the whole range of integration [0,1]. Use of a quadrature procedure which explicitly takes account of the $[1-x]^{-\frac{1}{2}}$ weight function allows such integrals to be evaluated extremely efficiently. For example, using the Gauss-Mehler quadrature procedure (see eq. 25.4.36 of ref. 21), Tellinghuisen (19) showed that 4 integration mesh points often yield results accurate to one part in 10^7!

The ordinary one-term RKR procedure described above has now become a standard spectroscopists' tool which is routinely used in the analysis of experimental data. However, although a general procedure for performing higher-order RKR calculations was reported by Vanderslice et al. (22) in 1965, it has not yet seen much use. In their approach, a single-term RKR calculation is first used to determine a realistic estimate of the potential energy function. This trial potential can then be used to evaluate the $I_{k,\ell}^{m,n}(E',J)$ integrals appearing in the integrands of eqs. (22) and (23) and the integration over v' repeated to give improved estimates of the turning points r_1 and r_2. The resulting improved potential may then be used to give more realistic estimates of the $I_{k,\ell}^{m,n}(E',J)$ factors and the process repeated until convergence is achieved.

One factor inhibiting the general adoption of this procedure was the lack, until relatively recently (3-5), of any simple and reliable procedure for evaluating the $I^{m,n}_{k,\ell}(E',J)$ integrals at *all* energies. In particular, the methods which had been devised (22,23,24) for calculating the "correction factors" in the integrands of eqs. (22) and (23) are all numerically unstable for vibrational levels lying near the dissociation limit. Another difficulty was the fact that two of these methods (23,24) appear to have neglected the contribution of the term $[E-U_0]^{\frac{1}{2}}\Delta(U_0,J)$ to eq. (23). However, with the techniques now available, it appears that reliable two-term or three-term RKR procedures should soon be replacing the familiar one-term method in routine spectroscopic analysis.

4. NEAR-DISSOCIATION BEHAVIOUR OF DIATOMIC MOLECULES

An extremely useful recent application of the JWKB quantization condition has been in the derivation of simple analytic expressions for the limiting near-dissociation behaviour of vibrational level spacings (25,26), rotational constants (4,27, 28) and other properties (29-31). While most of these expressions are based on the one-term quantization condition, analogous near-dissociation expressions have been obtained for the second and third terms of the higher-order JWKB quantization condition (32),

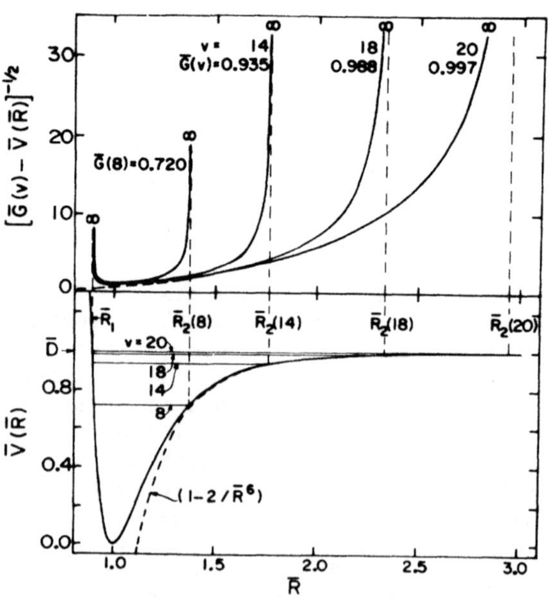

Figure 2. Lower: eigenvalues and turning points for 4 levels of a LJ(12,6) model potential. Upper: integrand of eq. (25) for these four levels. In both, energies and lengths are scaled by the potential well depth and equilibrium distance, respectively (Reproduced from J. Chem. Phys. 1970, 52, p. 3869).

and similar results can readily be obtained for any property defined in terms of $I_{k,\ell}^{m,n}$ type integrals corresponding to $k \geqslant 0$.

The most basic of these near-dissociation expressions is that for the distribution of vibrational energy levels, and its derivation illustrates the basic considerations which are involved in all such derivations. The natural starting point is the one-term JWKB expression for the density of vibrational levels (see eq. (8)) which may be rewritten as

$$\left(\frac{\partial v}{\partial E}\right)_J = \frac{1}{2\pi\beta}\int_{r_1}^{r_2} dr\,[E-V(r)]^{-\frac{1}{2}} \tag{25}$$

Fig. 2 shows the nature of the integrand of eq. (25) for four levels of a given model potential, where $\bar{R}=r/r_e$ and the potential energy function and vibrational eigenvalues are both scaled by the potential well depth. It is immediately clear that for the upper vibrational levels the dominant contribution to this integral comes from the region near the outer turning point. This suggests the fundamental approximation of this method, which involves replacing the exact potential by an approximate function which is accurate near the outer turning point though not necessarily highly reliable at shorter distances. As a justification of this approximation, the dashed curve seen in the upper portion of Fig. 2 shows the effect on the integrand of eq. (25) for the v=20 level of replacing the exact potential $V(r)=\varepsilon[(r_e/r)^6-1]^2$ by its long-range part: $\varepsilon[1-2(r_e/r)^6]$. While the extremely close agreement with the exact integrand (solid curve) for this case is partly due to the simplicity of the chosen model potential, it seems clear that for any potential this type of approximation will become increasingly exact the nearer the level in question lies to dissociation.

It has long been known that if two atoms lie sufficiently far apart that their electronic clouds overlap negligibly, their interaction energy may be expressed as a sum of inverse-power terms:

$$V(r) = D - \sum_{m \geqslant n} C_m/r^m \tag{26}$$

where D is the dissociation energy. The nature of the atoms to which the given molecular electronic state dissociates determines which (positive integer) values of m appear in eq. (26) (e.g., see ref. 33), and the perturbation theory has provided means by which many of these C_m coefficients may be calculated. Thus, for most molecular electronic states, the power n of the leading term in eq. (26) is known, and sometimes so is the associated C_n constant.

In view of the above, it seems appropriate to replace V(r) in eq. (25) by the asymptotically dominant contribution to the long-range interaction potential: $V(r)=D-C_n/r^n$. Doing this and changing the variable of integration to $y=r_2^n(v)/r$ transforms eq. (25) into

$$\left(\frac{\partial v}{\partial E}\right)_J = \frac{1}{2\pi\beta}(C_n)^{1/n}\,[D-E(v)]^{-(n+2)/2n}\int_1^{r_2/r_1} dy\,y^{-2}(y^n-1)^{-\frac{1}{2}} \tag{27}$$

For highly excited vibrational levels $r_2 >> r_1$ and it is a very minor additional approximation to replace r_2/r_1 by ∞; in this limit the integral is well known and eq. (27) becomes

$$\frac{dE(v)}{dv} = K_n[D-E(v)]^{(n+2)/2n} \tag{28}$$

where the constant K_n is an exactly-known (25,33) function of μ, n and C_n. In practise, it is often more convenient to use the integrated form of eq. (28), which for n≠2 is:

$$E(v) = D - X_0(n)[v_D-v]^{2n/(n-2)} \tag{29}$$

where (for n>2) v_D is the non-integer effective vibrational index associated with the dissociation limit, $X_0(n) = [K_n(n-2)/2n]^{2n/(n-2)}$ and

$$X_i(n) = \bar{X}_i(n)/[\mu^n(C_n)^2]^{1/(n-2)} \tag{30}$$

while the constant $\bar{X}_i(n)$ depends only on the value of the integer n (see Table 1). Another convenient form of these results is obtained on substituting eq. (29) into eq. (28) (or differentiating eq. (29)):

$$\frac{dE}{dv} = \left(\frac{2n}{n-2}\right)X_0(n)[v_D-v]^{(n+2)/(n-2)} \tag{31}$$

For practical applications of these results, it is useful to rewrite eqs. (28) and (31) in a form which facilitates graphical interpretation of the experimentally observed vibrational level spacings $\Delta G_{v+\frac{1}{2}}=E(v+1)-E(v)$. On making the approximation that

$$\Delta G_{v+\frac{1}{2}} \approx \left.\frac{dE(v)}{dv}\right|_{v+\frac{1}{2}} \tag{32}$$

Table 1. For energies in cm^{-1}, lengths in Å and mass in amu, numerical factors appearing in the limiting near-dissociation expressions for vibrational energies and rotational constants.

n	$\bar{X}_0(n)$	$\bar{X}_1(n)$	$\bar{X}_2(n)$	$\bar{X}_3(n)$	$\bar{X}_4(n)$
3	36410.	60221.	-24901.	1348.0	-73.367
4	13433.	4275.7	-205.65	-3.7691	-2.9067×10^{-1}
5	9170.9	1178.3	-15.377	-1.7742×10^{-1}	-5.3435×10^{-3}
6	7932.0	546.64	-2.7539	-1.9186×10^{-2}	-2.9942×10^{-4}

these expressions may be rewritten as

$$(\Delta G_{v+\frac{1}{2}})^{2n/(n+2)} = (K_n)^{2n/(n+2)} \; [D-E(v+\tfrac{1}{2})] \tag{33}$$

and

$$(\Delta G_{v+\frac{1}{2}})^{(n-2)/(n+2)} = [X_0(n)2n/(n-2)]^{(n-2)/(n+2)} \; (v_D-v-\tfrac{1}{2}) \tag{34}$$

They suggest that for vibrational levels lying sufficiently close to dissociation for eqs. (28) & (29) to be accurate, plots of the vibrational spacing, taken to the appropriate power, vs. $E(v+\frac{1}{2})$ or $v+\frac{1}{2}$ itself should be precisely linear and have intercepts corresponding, respectively, to the dissociation energy D and the vibrational index at dissociation, v_D. Moreover, the limiting slopes of these plots determine the constant $X_0(n)$ (or K_n), and hence define the value of the long-range potential constant C_n. In addition, once these three parameters (D, v_D and $X_0(n)$) are known, their substitution into eq.(29) yields reliable estimates of the energies of any vibrational levels lying above the highest one observed.

As an illustration of the utility of these results, Figs. 3 and 4 plot the vibrational spacings for the highest observed (prior to 1976) vibrational levels of the $B(^3\Pi^+_{0u})$ state of I_2 (for which case it is known (25,33) that n=5) in the manner suggested by eqs. (33) and (34). The linearity of these plots and the internal consistency of their slopes attests to the appropriateness of these equations for characterizing such data, and suggests that the ensuing dissociation energy, C_5 constant and predicted energies for levels v=73-87 should be fairly reliable.

When applying these equations, it is important to remember the two fundamental assumptions on which they are based: (i) that the vibrational level density of eq. (25) depends mainly on the nature of the potential near the outer turning point of the level in question, and (ii) that the long-range potential may be

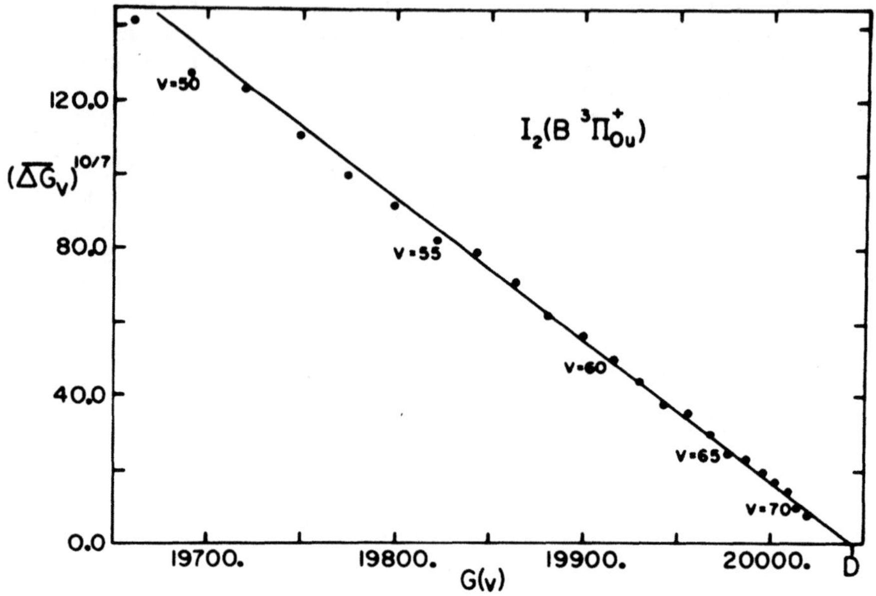

Figure 3. Data for B-state I_2 (34) plotted according to eq. (33) with n=5. (Reproduced from J. Mol. Spectrosc. 37 (1971), p. 109)

accurately represented by the single inverse-power term $V(r) \simeq D - C_n/r^n$. Both of these assumptions are exactly true only for levels lying virtually right at dissociation. While experimental data will sometimes include observations of levels lying sufficiently near this limit that the errors in eqs. (28) and (29) are small (e.g., see Figs. 3 and 4), for most molecular states this is not the case. However, in the latter case the knowledge of the expected limiting behaviour of the distribution of vibrational eigenvalues may still greatly facilitate extrapolations to estimate molecular dissociation energies and predict the energies of unobserved levels, expecially if a value of the appropriate C_n constant is provided by theory or some other experiment. In the long run, this transformation of an *extrapolation* beyond the observed vibrational levels *into an interpolation* between these data at low v and a known functional behaviour for $v \to v_D$ will probably prove to be the most significant feature of these results.

The approximations used in the derivation of eq. (28) may also be applied to any other property defined in terms of $I^{m,n}_{k,\ell}$ integrals corresponding to $k \geq 0$. While these approximations become relatively less accurate with increasing ℓ they will still be valid for vibrational levels lying extremely close to dissociation.

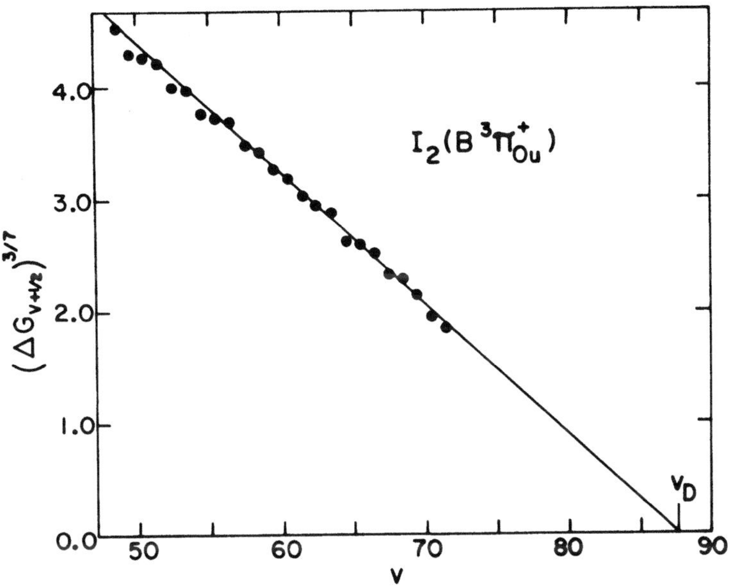

Figure 4. Data for B-state I_2 (34) plotted according to eq. (34) with n=5. (Reproduced from Molecular Spectroscopy I (1973), a Specialist Periodical Report of the Chemical Society (London), R.F. Barrow, D.A. Long and D.A. Millen editors).

For the rotational constants B_v, D_v,...etc. of eq. (9), the resulting expressions for their characteristic near-dissociation behaviour are (4,27,28):

$$B_v = X_1(n)(v_D - v)^{[2n/(n-2)]-2} \tag{35}$$

$$D_v = -X_2(n)(v_D - v)^{[2n/(n-2)]-4} \tag{36}$$

$$H_v = X_3(n)(v_D - v)^{[2n/(n-2)]-6} \tag{37}$$

$$L_v = X_4(n)(v_D - v)^{[2n/(n-2)]-8} \tag{38}$$

where $X_i(n)$ is given by eq. (30) and the associated $\bar{X}_i(n)$ constants are listed in Table 1. Analogous expressions for a number of other molecular properties have been derived by Stwally (29-31), and it seems clear that results of this type will play an increasingly important role in the interpretation of spectroscopic data in the years to come.

5. PROBLEMS

5.1 a) Derive equation (11).

b) Determine the generalized version of equation (11) which arises in the two-term (third-order) JWKB quantization condition.

5.2 Consider the potential $V(r) = -1/r + 1/4r^2$ where the energy and length units are chosen such that the factor $\beta = \sqrt{\hbar^2/2\mu} = 1$.

a) Making appropriate use of known integrals (from tables) show that the first-order quantization condition yields the eigenvalues:

$$E(v,J) = -1/4(v + J + 1)^2$$

b) At $J = 0$ the vibrational energy is clearly given by $E(v,0) = -1/4(v + 1)^2$ while the analogous expression for the rotational constant B_v is $B_v = 1/2(v + 1)^3$. Using these expressions as input to the first-order RKR equations (22) and (23), show that the expressions for (r_2-r_1) and $(1/r_1-1/r_2)$ so obtained agree with the starting potential.

5.3 For a potential whose long-range attractive tail has the form $V(r) \simeq D - A\times\exp(-br)$, derive the analog of eq. (28).

5.4 For the ground electronic state of $BeAr^+$, n=4, C_4=95340 cm^{-1} Å^6 and the four observed vibrational spacings are 344.98, 327.44, 310.06 and 292.91 cm^{-1}. Use the methods of section 4 to obtain estimates of the dissociation energy D and of v_D. (Hint: consider graphs analogous to Figs. 3 and 4).

5.5 Within the one-term quantization condition, show how the general expression for D_v given by eq. (11) gives rise to the limiting near-dissociation expression of eq. (36).

REFERENCES

1. Fröman, N.: 1979, in *Proceedings of the NATO Advanced Study Institute on Semiclassical Methods in Scattering and Spectroscopy* (this volume).

2. Dunham, J.L.: 1932, Phys. Rev. 41, p. 713 and 721.

3. Barwell, M.G., Le Roy, R.J., Pajunen, P., and Child, M.S.: 1979, J. Chem. Phys. 71 (in press).

4. Barwell, M.G.: 1976, M.Sc. Thesis, University of Waterloo.

5. Pajunen, P.: 1978, D. Phil. Thesis, Oxford University.

6. Beckel, C.L.: 1969, J. Chem. Phys. 50, p. 2372.

7. Albritton, D.L., Harrop, W.J., Schmeltekopf, A.L., and Zare, R.N.: 1973, J. Mol. Spectrosc. 46, p. 25.

8. Tellinghuisen, J.: 1973, Chem. Phys. Lett. 18, p. 544.

9. Wei, J., and Tellinghuisen, J.: 1974, J. Mol. Spectrosc. 50, p. 317.

10. Kirschner, S.M., and Watson, J.K.G.: 1973, J. Mol. Spectrosc. 47, p. 234.

11. Kirschner, S.M., and Watson, J.K.G.: 1974, J. Mol. Spectrosc. 51, p. 321.

12. Brown, J.D., Burns, G., and Le Roy, R.J.: 1973, Can. J. Phys. 51, p. 1664.

13. Simons, G., Parr, R.G., and Finlan, J.M.: 1973, J. Chem. Phys. 59, p. 3229.

14. Thakkar, A.: 1975, J. Chem. Phys. 62, p. 1693.

15. Rydberg, R.: 1931, Z. Physik 73, p. 326; ibid 1933, Z. Physik 80, p. 514.

16. Klein, O.: 1932, Z. Physik 76, p. 226.

17. Rees, A.L.G.: 1947, Proc. Phys. Soc. (London) 59, p. 998.

18. Vanderslice, J.T., Mason, E.A., and Maisch, W.G.: 1959, J. Mol. Spectrosc. 3, p. 17; ibid 1960, J. Mol. Spectrosc. 5, p. 83.

19. Tellinghuisen, J.: 1972, J. Mol. Spectrosc. 44, p. 194; Tellinghuisen, J.: 1974, Computer Phys. Comm. 6, p. 221.

20. Dickinson, A.S.: 1972, J. Mol. Spectrosc. 44, p. 183.

21. Abramowitz, M., and Stegun, I.A.: *Handbook of Mathematical Functions* (Natl. Bur. Std. Appl. Math. Ser. 55 (1964); also Dover, New York (1965)).

22. Vanderslice, J.T., Davies, R., and Weissman, S.: 1965, J. Chem. Phys. 43, p. 1075; Davies, R.H., and Vanderslice, J.T.: 1966, J. Chem. Phys. 45, p. 95.

23. Kirschner, S.M., and Watson, J.K.G.: 1974, J. Mol. Spectrosc. 51, p. 321.

24. Huffaker, J.N.: 1977, J. Mol. Spectrosc. 65, p. 1.

25. Le Roy, R.J., and Bernstein, R.B.: 1970, Chem. Phys. Lett. 5, p. 42; ibid; 1970, J. Chem. Phys. 52, p. 3869.

26. Stwalley, W.C.: 1970, Chem. Phys. Lett. 6, p. 241.

27. Le Roy, R.J.: 1972, Can. J. Phys. 50, p. 953.

28. Le Roy, R.J., and Barwell, M.G.: 1975, Can. J. Phys. 53, p. 1983.

29. Stwalley, W.C.: 1973, J. Chem. Phys. 58, p. 3867.

30. Stwalley, W.C., 1973, in Smith, D.W. and McRae, W.B. (Eds.), *Energy, Structure and Reactivity* (Wiley), p. 259.

31. Stwalley, W.C.: 1978, Contemp. Phys. 19, p. 65.

32. Kirschner, S.M., and Le Roy, R.J.: 1978, J. Chem. Phys. 68, p. 3139.

33. Le Roy, R.J.: 1973, in *Molecular Spectroscopy*, Vol. 1, Barrow, R.F., Long, D.A., and Millen, D.J. (Eds.), p. 113.

34. Brown, W.G.: 1931, Phys. Rev. 38, p. 709.

SEPARABLE SPECTROSCOPIC APPLICATIONS

M. S. Child

Theoretical Chemistry Department,
1 South Parks Road, Oxford.

ABSTRACT: Primitive semiclassical and uniform approximations to Franck-Condon integrals and transition matrix elements are derived, and applications to spectroscopic intensity patterns and to diatomic predissociation are discussed. The use of semiclassical connection formulae for the general treatment of tunneling and curve-crossing problems is introduced. Applications to quantisation in a double minimum potential and to predissociation are used to illustrate the technique.

1. FRANCK-CONDON FACTORS

1.1 Semiclassical Normalisation

The general JWKB wavefunction in any classically accessible region may be written

$$\psi(r) \simeq C\,[k(r)]^{-\frac{1}{2}} \sin\,[\phi(r) + \pi/4] \tag{1}$$

where

$$k(r) = [2\mu(E-V(r))]^{\frac{1}{2}}/\hbar \tag{2}$$

$$\phi(r) = \int_a^r k(r)\,dr \tag{3}$$

and a is the left hand classical turning point. Thus the essential difference between bound and continuum wavefunctions in the range covered by (1) lies in the normalisation coefficient C.

M. S. Child (ed.), Semiclassical Methods in Molecular Scattering and Spectroscopy, 127–154.

Fröman shows in Chapter 1 that for a bound state

$$C_{bound} = [2\mu\, \hbar\bar{\omega}/\pi\hbar^2]^{\frac{1}{2}} \tag{4}$$

where $\hbar\bar{\omega}$ denotes the local level spacing

$$\hbar\bar{\omega} = (dE/dv) \tag{5}$$

In spectroscopic applications the continuum normalisation must be to a delta function of energy

$$\int_0^\infty \psi_E(r)\ \psi_{E'}(r)dr = \delta(E - E') \tag{6}$$

which is shown in standard texts [1] to require that in the limit $r \rightarrow \infty$

$$\psi(r) \overset{r \rightarrow \infty}{\sim} [2\mu/\pi\hbar^2 k(\infty)]^{\frac{1}{2}} \sin [k(\infty)r + \eta] \tag{7}$$

Thus by comparison with equation (1)

$$C_{cont} = [2\mu/\pi\hbar^2]^{\frac{1}{2}} \tag{8}$$

Notice that even if the continuum wavefunction is determined by exact numerical integration, normalisation by equation (1) and (8) is frequently much more convenient than by equation (7), because in many spectroscopic applications the tail of the wavefuntion required for comparison with (7) is irrelevant to the problem

Equations (4) and (8) show that the essential difference between the bound and contimuum wavefunction lies in the factor $(\hbar\bar{\omega})^{\frac{1}{2}}$ in equation (4), and this is easily understood. The continuum probability density $|\psi_E(r)|^2$ is naturally continuous in energy at any given r, whereas the discrete density $|\psi_v(r)|^2$ is concentrated at E_v, but the mean discrete density between $E_{v-\frac{1}{2}}$ and $E_{v+\frac{1}{2}}$ given by $|\psi_v(r)|^2/\hbar\bar{\omega}$ is identical with that for a continuum state. Put another way if the bound state normalisation to unity is interpreted as defining a single quantum state, the continuum normalisation ensures a unit density of states per unit energy.

An immediate consequence of this normalisation difference is that the differential oscillator strength per unit energy may be expected to vary smoothly through any dissociation limit [2,3]. Fig 1 shows a molecular example of this behaviour, but the argument is not restricted to molecules and a similar picture applies at the atomic ionisation limit [6]

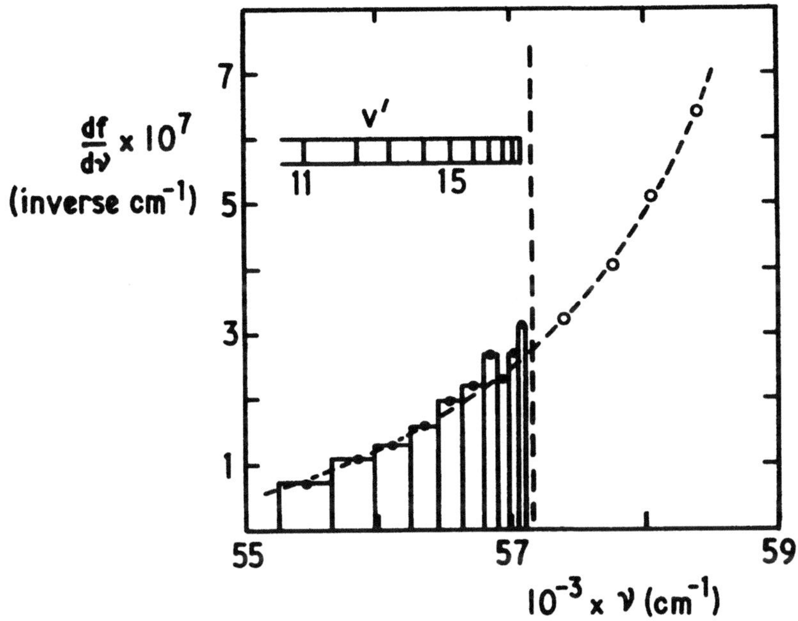

Fig 1. Differential oscillator strengths for the v'' = 0 progression of the B ← X system of O_2. Continuous [4] and discrete [5] data points represented by open and closed circles respectively.

1.2 Primitive and Airy Uniform Approximation

A primitive approximation to the Franck-Condon overlap

$$S_{12} = \int_0^\infty \psi_1(r)\ \psi_2(r)\ dr \tag{9}$$

may be obtained by substituting the JWKB forms for the $\psi_i(r)$ and approximating the integral by stationary phase [7]. The dominant contribution to this integral may be written

$$S_{12} \simeq (C_1C_2/4) \int_0^\infty [k_1(r)k_2(r)]^{-\frac{1}{2}} \{\exp[i\Delta\phi(r)] + \exp[-i\Delta\phi(r)]\}dr \tag{10}$$

where

$$\Delta\phi(r) = \phi_2(r) - \phi_1(r) \tag{11}$$

Corresponding terms in $\phi_2(r) + \phi_1(r)$ have been omitted from (10) because the $\phi_i(r)$ defined by (3) increase monotonically with r; hence the phase sum can never be stationary.

The stationary phase condition requires according to (3) that

$$\frac{d\Delta\phi}{dr} = k_2(r) - k_1(r) = 0 \tag{12}$$

in other words that the dominant contribution to the integral comes from the pseudo crossing point r_x at which

$$E_1 - V_1(r) = E_2 - V_2(r) \tag{13}$$

The location of this point is illustrated in relation to the potential curves and in phase space in Fig. 2.

It is assumed in what follows that this equation has only one root, but the generalisation to include more roots is trivial at the stationary phase level

The approximation is completed by performing the expansion

$$\Delta\phi(r) \simeq \Delta\phi_x + \tfrac{1}{2}\Delta\phi''(r - r_x)^2 \tag{14}$$

by formally extending the integration range to $(-\infty,\infty)$, and by replacing the $k_i(r)$ by their values at r_x. The justification is that the extent of the integration range and the form of $\Delta\phi(r)$ is immaterial in any rapidly oscillatory region. The terms in (10) therefore reduce to standard integrals of the form

$$\int_{-\infty}^{\infty} \exp\,(\pm i|\alpha|x^2)\,dx = (\pi/|\alpha|)^{\frac{1}{2}} \exp\,(\pm i\pi/4) \tag{15}$$

It is convenient to adopt a convention in which $\Delta\phi''$ is negative, in which case

$$S_{12} \simeq \frac{C_1C_2}{2}\left[\frac{2\pi}{(-\Delta\phi'')k_1(r_x)k_2(r_x)}\right]^{\frac{1}{2}} \sin\,(\Delta\phi_x + \pi/4) \tag{16}$$

It remains to note according to (2), (3) and (12) that

$$k_1(r_x) = k_2(r_x) = \mu\, u_x/\hbar \tag{17}$$

and

$$\Delta\phi'' = k_2'(r) - k_1'(r) = -\,2(F_1 - F_2)/\hbar u_x \tag{18}$$

where u_x is the classical velocity and F_i are the classical forces

$$F_i = -(dV_i/dr) \tag{19}$$

all evaluated at r_x. Thus on substituting for the coefficients C_i from (4), the bound-bound Franck-Condon integral becomes

$$S_{12} = P_x^{\frac{1}{2}} \sin(\Delta\phi_x + \pi/4) \tag{20}$$

where

$$P_x = [2\hbar\bar{\omega}_1\hbar\bar{\omega}_2/\pi\hbar u_x(F_1 - F_2)] \tag{21}$$

$$\Delta\phi_x = \int_{a_2}^{r_x} k_2(r)dr - \int_{a_1}^{r_x} k_1(r)dr \tag{22}$$

The corresponding expressions for bound-continuum and continuum -continuum overlaps are obtained by omission of one and both factors $\hbar\bar{\omega}$ respectively.

Notice that the convention that $\Delta\phi''$ is negative implies by (18) that $(F_1 - F_2)$ is positive; it also ensures that $\Delta\phi_x$ is a (necessarily positive) maximum of $\Delta\phi(r)$ because both component integrals in (11) are increasing functions of r.

If the motion in $V_2(r)$ is bound, equation (20) may also be written in terms of

$$\Delta\bar{\phi}_x = \int_{a_1}^{r_x} k_1(r)dr + \int_{r_x}^{b_2} k_2(r)dr, \tag{23}$$

where b_2 is the right hand turning point, because

$$\Delta\phi_x + \Delta\bar{\phi}_x = \int_{a_2}^{b_2} k_2(r)dr = (v_2 + \tfrac{1}{2})\pi \tag{24}$$

Thus on substitution for $\Delta\phi_x$ in (20)

$$S_{12} \simeq (-1)^{v_2} P_x^{\frac{1}{2}} \sin(\Delta\bar{\phi}_x + \pi/4) \tag{25}$$

The limited validity of equation (1) implies that these primitive semiclassical results (20) and (25) apply only to situations in which the pseudo-crossing point lies in the classically accessible range, $a_2 < r_x < b_2$ (see Fig 2). The aim of the uniform approximations given below is to extend the theory to situations in which the roots of equation (13) are classically inaccessible (as shown by r_x' and r_x'' in Fig 2). The method adopted also removes the singularity due to divergence of P_x when velocity u_x in the denominator of (21) falls to zero at the classical turning points. The resulting formulae depend

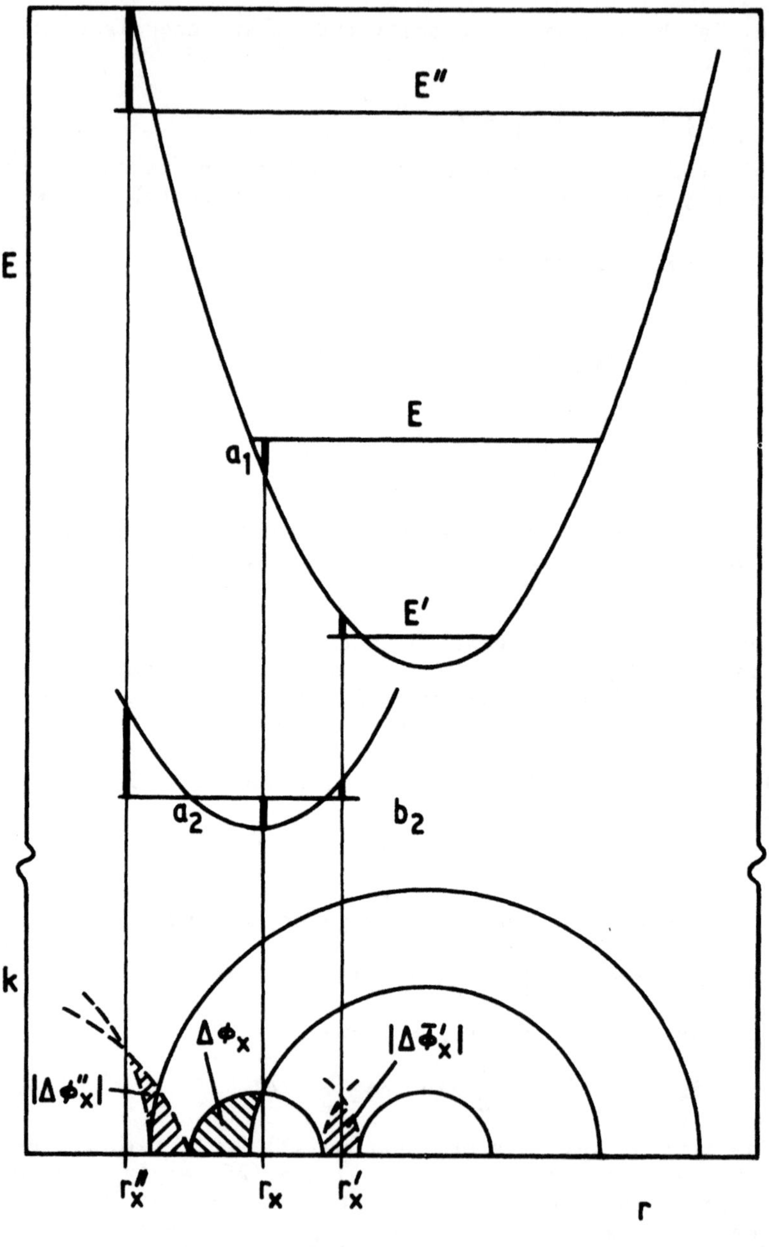

Fig 2. Psuedo crossing points (r_x, r_x',r_x'') and semiclassical phases in classically allowed (E) and forbidden (E',F'') cases.

on the same quantities P_x and $\Delta\phi_x$ or $\Delta\overline{\phi}_x$ as those defined by (21) (22) and (23), taking note that all these quantities become purely imaginary in classically inaccessible cases. This is indicated in the lower part of Fig 2, by using dashed lines to indicate the magnitude of the imaginary reduced momenta

$$|k(r)| = [2\mu(V(r) - E)]^{\frac{1}{2}}/\hbar \tag{26}$$

The simplest uniform approximations [8] is given in terms of the Airy function (see appendix).

$$S_{12} = (\pi\,|P_x|)^{\frac{1}{2}}\,|\xi|^{\frac{1}{4}}\,\mathrm{Ai}(-\xi) \tag{27a}$$

$$= (-1)^{v_2}(\pi\,|P_x|)^{\frac{1}{2}}\,|\overline{\xi}|^{\frac{1}{4}}\,\mathrm{Ai}(-\overline{\xi}) \tag{27b}$$

with the arguments ξ and $\overline{\xi}$ given by

$$\begin{aligned} \xi &= (3\,\Delta\phi_x/2)^{2/3}, \quad a_2 < r_x < b_2 \\ &= -\,(3\,|\Delta\phi_x|/2)^{2/3}, \quad r_x < a_2 \end{aligned} \tag{28a}$$

$$\begin{aligned} \overline{\xi} &= (3\,\Delta\overline{\phi}_x/2)^{2/3}, \quad a_2 < r_x < b_2 \\ &= -\,(3\,|\Delta\overline{\phi}_x|/2)^{2/3}, \quad r_x > b_2 \end{aligned} \tag{28b}$$

The choice between (27a) and (27b) in the classical case, $a_2 < r_x < b_2$, is made such that the smaller of the two phase differences $\Delta\phi_x$ and $\Delta\overline{\phi}_x$ is adopted. The derivation of these results is far from simple [9,10] but that of a related transitional approximation is set as example 1 at the end of this chapter.

Fig. 3 shows the level of agreement between exact $n \to 0$ overlap integrals and the uniform Airy approximation for the model defined by

$$\begin{aligned} H_1 &= \tfrac{1}{2}p^2 + \tfrac{1}{2}\,r^2 \\ H_2 &= \tfrac{1}{2}p^2 + \tfrac{1}{2}\,(r - c)^2 \end{aligned} \tag{29}$$

with $c = 2.5$ scaled units. The error in every case is less than 5%. The primitive approximation on the other hand is seriously inaccurate over all but a very small part of the classically accessible range. Substantial improvements in the primitive approximation are expected for transitions from levels with $n \neq 0$, but the Airy approximation, which requires little additional effort, must still be superior.

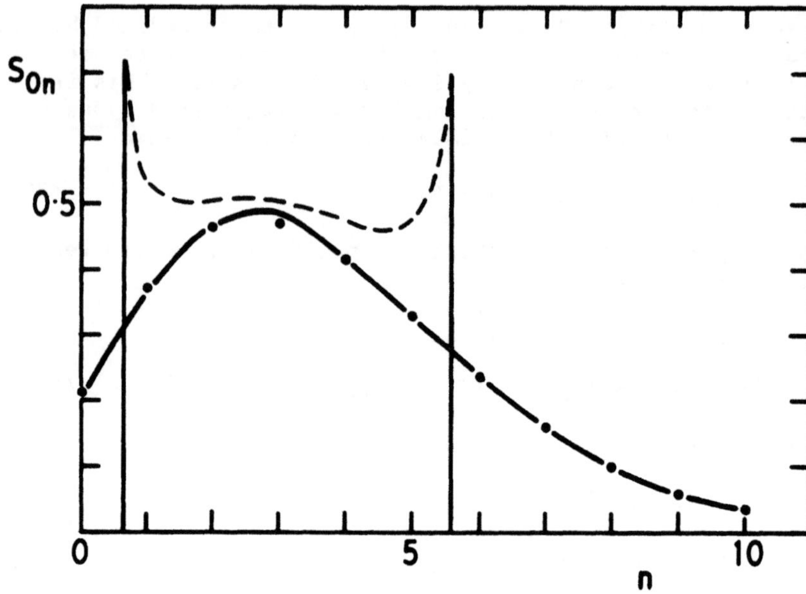

Fig 3. Comparison between exact (points), Airy uniform (solid line) and primitive (dashed line) $n \to 0$ overlap integrals for the model defined by (29). Verticals indicate the classical thresholds.

The above results are readily extended to the calculation of matrix elements, using the formula

$$M_{12} = \int_0^\infty \psi_1(r)\; M(r)\; \psi_2(r)dr \simeq M(r_x)\; S_{12} \tag{30}$$

with S_{12} given by (22) or (27), provided that the function $M(r)$ is essentially constant over the stationary phase region, in other words provided that

$$(d\; \ell nM/dr)/(\Delta\phi'')^{\frac{1}{2}} = (d\; \ell nM/dr)[\hbar u_x/(F_1-F_2)]^{\frac{1}{2}} \ll 1 \tag{31}$$

One important restriction on the use of equations (16), (22) and (27) in any application is however that equation (13) for the pseudo-crossing point should have only a single root. Situations involving one or other potential with a very shallow branch can give rise to multiple roots. The exact Franck-Condon overlap then shows complicated higher interference structure [11], and even the present uniform expressions diverge for critical transitions for which the force difference $(F_1 - F_2)$ in equation (21) goes to zero [12].

1.3 Condon Reflection Characteristics

A different type of uniform approximation, suggested by analogy between equations (1) and (20), may be written [11,13]

$$S_{12} = [u(\xi)/u_x(F_1 - F_2)]^{\frac{1}{2}} \psi_{n_2}(\xi) \qquad (32)$$

for a bound-continuum transition, with an additional factor $\hbar\bar{\omega}_1$ in the bound-bound case. Here $\psi_{n_2}(\xi)$ is the n_2th wavefunction for any convenient potential $V_2(\xi)$; the argument ξ is implicitly y defined by the equation

$$\frac{1}{\hbar}\int_{a_2}^{\xi} [2\mu(E_2 - V_2(\xi))]^{\frac{1}{2}} d\xi = \Delta\phi_x \qquad (33)$$

and $u(\xi)$ is the classical velocity

$$u(\xi) = [\mu(E_2 - V_2(\xi))/2]^{\frac{1}{2}} \qquad (34)$$

Choice of n_2 as the index in (32) ensures that real values of the integral in (33) and of $\Delta\phi_x$ lie in the common range $(0,(2n_2 + 1)\pi/2)$.

The simplest reference model is the scaled harmonic oscillator for which

$$\psi_n(\xi) = \pi^{-\frac{1}{4}}(2^n n!)^{-\frac{1}{2}} H_n(\xi) \exp(-\xi^2/2) \qquad (35)$$

$$\Delta\phi_x = \tfrac{1}{2}\{\xi(2n+1 - \xi^2)^{\frac{1}{2}} - (2n+1)[\pi - \cos^{-1}(\xi/(2n+1)^{\frac{1}{2}})]\} \qquad (36)$$

for $a_2 < r_x < b_2$, and

$$\Delta\phi_x = \frac{i}{2}\{\xi(\xi^2 - 2n+1)^{\frac{1}{2}} - (2n+1)\ln[(\xi + (\xi^2 - 2n+1)^{\frac{1}{2}}/(2n+1)^{\frac{1}{2}}]\} \qquad (37)$$

for $r_x < a_2$, $\Delta\phi_x$ being replaced by $\Delta\bar{\phi}_x$ in (35) when $r_x > b_2$

$$u(\xi) = [(2n+1) - \xi^2]^{\frac{1}{2}} \qquad (38)$$

Fig 4 gives a revealing illustration of the difference between this harmonic uniform approximation and the previous Airy result (27). The former is seen as a phase space mapping between the actual phase overlap and the phase generated at the point ξ by the harmonic motion, whereas the comparison in the latter case is with motion under a linear potential. This also explains why it is necessary to switch from one form of equation (27) to the other.

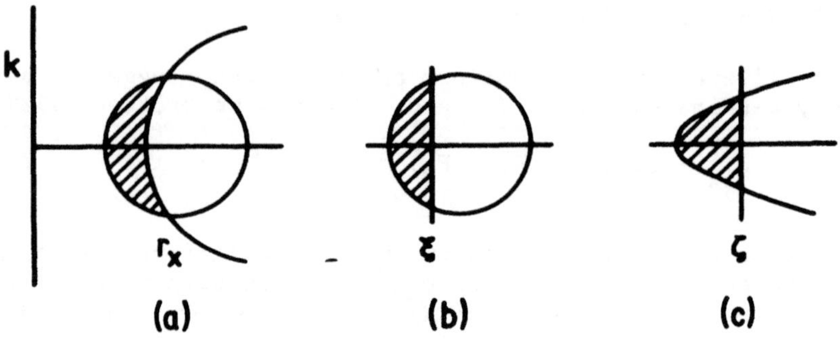

Fig 4. Phase space mappings between (a) the real problem, (b) the harmonic, and (c) the Airy model. The shaded areas are defined by equations (28) and (36) to be equal.

The transcendental nature of (31),(34) and (35) complicates the determination of ξ, but the uniform harmonic approximation defined by (32) and (35)-(38) may be marginally more accurate than the uniform Airy result. The rms. errors for the model illustrated in Fig 3 are 1% and 2% respectively for example.

The main advantage of (32) over (27) is however that it emphasises the relation between the forms of the Franck-Condon overlap function and the parent wavefunction. Thus provided r_x varies along a relevant progression over the full range $r_x < a_2$ to $r_x > b_2$ as illustrated in Fig. 2, equation (33) shows that ξ will also run from one classically forbidden region to the other and hence S_{12} will have the same number of nodes (as a function of E or n_1) as $\psi_{n2}(\xi)$. This condition is always satisfied for bound-continuum transitions, but it requires an adequate displacement between the potential minima in the bound-bound case.

This qualitative connection was originally expressed in quantitative form by the reflection formula [14].

$$S_{12}(E) = (dV_1/dr)^{-\frac{1}{2}} \psi_{n_2}(r(E)) \tag{39}$$

with r(E) given by

$$V_1(r) = E \tag{40}$$

The derivation of (39) assumes that the continuum wavefunction $\psi_1(r)$ behaves as a delta function centred at the turning point of the upper potential $V_1(r)$. Comparison of equations (32) and (33) shows that the validity of (39) requires that the potential $V_1(r)$ is so steep that the second contribution to $\Delta\phi_x$ is negligible in equation (22) because (33) then becomes

$$\frac{1}{\hbar}\int_{a_2}^{\xi}[2\mu(E_2 - V_2(\xi))]^{\frac{1}{2}}\,d\xi = \frac{1}{\hbar}\int_{a_2}^{r_x}[2\mu[E_2 - V_2(r)]^{\frac{1}{2}}dr \quad (41)$$

with solution

$$\xi = r_x \quad (42)$$

It then follows that $u(\xi) = u_x$ in (32). Finally neglect of the second term in (22) implies that $F_1 >> F_2$, so only difference between (32) and (39) in this limit lies in the appearence of r_x given by (13) rather than $r(E)$ given by (40).

1.4 Analysis of Predissociation level widths

The type I (curve-crossing) predissociation level width, given by the golden rule formula

$$\Gamma = 2\pi\left|\int_0^{\infty} \psi_1(r)\ H_{12}(r)\ \psi_2(r)\ dr\right|^2 \quad (43)$$

may be approximated in the light of (21), (27) and (30) as

$$\Gamma = [4\pi|H_{12}(r_x)|^2\ \hbar\bar{\omega}_2/\hbar u_x(F_1 - F_2)]\ \xi^{\frac{1}{2}}\ Ai^2(-\xi) \quad (44)$$

with ξ given by

$$\xi = (3\ \Delta\phi_x/2)^{2/3} \quad (45)$$

$$\Delta\phi_x = \int_{a_2}^{r_x} k_2(r)dr - \int_{a_1}^{r_x} k_1(r)dr \quad (46)$$

where r_x is the actual crossing point between the two curves,

$$V_1(r_x) = V_2(r_x) = E_x \quad (47)$$

because the two energies E_1 and E_2 are equal in a predissociation case. Equations (44) - (46) assume that $\Delta\phi_x < \Delta\bar{\Phi}_x$ otherwise ξ and $\Delta\phi_x$ would be replaced by $\bar{\xi}$ and $\Delta\bar{\Phi}_x$.

The dominant variation in equation (44) comes from the Airy function itself, and the form of this function in the Appendix predicts an initial exponential rise in Γ for energies below the crossing point (corresponding to $\xi < 0$) followed by an oscillatory variation for $E > E_x$ and $\xi > 0$. This means that an initial estimate of the crossing point energy is immediately available whenever this transition from exponential to oscillatory behaviour is observed. Subsequent analysis is concerned with refining this estimate, and with determining $H_{12}(r_x)$ and the repulsive potential $V_1(r)$. It is assumed that the bound state potential is known. A systematic procedure based on the following observations has been developed [15].

(i) Given E_x, the velocity u_x in (44) is known, and the mean vibrational spacing $\hbar\omega_2$ is available from the spectrum. A convenient reduced linewidth is therefore given by

$$\gamma = \Gamma(E-E_x)^{\frac{1}{2}}/\hbar\bar{\omega} = \pi A\ \xi^{\frac{1}{2}}\ Ai^2(-\xi)$$
$$\overset{\xi>1}{\sim}\ A\sin^2(\Delta\phi_x + \pi/4) \tag{48}$$

$$A = 2(2\mu)^{\frac{1}{2}}H_{12}^2(r_x)/\hbar(F_1-F_2) \tag{49}$$

The maximum value of γ in the sinusoidal region therefore fixes the ratio $H_{12}^2(r_x)/(F_1-F_2)$, and the remaining values are used to assign values of $\Delta\phi_x$ to the observed levels ($\Delta\phi_x$ must be deduced from ξ for levels near the crossing point).

(ii) Knowledge of the bound state curve $V_2(r)$ fixes

$$\phi_2(r_x) = \int_{a_2}^{r_x} k_2(r)dx \tag{50}$$

for any assumed value of the crossing point. Thus the purely repulsive component is given by

$$\phi_1(r_x) = \phi_2(r_x) - \Delta\phi_x \tag{51}$$

and this should vary linearly with $(E - E_x)^{3/2}$ for energies close to the crossing point, an observation which is used to refine E_x.

(iii) Knowledge of $\phi_1(r_x)$ as a function of E may be inverted by the RKR method to give the repulsive turning point $a_1(U)$ as a function of energy.

$$a_1(U) = r_x - \frac{\hbar}{\pi}\left(\frac{2}{\mu}\right)^{\frac{1}{2}}\int_{E_x}^{U}\frac{d\phi_1}{dE}\frac{dE}{(U-E)^{\frac{1}{2}}} \tag{52}$$

An example of the application of this procedure to the magnetically induced predissociation of the B state of I_2 is illustrated in Fig. 4.

2. TUNNELING AND CURVE-CROSSING

2.1 Semiclassical Connection Formulae

A wide variety of tunneling and curve-crossing problems may be handled by use of semiclassical connection formulae [1,16]

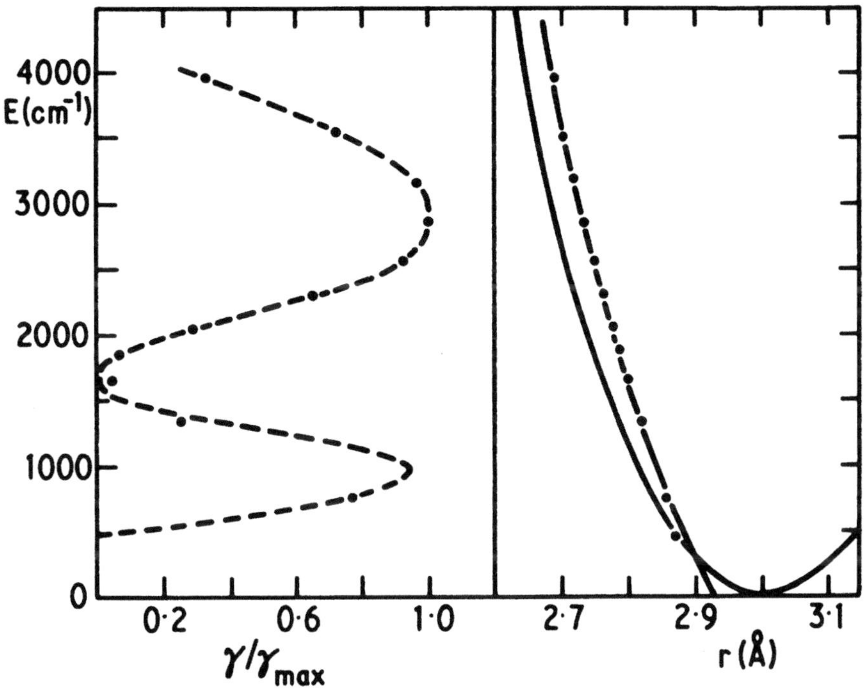

Fig 5. Analysis of the magnetically induced predissociation of I_2 [15]. Data points plotted as (γ/γ_{max}) compared with $\pi\xi^{\frac{1}{2}}Ai^2(-\xi)$ to the left. Recovered turning points compared with a published curve [29] to the right.

These formulae link the coefficients (P',P'') in the WKB expansion

$$\psi(r) \simeq [\mu/2\pi\hbar^2 k(r)]^{\frac{1}{2}} \{P' \exp [i\int_p^r k(r)dr] + P'' \exp [-i\int_p^r k(r)dr]\} \quad (53)$$

in one classically accessible region of space with those in another. Here and elsewhere prime and double prime apply to positive and negative momentum components respectively

Clearly the phases of (P',P'') depend on the choice of the

phase reference point p, and the same function $\psi(r)$ could be defined with reference to a new point q and new coefficients (Q',Q'') related to the old by

$$\begin{pmatrix} Q' \\ Q'' \end{pmatrix} = \begin{pmatrix} \exp(i\sigma) & , \quad 0 \\ 0 & , \exp(-i\sigma) \end{pmatrix} \begin{pmatrix} P' \\ P'' \end{pmatrix} \tag{54}$$

where

$$\sigma = \int_p^q k(r)dr, \tag{55}$$

provided p and q lie in the same semiclassical segment. This is the simplest connection formula.

Another familiar example is the relation between the coefficients (A',A'') defined with respect to a left hand turning point, because equation (1) implies in this case that

$$\begin{pmatrix} A' \\ A'' \end{pmatrix} = A \begin{pmatrix} \exp(-i\pi/4) \\ \exp\ (i\pi/4) \end{pmatrix} \tag{56}$$

Similarly at a right hand turning point

$$\begin{pmatrix} B' \\ B'' \end{pmatrix} = B \begin{pmatrix} \exp(i\pi/4) \\ \exp(-i\pi/4) \end{pmatrix} \tag{57}$$

Passage from A'' to A' and from B' to B'' therefore involves a phase change of $(-\pi/2)$ as indicated in Fig 6.

The presence of a potential barrier is handled by mapping the general problem onto that of a quadratic barrier [17]. The resulting connection between the right hand (C_1',C'') and left hand (B',B'') coefficients shown in Fig 6(b) may be written

$$\begin{pmatrix} C' \\ C'' \end{pmatrix} = \begin{pmatrix} (1+\kappa^2)^{\frac{1}{2}}\exp(-i\phi), & -i\kappa\ \exp(i\theta) \\ i\kappa\ \exp(-i\theta) & , \quad (1+\kappa^2)^{\frac{1}{2}}\ \exp\ (i\phi) \end{pmatrix} \begin{pmatrix} B' \\ B'' \end{pmatrix} \tag{58}$$

or

$$\underset{\sim}{C} = \underset{\sim}{M}\ \underset{\sim}{B} \tag{59}$$

where

$$\kappa = \exp\ (-\pi\varepsilon) \tag{60}$$

$$\pi\varepsilon = -\int_b^c |k(r)|dr \quad \text{for} \quad E < V_{max}$$

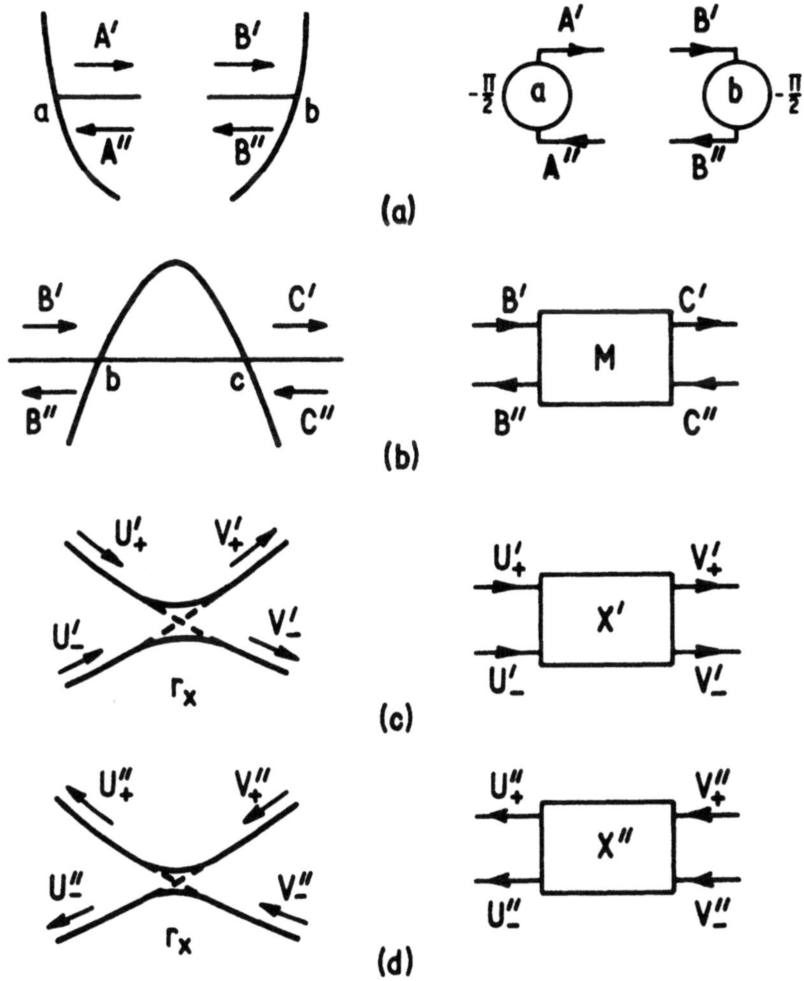

Fig 6. Diagrammatic connections (a) turning points, (b) barrier maximum, (c) curve-crossing, positive momentum, (d) curve-crossing negative momentum.

$$= -i \int_{r_+}^{r_-} k(r)\,dr \quad \text{for } E > V_{max} \tag{61}$$

$$\Theta = 0 \qquad \text{for } E < V_{max}$$

$$= \int_{r_+}^{r_{max}} k(r)dr + \int_{r_-}^{r_{max}} k(r)dr, \qquad E > V_{max} \tag{62}$$

$$\phi = \arg \Gamma\,(i\varepsilon+\tfrac{1}{2}) - \varepsilon \ln|\varepsilon| + \varepsilon \tag{63}$$

and $r_\pm$ are the complex points at which $k(r)$ vanishes for $E > V_{max}$. Equations (14) and (15) reduce to

$$\varepsilon = (E-V_{max})/\hbar\omega^* \quad ; \qquad \theta = 0 \tag{64}$$

in the case of a quadratic barrier with imaginary frequency $i\omega^*$. The coefficients (B',B'') and (C',C'') are defined with respect to the phase reference points b and c in Fig. 6(b) respectively. Numerical values of the phase correction $\phi(\varepsilon)$ are given in the table below.

The curve-crossing case is most conveniently handled with reference to the adiabatic curves.

$$V_\pm(r) = \tfrac{1}{2}\,[V_1(r) + V_2(r)] \pm \tfrac{1}{2}\{[V_1(r) - V_2(r)]^2 + 4V_{12}^2(r)\}^{\frac{1}{2}} \tag{65}$$

and with the corresponding functions $k_\pm(r)$ in the analogue of (51) given by

$$k_\pm(r) = [2\mu(E - V_\pm(r)]^{\frac{1}{2}}/\hbar \tag{66}$$

The amplitude coefficients {U} and {V} in figs 6(c) and 6(d) are defined with respect to the curve-crossing point r_x. The connection formulae may be shown[18,20] to be.

$$\begin{pmatrix} V'_+ \\ V'_- \end{pmatrix} = \begin{pmatrix} [1-\lambda^2]^{\frac{1}{2}} \exp(i\chi), & -\lambda \\ \lambda\ , & [1-\lambda^2]^{\frac{1}{2}} \exp(-i\chi) \end{pmatrix} \begin{pmatrix} U'_+ \\ U'_- \end{pmatrix} \tag{67}$$

$$\begin{pmatrix} U_+'' \\ U_-'' \end{pmatrix} = \begin{pmatrix} [1-\lambda^2]^{\frac{1}{2}} \exp(i\chi) , & \lambda \\ -\lambda, & [1-\lambda^2]^{\frac{1}{2}} \exp(-i\chi) \end{pmatrix} \begin{pmatrix} V_+'' \\ V_-'' \end{pmatrix} \tag{68}$$

or $\underset{\sim}{V}' = \underset{\sim}{X}' \, \underset{\sim}{U}'; \; \underset{\sim}{U}'' = \underset{\sim}{X}'' \, \underset{\sim}{V}''$ (69)

where

$$\lambda = \exp(-\pi\nu) \tag{70}$$

$$2\pi\nu = \text{Im} \left\{ \int_{r+}^{r-} [k_-(r) - k_+(r)]dr \right\} \tag{71}$$

$$\chi = \arg\Gamma(i\nu) - \nu \ell n \nu + \nu\pi/4 \tag{72}$$

with the integration limits $r_\pm$ in (71) taken to be the complex transition points at which the adiabatic terms $V_\pm(r)$ intersect, in other words at which

$$[V_1(r) - V_2(r)]^2 + 4V_{12}^2(r) = 0 \tag{73}$$

Numerical values of the phase term χ are given in the table below. Note that equations (67)-(69) are identical to corresponding forms in reference [16], but with the previous coefficients (U_1, U_2, V_1, V_2) now replaced by (U_+, $-U_-$, V_-, V_+) respectively for both primed and unprimed sets.

As presented equation (71) gives the Stückelberg [21] form for the switching parameter ν, but the integral is not always readily evaluated. The Landau-Zener [22,23] approximation is

$$\nu = V_{12}^2/\hbar u_x(F -F_2) \tag{74}$$

where

$$V_{12} = V_{12}(r_x) \tag{75}$$

$$u_x = [2(E - E_x/\mu]^{\frac{1}{2}} \tag{76}$$

$$F_i = -(dV_i/dr)_{r_x}; \quad i = 1,2 \tag{77}$$

Note that V_{12} and F_1 and F_2 are defined in the diabatic representation. Note also that (67)-(77) are inapplicable close to and below the crossing point [1].

Table 1. Values and properties of the phase functions $\phi(x)$, $\chi(x)$ given by equations (63) and (72)

x	$\phi(x)$	$\chi(x)$	x	$\phi(x)$	$\chi(x)$
0	0.000	-0.785	0.6	0.080	-0.152
0.1	0.137	-0.512	0.7	0.068	-0.129
0.2	0.150	-0.376	0.8	0.058	-0.111
0.3	0.135	-0.287	0.9	0.051	-0.098
0.4	0.115	-0.226	1.0	0.045	-0.087
0.5	0.096	-0.183			

$$\phi(x) = -\phi(-x) \overset{x>1}{\sim} 1/24x + 7/2880x^3$$

$$\chi(x) \overset{x>1}{\sim} -1/12x - 1/360x^3 - 1/1260x^5$$

2.2 Double minimum potential

The problem of quantisation in a double minimum potential will be taken as a first example of the use of connection formulae. The relevant linkage of elementary components from Fig. 6 is illustrated in Fig. 7.

Here we note that equation (58) connects (B',B'') to (C', C'') provided the phase reference points are taken as b and c. Equations (54),(56) and (57) also imply a connection between (B',B'') and the amplitude coefficient A at turning point a, and similarly for (C',C'') and a coefficient D at d.

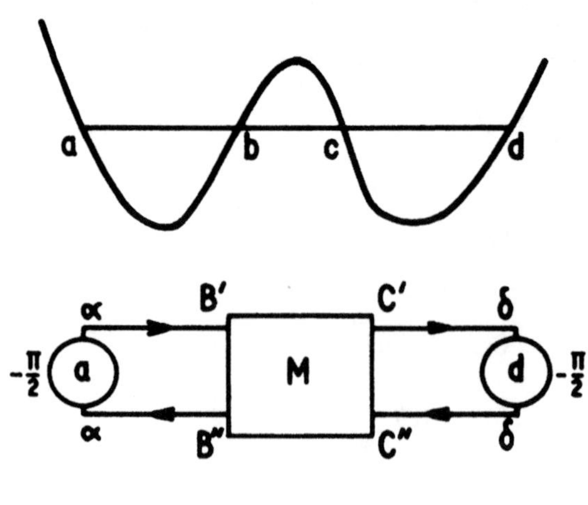

Fig 7. Diagram and potential curve for the double minimum potential.

These connections may be expressed in the form

$$\underset{\sim}{B} = A \underset{\sim}{L}(\alpha)\underset{\sim}{u} \quad ; \quad \underset{\sim}{C} = D \underset{\sim}{L}^{-1}(\delta)\underset{\sim}{u} \tag{78}$$

where A and D are real numbers and

$$\underset{\sim}{L}(\alpha) = \begin{pmatrix} \exp(i\alpha - i\pi/4), & 0 \\ 0, & \exp(-i\alpha + i\pi/4) \end{pmatrix} \tag{79}$$

$$\underset{\sim}{u} = \begin{pmatrix} 1 \\ 1 \end{pmatrix} \tag{80}$$

Thus the overall connection implied by the diagram in Fig 6. may be written

$$D \underset{\sim}{L}^{-1}(\delta)\underset{\sim}{u} = A \underset{\sim}{M} \underset{\sim}{L}(\alpha)\underset{\sim}{u} \tag{81}$$

or

$$\underset{\sim}{L}(\delta) \underset{\sim}{M} \underset{\sim}{L}(\alpha)\underset{\sim}{u} = (D/A)\underset{\sim}{u} \tag{82}$$

This means that the energy must be chosen such that the composite matrix on the left of (82) has $\underset{\sim}{u}$ as an eigenvector, with the implication, given the form of $\underset{\sim}{u}$ that the sums of the elements in the two rows of the product matrix must be equal;

$$-i\{[1 + \kappa^2]^{\frac{1}{2}} \exp(i\alpha+i\delta-i\phi) + \kappa \exp(-i\alpha+i\delta+i\theta)\}$$
$$= i\{\kappa \exp(i\alpha-i\delta-i\theta) + [1 + \kappa^2]^{\frac{1}{2}} \exp(-i\alpha-i\delta+i\phi)\} \tag{83}$$

It follows by simple rearrangement that the quantisation condition may be written

$$[1 + \kappa^2]^{\frac{1}{2}} \cos(\alpha + \delta - \phi) = -\kappa \cos(\alpha - \delta - \theta) \tag{84}$$

As derived this is a first order result, but the structure of the equation is retained in all orders [24].

Various limiting results applicable to symmetric and non-symmetric double wells at high and low energies may be derived from (84) [16]. Present attention is restricted to the symmetric situations, in which case $\theta = 0$. It is convenient to define the modified phase integral

$$\alpha' = \alpha - \phi/2 \tag{85}$$

and to express (84) in the form

$$\cos\ 2\alpha' = -\ \kappa/[1 + \kappa^2]^{\frac{1}{2}} \tag{86}$$

A graphical solution of this equation si shown in Fig 8. Clearly at low energies, $E \ll V_{max}$ when $\kappa \to \infty$, the roots lie close to the single-well energies E_v at which

$$\alpha' = \int_a^b k(r)dr - \tfrac{1}{2}\phi = (v + \tfrac{1}{2})\pi \tag{87}$$

and the expansion

$$\alpha'(E) = (v + \tfrac{1}{2})\pi + (E - E_v)\pi/\hbar\omega \tag{88}$$

is appropriate,where $\hbar\omega$ is the mean separation between the levels; given by (87). It follows on substitution in (86) that

$$E \simeq E_v \pm (\hbar\omega/2\pi\kappa)$$

$$= E_v \pm (\hbar\omega/2\pi)\ \exp\ (-\int_b^c |k(r)|dr) \tag{89}$$

This is the standard primitive semiclassical result [7].

Equation (86) may also be used to analyse the high energy limit, $E \gg V_{max}$, for which $\kappa \to 0$. The roots lie close to energies E' at which

$$2\alpha' = \int_a^d k(r)dr - \quad = (v' + \tfrac{1}{2})\pi \tag{90}$$

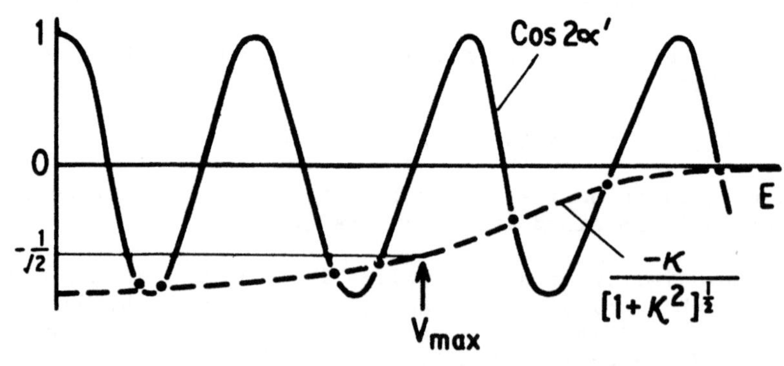

Fig 8. Graphical quantisation for the symmetric double well potential.

These are levels of the full double well, and the expansion

$$2\alpha'(E) \simeq (v' + \tfrac{1}{2})\pi + (E-E_v')\pi/\hbar\omega' \tag{91}$$

now yields

$$F = E_v' - (-1)^{v'} \kappa\hbar\omega'/\pi$$

$$\simeq E_v' - (-1)^{v'} (\hbar\omega'/\pi) \exp[-\pi(E-V_{max})/\hbar\omega^*] \tag{92}$$

where the quadratic approximation (64) has been applied. The corrections implied by the second term in (92) are less easy than those in (89) to detect experimentally but the location of V_{max} in Fig 7 (at which $\kappa=1$ and $\kappa[1+\kappa^2]^{-\frac{1}{2}} = 2^{-\frac{1}{2}}$) shows that perturbations above the barrier are as important in any full analysis as the level doubling below it. Such an analysis will yield both the 'barrier' and 'well' phase integrals which may be separately inverted by the RKR technique to determine the potential function [25]

Equation (84) has also been applied to quantisation problems in more than one degree of freedom [25,26].

2.2 Predissociation

The case of predissociation at arbitrary coupling strengths [20,27,28] will be taken to illustrate application of the curve-crossing connection formulae. The appropriate diagram is given in Fig. 9.

In view of the normalising factor in (53) the boundary conditions require that $|V_-'| = |V_-''| = 1$, because the wavefunction always belongs to the continuum, and the problem is to determine the amplitudes (A_+, A_- and B_+) at the turning points (a_+, a_- and b_+) and the asymptotic phase (say γ) consistent with this normalisation, with the known phases

$$\alpha_\pm = \int_{a_\pm}^{r_x} k(r)dr \; ; \quad \beta_+ = \int_{r_x}^{b_+} k_+(r)dr \tag{93}$$

and with the coupling strength parameter λ in (67). Fortunately it is sufficient to restrict attention to the upper part of Fig. 8

The behaviour at the turning points requires that

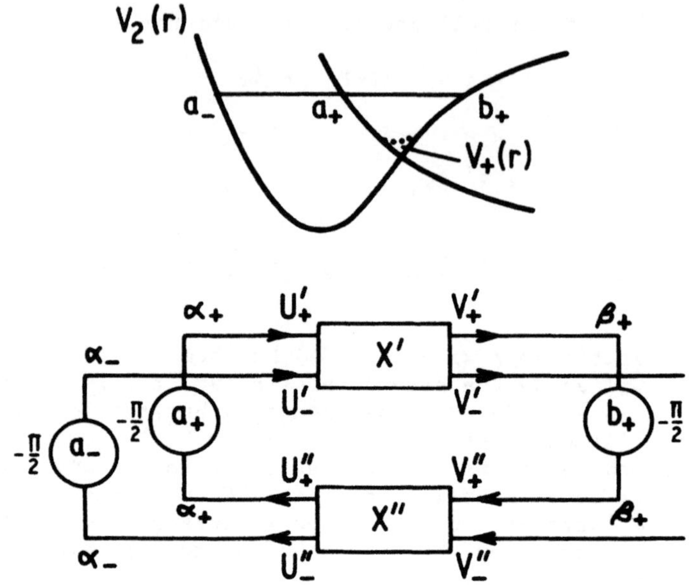

Fig 9. Diagram and potential curves for predissociation.

$$\begin{pmatrix} U'_+ \\ U'_- \end{pmatrix} = \begin{pmatrix} A_+ \exp(i\alpha_+ - i\pi/4) \\ A_- \exp(i\alpha_- - i\pi/4) \end{pmatrix} \tag{94}$$

$$\begin{pmatrix} V'_+ \\ V'_- \end{pmatrix} = \begin{pmatrix} B_+ \exp(-i\beta_+ + i\pi/4) \\ \exp(i\gamma) \end{pmatrix} \tag{95}$$

Thus the connection implied by (67) means that

$$A_+[1-\lambda^2]^{\frac{1}{2}} \exp(i\chi+i\alpha_+ - i\pi/4) - A_-\lambda \exp(i\alpha_- - i\pi/4)$$
$$= B_+ \exp(-i\beta_+ + i\pi/4) \tag{96}$$

$$A_+\lambda \exp(i\alpha_+ - i\pi/4) + A_-[1-\lambda^2]^{\frac{1}{2}} \exp(i\alpha_- - i\chi - i\pi/4)$$
$$= \exp(i\gamma) \tag{97}$$

It follows on equating the phases on both sides of (96) and rearranging the result that

$$A_+[1-\lambda^2]^{\frac{1}{2}}\cos(\alpha_+ + \beta_+ \chi) = \lambda A_- \cos(\alpha_- + \beta_+) \tag{98}$$

A second equation connecting A_+ and A_- may be obtained by equating moduli on both sides of (97)

$$A_+^2\lambda^2 + A_-^2(1-\lambda^2) + 2A_+A_-\lambda[1-\lambda^2]^{\frac{1}{2}}\cos(\alpha_+ - \alpha_- + \chi) = 1 \tag{99}$$

It follows on combining (98) and (99) that the amplitude of the wavefunction at the turning point a_- is determined by

$$A_-^2 = \frac{u(1+u)\cos^2\theta_+'}{[\cos^2\theta_2' + 2u\cos(\theta_+'-\theta_2')\cos\theta_+'\cos\theta_2' + u^2\cos^2\theta_+']^2} \tag{100}$$

where

$$u = \lambda^{-2} - 1 = \exp(2\pi\nu) - 1 \tag{101}$$

$$\theta_+' = \alpha_+ + \beta_+ + \chi \qquad \theta_2' = \alpha_- + \beta_+ \tag{102}$$

The reason for this notation is that the phases θ_+' and θ_2' closely approximate to the complete phase integrals for the adiabatic and diabatic potential wells $V_+(r)$ and $V_2(r)$ in Fig 9. respectively.

Attention is focussed on A_-^2 because it is the energy variation of this quantity which determines the spectroscopic lineshape. Consider the weak interaction limit, u<<1, for example. Equation (100) shows that A_-^2 is of order u except at energies E_2 at which

$$\theta_2' = (v_2 + \tfrac{1}{2})\pi \tag{103}$$

when A_-^2 becomes of order u^{-1}. The expansion

$$\theta_2'(E) \simeq (v_2 + \tfrac{1}{2})\pi + (E-E_2)\pi/\hbar\omega_2 \tag{104}$$

leads to the following Lorentzian approximation to A_-^2 in this case

$$A_-^2 = (\hbar\omega_2 \; \Gamma/2\pi)/[(E-E_2-\Delta)^2 + \Gamma^2/4] \tag{105}$$

$$\Gamma = 2(\hbar\omega_2/\pi)\, u \cos^2\theta_+' \tag{106}$$

$$\Delta = (\hbar\omega_2/\pi)\, u \sin\theta_+' \cos\theta_+' \tag{107}$$

It is readily verified that (106) is identical with the stationary phase approximation to (44), but the present result does not extend to the crossing point. Equation (107) indicates a level shift of the same order of magnitude.

The advantage of the present technique over (44) is that it is not restricted to the weak interaction case. A simple general theory may be developed by noting that the denominator vanishes identically if by chance.

$$\theta_2' = (v_2 + \tfrac{1}{2})\pi \quad \text{and} \quad \theta_+' = (v_+ + \tfrac{1}{2})\pi \tag{108}$$

at any energy; such levels give rise to completely sharp lines in the spectrum. The positions and widths of neighbouring levels may be estimated by use of the approximations

$$\theta_2'(E) \simeq (v_2 + \tfrac{1}{2})\pi + (E-E_2)\pi/\hbar\bar{\omega}_2$$

$$\theta_+'(E) \simeq (v_+ + \tfrac{1}{2})\pi + (E-E_+)\pi/\hbar\bar{\omega}_+ \tag{109}$$

where E_2 and E_+ denote energies at which the two parts of (108) are separately satisfied. It is found after substitution in (100) followed by expansion of the trigonometric functions in appropriate small terms that

$$A_-^2 \simeq \left(\frac{\hbar\bar{\omega}_2}{1+x}\right) \frac{(\Gamma/2\pi)}{(E-\bar{E})^2 + \Gamma^2/4} \tag{110}$$

where

$$\bar{E} = (E_2 + xE_+)/(1 + x) \tag{111}$$

$$\Gamma = 2\pi\, x(\gamma + x)\, (E_2 - E_+)^2/\hbar\bar{\omega}_2(1 + x)^3 \tag{112}$$

$$x = u\hbar\bar{\omega}_2/\hbar\bar{\omega}_+ \tag{113}$$

$$\gamma = \hbar\bar{\omega}_2/\hbar\bar{\omega}_+ \tag{114}$$

with the Landau-Zener coupling strength parameter u given by (70), (71), (74) and (101).

Fig. 10 illustrates the physical significance of these results. Here E_2 and E_+ are visualised as hypothetical bound vibration-rotation levels in the diabatic and adiabatic wells, $V_2(r)$ and $V_+(r)$ respectively. Equation (111) shows that the observed level $\bar{E}$ is a simple average of E_2 and E_+ determined by the coupling parameter u and the frequency ratio $(\hbar\bar{\omega}_2/\hbar\bar{\omega}_+)$. Secondly the quadratic dependence of the level width Γ on (E_2-E_+) implies a completely sharp level at any coincidence between E_2 and E_+ brought about by the two different rotational constants.

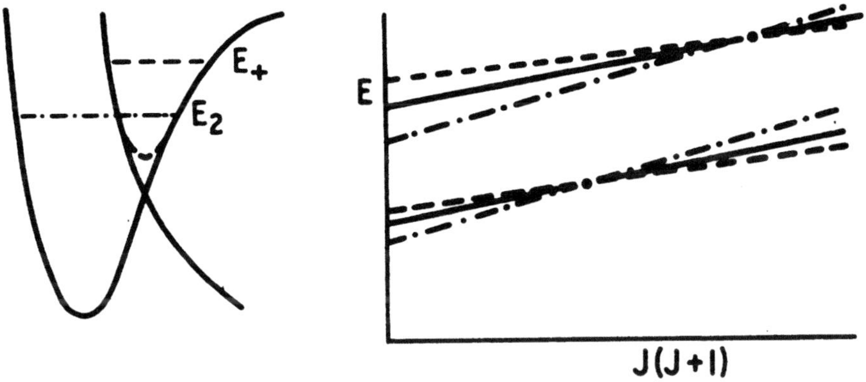

Fig 10. Diabatic (-.-.-.-), adiabatic (----), and actual (——) term values.

Finally in the near diabatic limit ($x \to 0$) $\bar{E} \simeq E_2$ and Γ is of order x, while in the near adiabatic limit ($x \to \infty$), $\bar{E} \simeq E_+$ and Γ is of order x^{-1}. Thus equations (110)-(114) cover the full range of coupling strengths in a physically acceptable manner. The B' ← X spectrum of IBr has been interpreted in detail using these equations in the intermediate coupling regime [27]. Excellent agreement with exact numerical calculations has also been reported [28].

3. PROBLEMS

1. At energy E, the potential

$$V_i(x) = E_x - F_i x$$

supports the continuum wavefunction

$$\psi_i(x) = D_i\, A_i(-\beta_i\,(x - a_i))$$

where

$$\beta_i = (2\mu\, F_i/\hbar^2)^{1/3}, \quad a_i = -(E - E_x)/F_i$$

$$D_i = (2\mu\,/\beta_i\,\hbar^2)^{\frac{1}{2}}$$

(i) Use the equations (1) and (8) in the text and the asymptotic form

$$Ai(-z) \overset{z>>1}{\sim} \pi^{-\frac{1}{2}}\, z^{-\frac{1}{4}} \sin\,(2/3\, z^{3/2} + \pi/4)$$

to verify the above normalisation.

(ii) Use the identities

$$Ai(-z) = \frac{1}{2\pi}\int_{-\infty}^{\infty} \exp\left(\frac{iu^3}{3} - iuz\right) du$$

$$\delta(\alpha) = \frac{1}{2\pi}\int_{-\infty}^{\infty} \exp(-i\alpha x)\, dx$$

$$\int_{-\infty}^{\infty} f(u)\ \delta(u\beta_1 + v\beta_2)\, du = f(-v\beta_2/\beta_1)/\beta_1$$

to prove that

$$S_{12} = D_1 D_2 (\beta_1^3 - \beta_2^3)^{-1/3}\, Ai(-\beta_1\beta_2(a_1 - a_2)(\beta_1^3 - \beta_2^3)^{-1/3})$$

(iii) Verify that this result is consistent with equations (27a) and (28a) of the text.

2. Calculate the dependence on $K = J(J + 1)$ of the phase term $\Delta\phi_x$ which governs the predissociation linewidth in terms of the properties of the artificial potential well.

$$V_+(r) = V_1(r) \qquad r < r_x$$

$$= V_2(r) \qquad r > r_x$$

(i) Write

$$\Delta\phi_x(v_2,K) = \frac{\sqrt{2\mu}}{\hbar}\int_{a_+}^{b_+} [G(v_2) + B(v_2)\, K - V_+(r) - K\hbar^2/2\mu r^2]^{\frac{1}{2}} dr$$

(ii) Show that at $K = 0$

$$\left(\frac{\partial\Delta\phi_x}{\partial K}\right)_{v_2} = \frac{\pi[B(v_2) - B_+]}{\hbar\bar{\omega}_+}$$

where

$$B_+ = \frac{\hbar^2}{2\mu}\int_{a_+}^{b_+} \frac{1}{r^2}\frac{dr}{[G(v_2) - V_+(r)]^{\frac{1}{2}}} \Bigg/ \int_{a_+}^{b_+} \frac{dr}{[G(v_2) - V_+(r)]^{\frac{1}{2}}}$$

and $\hbar\bar{\omega}_+$ is the energy over which $\Delta\phi_x(v_2,0)$ varies by π

3. Derive the quantisation condition for restricted rotation by a symmetric three fold barrier. Show first that the amplitude coefficients (A_i', A_i'') and (A_j', A_j'') immediately to the left of successive barriers are related by

$$\underset{\sim}{A}_i = \underset{\sim}{K}(\alpha)\, \underset{\sim}{M}\, \underset{\sim}{A}_j$$

$$\underset{\sim}{K}(\alpha) = \begin{pmatrix} \exp(i\alpha), & 0 \\ 0, & \exp(-i\alpha) \end{pmatrix}$$

where α is the phase integral for a single well and $\underset{\sim}{M}$ is the barrier transmission matrix (58). Deduce that the solution is single valued provided that

$$(\underset{\sim}{K}(\alpha)\, \underset{\sim}{M})^3\, \underset{\sim}{A}_i = \underset{\sim}{A}_i$$

This means that the product $\underset{\sim}{K}\,\underset{\sim}{M}$ should have one eigenvalue equal to the cube root of unity. Hence obtain the quantisation condition

$$\cos(\alpha - \phi) = (1 + \kappa^2)^{-\frac{1}{2}} \cos(2\pi k/3), \; k = 0, \pm 1$$

and sketch this result graphically.

4. Draw the connection formula diagram, appropriate to two potential wells, which cross at a point r_x in the classically accessible region, and deduce that, in the notation of equations (101) and (102), the quantisation condition may be written

$$\cos\theta_1' \cos\theta_2' = u \cos\theta_+' \cos\theta_-'$$

Discuss the limiting behaviour when $u \to 0$ and $u \to \infty$. Use the notation.

$$\theta' = \alpha_- + \beta_- - X$$

REFFRENCES

1. Child M.S.: 1974, *Molecular Collision Theory*,(Academic Press)
2. Smith, A.L.: 1971, J. Chem Phys. 55, p 4344.
3. Allison, A.C., and Dalgarno, A.: 1971, J. Chem Phys 55, p 4342.
4. Hudson, R.D., Carter, V.L., and Stein, J.A.: 1966, J. Geophys. Res. 7 , p 2295
5. Farmer, A.J.D., Fabian, W., Lewis, B.R., Lokan, K.H., and Haddad, G.N.: 1969, J. Quant. Spect. Radiative Transfer 8, p 1969.
6. Fano, U., Cooper, J.W.: 1968, Rev. Mod. Phys. 40, p 441
7. Landau, L.D., and Lifshitz, E.M.,: 1965, *Quantum Mechanics* 2nd edition, (Pergamon).
8. Miller, W.H.: 1968, J. Chem. Phys. 48, p 464.
9. Child, M.S.: 1975, Mol. Phys. 29, p 1421
10. Bieniek, R.J.: 1977, Phys. Rev. A. 15, p 1513.
11. Hunt, P.M., and Child, M.S.,: 1978, Chem. Phys. Lett. 58, p 202.
12. Atabek, O., and Lefebvre, R.: 1977, J. Chem,Phys, 67, p. 4983
13. Child, M.S.: 1978, Mol. Phys. 35, p 759
14. Winans, J.G., and Stückelberg, E.C.G.,: 1928, Proc. Natl, Acad, Amer. 14, p 867
15. Child, M.S.: 1974 in *Molecular Spectroscopy* vol 2 ed. Barrow R.F., Long, D.A., and Millen D.J., (Chemical Society Specialist Periodical Report).
16. Child, M.S.: 1974, J. Mol. Spect. 53, p 280.
17. Miller, S.C., and Good, R.H., 1953, Phys. Rev. 91, p. 174.
18. Kotova, L.P.,: 1969, , Sow, Phys. J.E.T.P. 28, p 719.
19. Child, M.S.: 1969, Mol. Phys. 16, p 313.
20. Bandrauk, A.D., and Child, M.S.: 1970, Mol. Phys. 19, p 95
21. Stückelberg, E.C.G.: 1932, Hew. Phys. Acta 5, p 369.
22. Landau, L.D.: 1932, Phys. Z. Sow 2, p 46.
23. Zener, C: 1932, Proc. Roy. Soc. A137, p 696.
24. Fröman, N., Fröman, P.O., Myhrman, U., and Paulsson, R.: 1972, Ann. Phys. (N.Y.),
25. Pajunen, P.: 1978, D.Phil thesis, Oxford University.
26. Colwell, S.M., Handy, N.C., and Miller, W.H.,: 1978, J. Chem
27 Phys. 68, p 745.
27. Child, M.S.,: 1976, Mol. Phys. 32, p 1495
28. Child, M.S., and Lefebvre, R.: 1978, Chem. Phys. Lett. 55, p. 213.
29. Chapman, G.D., and Bunker, P.R.: 1972, J. Chem. Phys. 57, p 2951.

THE CLASSICAL S MATRIX

M.S. Child

Theoretical Chemistry Department
1 South Parks Road, Oxford OX1 3TG, U.K.

ABSTRACT: The classical theory of canonical transformations is briefly reviewed and applied to the generation of angle-action variables for one vibrational degree of freedom. Analytical aspects of classical inelastic scattering are discussed in a modified angle-action representation. The origins of the primitive semiclassical and various uniform approximations for the S matrix are outlined. Numerical comparisons between exact and semiclassical transition probabilities are given for collinear vibrationally inelastic and reactive models.

1. CLASSICAL CONSIDERATIONS

1.1 Canonical transformations

A transformation from one system of coordinates and momenta (p_1,q_1) to another system (p_2,q_2) is canonical [1] if Hamilton's equations

$$\dot{q}_i = (\partial H_i/\partial p_i)_{q_i}, \quad \dot{p}_i = -(\partial H_i/\partial q_i)_{p_i} \tag{1}$$

apply both in the old system with Hamiltonian $H_1(p_1,q_1)$ and in the new system with $H_2(p_2,q_2)$ such that

$$H_1(p_1,q_1) = H_2\,[p_2(p_1,q_1),\, q_2(p_1,q_1)] \tag{2}$$

The following argument shows that a sufficient condition is that the transformation should have unit Jacobian,

M. S. Child (ed.), Semiclassical Methods in Molecular Scattering and Spectroscopy, 155–177.

$$J = \partial(p_1,q_1)/\partial(p_2,q_2) = 1. \tag{3}$$

Suppose that (1) applies with $i = 1$. It follows from (2) that

$$\dot{q}_1 = (\partial H_1/\partial p_1)_{q_1} = (\partial p_2/\partial p_1)_{q_1}(\partial H_2/\partial p_2)_{q_2} + (\partial q_2/\partial p_1)_{q_1}(\partial H/\partial q_2)_{p_2}$$

$$\dot{p}_1 = -(\partial H_1/\partial q_1)_{p_1} = -(\partial p_2/\partial q_1)_{p_1}(\partial H_2/\partial p_2)_{q_2} - (\partial q_2/\partial q_1)_{p_1}(\partial H/\partial q_2)_{p_2} \tag{4}$$

and these equations may be inverted to give

$$(\partial H_2/\partial p_2)_{q_2} = +J[(\partial q_2/\partial p_1)_{q_1}\dot{p}_1 + (\partial q_2/\partial q_1)_{p_1}\dot{q}_1] = J\dot{q}_2$$

$$(\partial H_2/\partial q_2)_{p_1} = -J[(\partial p_2/\partial p_1)_{q_1}\dot{p}_1 + (\partial p_2/\partial q_1)_{p_1}\dot{q}_1] = -J\dot{p}_2 \tag{5}$$

because the terms in parentheses are the total time derivatives of $q_2(p_1,q_1)$ and $p_2(p_1,q_1)$. Thus the validity of (3) implies that (1) also applies with $i = 2$.

A standard technique for the construction of canonical transformation involves the use of generating functions, involving one new and one old variable [1]. These are conventionally denoted $F_1(q_1,q_2)$, $F_2(q_1,p_2)$, $F_3(p_1,q_2)$, $F_4(p_1,p_2)$, with the transformation equations defined by one or other of the forms.

$$p_1 = (\partial F_1/\partial q_1)_{q_2}, \quad p_2 = -(\partial F_1/\partial q_2)_{q_1} \tag{6}$$

$$p_1 = (\partial F_2/\partial q_1)_{p_2}, \quad q_2 = (\partial F_2/\partial p_2)_{q_1} \tag{7}$$

$$q_1 = -(\partial F_3/\partial p_1)_{q_2}, \quad p_2 = -(\partial F_3/\partial q_2)_{p_1} \tag{8}$$

$$q_1 = -(\partial F_4/\partial p_1)_{p_2}, \quad q_2 = (\partial F_4/\partial p_2)_{p_1} \tag{9}$$

The mixed nature of these equations means that it is necessary to rearrange (6),(7),(8) or (9) to obtain explicit expressions for $q_2(p_1,q_1)$ and $p_2(p_1,q_1)$. On the other hand equation (3) is always satisfied because it follows from (6) for example that for any $F_1(q_1,q_2)$

$$(\partial p_1/\partial q_2)_{q_1} = (\partial^2 F/\partial q_1 \partial q_2) = -(\partial p_2/\partial q_1)_{q_2} \tag{10}$$

which may be rearranged to give

$$-\left(\frac{\partial p_1}{\partial q_2}\right)_{q_1} \Big/ \left(\frac{\partial p_2}{\partial q_1}\right)_{q_2} = -\frac{\partial(p_1,q_1)}{\partial(q_2,q_1)} \; \frac{\partial(q_1,q_2)}{\partial(p_2,q_2)} = \frac{\partial(p_1,q_1)}{\partial(p_2,q_2)} = 1 \tag{11}$$

Finally the arbitrary choice of any one of the generating functions defines the transformation, but for any given transformation the validity of (6) - (9) requires the following interrelations between them.

$$F_2(q_1,p_2) = F_1[q_1,q_2(q_1,p_2)] + p_2\, q_2(q_1,p_2) \tag{12}$$

$$F_3(p_1,q_2) = F_1[q_1(p_1,q_2),q_2] - p_1\, q_1(p_1,q_2) \tag{13}$$

$$F_4(p_1,p_2) = F_1[q_1(p_1,p_2),q_2(p_1,p_2)] + p_2\, q_2(p_1,p_2) - p_1\, q_1(p_1,p_2). \tag{14}$$

1.2 Angle Action Variables

Problems involving periodic or quasi-periodic [1,2] motion are greatly simplified by transformation from cartesian variables (p,q) to angle-action variables, denoted here by (N,θ), such that the transformed Hamiltonian $\hat{H}(N)$ is independent of the angle θ. Attention is restricted to the simple case of vibration in one degree of freedom, for which the classical motion defines a closed orbit in phase space as illustrated in Fig 1.

Each such orbit may be regarded as an energy contour in phase space, and the object is to find a function $N(p,q)$ the contours of which coincide with these orbits. A second condition is that the conjugate variable $\theta(p,q)$ which defines a given point on the orbit should increase by 2π for each complete period of the motion. This angle need not however have a simple geometrical interpretation.

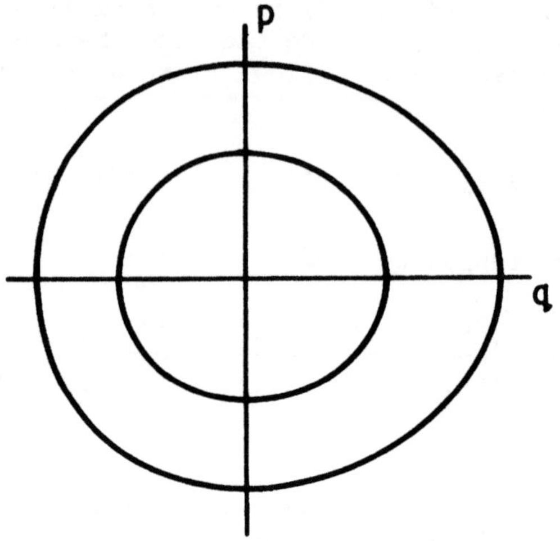

Fig 1. Phase space orbits for the Morse oscillator.

The transformation from (p,q) to (N,θ) is most readily achieved by means of an F_2 type generator, known as Hamilton's characteristic function $W(q,N)$. The necessary conditions on $W(q,N)$ are therefore that

$$p = (\partial W/\partial q)_N, \quad \theta = (\partial W/\partial N)_q \tag{15}$$

subject to the condition that

$$\oint \left(\frac{\partial\theta}{\partial q}\right)_N dq = 2\pi \tag{16}$$

Now the equality between the mixed partial derivatives of $W(q,N)$ allows us to rewrite (16) as

$$\oint \left(\frac{\partial p}{\partial N}\right)_q dq = \frac{\partial}{\partial N}\left[\oint p\ (N,q)\ dq\right] = 2\pi \tag{17}$$

from which it follows by integration with respect to N that

$$N = \frac{1}{2\pi}\oint p(N,q)\ dq \tag{18}$$

It now remains to express $p(N,q)$ in terms of the unknown transformed Hamiltonian $\tilde{H}(N)$ by rearranging the equation

$$H(p,q) = p^2/2\mu + V(q) = \tilde{H}(N) \tag{19}$$

and then to substitute in equation (18) to obtain the following implicit equation for $\tilde{H}(N)$

$$N = \frac{1}{\pi} \int_a^b [2\mu(\tilde{H}(N) - V(q))]^{\frac{1}{2}} \, dq \tag{20}$$

where a and b denote the turning points.

Finally, after solution of (20) for $\tilde{H}(N)$, the Hamilton-Jacobi equation [1,2]

$$H\,[(\partial W/\partial q)_N, q] = \tilde{H}(N) \tag{21}$$

is solved for the generator $W(q,N)$, and equations (15) are rearranged to obtain explicit expressions for (N,θ) in terms of (p,q) or vice versa. Application of this procedure to the simple harmonic oscillator is set as problem 1 at the end of the chapter.

The simplicity of the angle-action representation lies in the independence of $\tilde{H}(N)$ on the angle θ, because Hamilton's equations then become

$$\dot{N} = -(\partial\tilde{H}/\partial\theta)_N = 0, \quad \dot{\theta} = (\partial\tilde{H}/\partial N)_\theta = \omega(N) \tag{22}$$

with solution

$$N = \text{const}, \quad \theta = \omega(N)t + \text{const} \tag{23}$$

Thus N is a constant of the motion and θ increases linearly with time, at a frequency which may depend on N.

1.3 Classical Inelastic Scattering

Inelastic scattering arises from coupling between internal and translational degrees of freedom, expressed in the simplest form by the Hamiltonian

$$H = P^2/2m + H_{int}(p,q) + V(q,R) \tag{24}$$

or if angle action variables are employed for the internal motion

$$H = P^2/2m + \tilde{H}(N) + V(N,\theta,R) \tag{25}$$

Here $V(q,R)$ is defined to vanish as $R \to \infty$. Thus in the asymptotic limits equation (23) apply for the internal motion, the translational momentum in P is conserved and the coordinate R varies linearly with time;

$$R = R_o + Pt/m \tag{26}$$

Taken in conjunction with (23) this means that the modified angle variable

$$\bar{\theta} = \theta - m\omega(N)R/P \tag{27}$$

is a constant of the asymptotic motion.

It is therefore frequently convenient to visualise changes of the internal system in $(N,\bar{\theta})$ rather than (N,θ) space.

The modified angle $\bar{\theta}$ also has formal significance as the conjugate variable to N in the $(N,\bar{\theta},E,t)$ rather than the $(N,\theta,P,R,)$ asymptotic system [3]. Another aspect of this situation relevant to the scattering context is that the line integral

$$\Lambda\ (N_1,N_2) = \int_{N_1}^{N_2} \theta\ dN + \int_{R_1}^{R_2} PdR \tag{28}$$

taken along a trajectory from (N_1,R_1) to (N_2,R_2), with R_1 and R_2 large, acts as an F_4 type generator of the transformation from $(N_1\bar{\theta}_1)$ to $(N_2,\bar{\theta}_2)$ in the sense that [3,4]

$$(\partial\Lambda/\partial N_1)_{N_2} = -\ \bar{\theta}_1\ ,\ (\partial\Lambda/\partial N_2)_{N_1} = \bar{\theta}_2 \tag{29}$$

This follows because the momentum P is an explicit function of N in the asymptotic region

$$P = [2m\ (E - \tilde{H}(N))]^{\frac{1}{2}} \tag{30}$$

Thus on differentiation of (28) with respect to N_2 for example, followed by comparison with (22) and (27)

$$(\partial\Lambda/\partial N_2)_{N_1} = \theta_2 + R_2(dP_2/dN_2) = \theta_2 - m\omega(N_2)R_2/P_2 = \bar{\theta}_2 \tag{31}$$

The canonical nature of the transformation in $(N,\bar{\theta})$ space implied by (29) means that volume is conserved in the internal part of the phase space, whereas Liouville's theorem [1]normally implies that only the total volume of phase space is conserved in a dynamical transformation. The present reduced conservation law is a consequence of the fact that it is always possible formally to replace one pair of variables, in this case (P,R) by (E,t) in a conservative system [1].

The above discussion points to the advantages of the $(N,\bar{\theta})$ system in visualising the effect of a collision, but the practical solution of Hamilton's equations is more conveniently obtained in the cartesian system.

$$\dot{P} = -(\partial V/\partial R) \qquad \dot{R} = P/m$$

$$\dot{P} = -(\partial H/\partial q) \qquad \dot{q} = (\partial H/\partial p) \tag{32}$$

or by solving the equivalent equations involving (N,θ) in place of (p,q) subject to the boundary conditions applicable to a suitably prepared asymptotic state. The time evolution of the internal state may be pictured in terms of the motion of a particle over the surface of a distorted cylinder as illustrated in Fig 2.

In the absence of any interaction the regular orbital motion at frequency ω appears at rest in the $(N,\bar{\theta})$ or equivalent $(\bar{p},\bar{q})$ space leaving the representative point to follow a straight line; the family of such trajectories defines the cylinder as seen in Fig 2. A similar rotation has been applied in the inelastic case, but the twisting of the curves now distorts the cylinder. The projections to the right of the diagram show that action is conserved in the elastic case, but that the distortion due to $V(q,R)$ in the inelastic case leads to a range of possible final action values.

This may be seen in another way by plotting the final quantum number (related to the action by $n = N/\hbar - \frac{1}{2}$) as a function of initial angle in Fig 3. As normally found in practice, this shows two initial angles leading to any given final action n_2 within the classically accessible range $n_{min} < n_2 < n_{max}$. Since the quantum numbers have unit spacing along the axis and the initial angle has a uniform probability, $d\theta/2\pi$, of lying within any range $\theta \to \theta + d\theta$, the quantised probability for the observation of final quantum number n_2 is given by the equation

$$P_{n_1 n_2} = \sum_{\nu} \left[\frac{1}{2\pi} \frac{d\theta_1}{dn_2} \right] \tag{33}$$

the sum being taken over the points marked $\theta_1^{(a)}$ $\theta_1^{(b)}$ in Fig 3.

2. DERIVATION OF THE S MATRIX

Two routes to the semiclassical S matrix have been suggested; one, due to Marcus et al [5,7], employs a semiclassical approx-

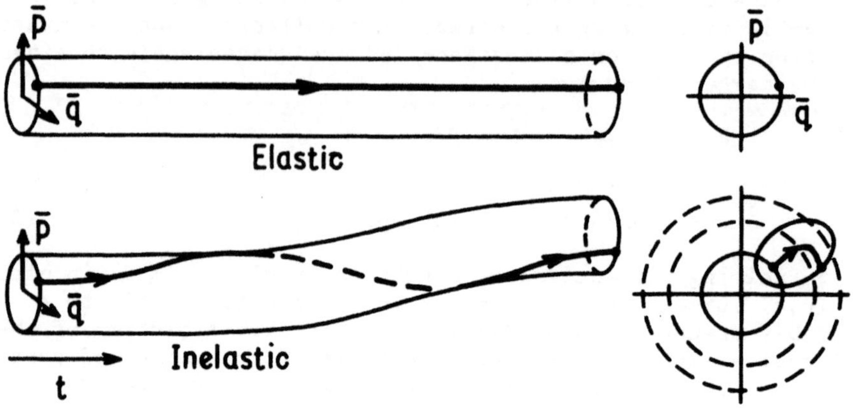

Fig 2. Elastic and inelastic scattering seen as time evolutions in modified phase space.

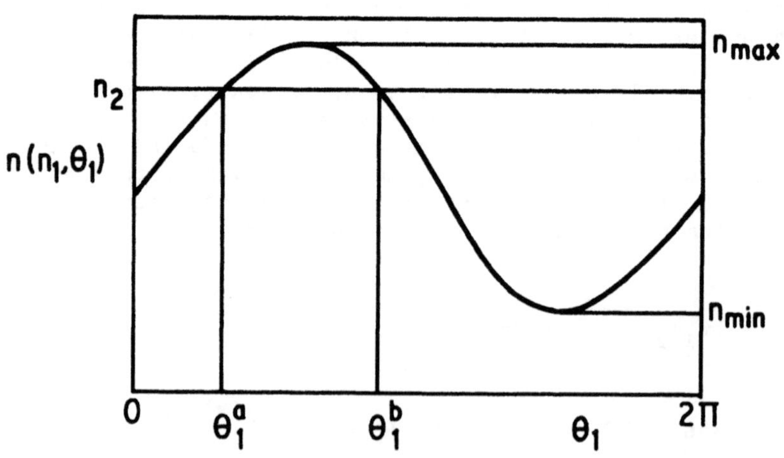

Fig 3. Final quantum number $n(n_1,\theta_1)$ as a function of initial angle θ_1

imation to the wavefunction, while the other due to Pechukas [8] and Miller [4,9] starts from the Feynman path integral form for the propagator [10]. The former approach is adopted below. Extensive reviews from both points of view have been published [3,11 -14].

2.1 The Semiclassical Wavefunction

The form of the semiclassical wavefunction is most readily obtained by expressing the Hamiltonian in angle-action variables

$$\hat{H} = -\frac{\hbar^2}{2m}\frac{\partial^2}{\partial R^2} + H_{int}(\hat{N}) + V(\hat{N},\theta,R), \tag{34}$$

with the action operator given by,

$$\hat{N} = (-i\,\partial/\partial\theta + \tfrac{1}{2})\hbar \tag{35}$$

in order to be consistent with the Bohr quantisation condition and with the commutation relation $[N,\theta] = -i\hbar$. Since N commutes with $H_{int}(N)$, they have common eigenfunctions

$$\hat{N}\ e^{in\theta} = (n + \tfrac{1}{2})\hbar\ e^{in\theta}, \tag{36}$$

and the initial asymptotic wavefunction may be written

$$\Psi(\theta,R) \sim (2\pi k_1)^{-\frac{1}{2}} \exp\,(in_1\theta + i\,k_1R), \tag{37}$$

where

$$k_1^2\,\hbar^2/2m + \tilde{H}_{int}\,[(n_1 + \tfrac{1}{2})\hbar] = E \tag{38}$$

In seeking a WKB like wavefunction

$$\Psi(\theta,R) = \exp\,[i\,W(\theta,R)/\hbar] \tag{39}$$

which goes over to (41) in the asymptotic region, it is convenient to consider the effect of a typical term $(-i\hbar)^r\,(\partial/\partial\theta)^r$ in the Hamiltonian

$$(-i\hbar)^r\frac{\partial^r\Psi}{\partial\theta^r} = \left\{\left(\frac{\partial W}{\partial\theta}\right)^r - \frac{i\hbar}{2}\frac{\partial}{\partial\theta}\left[r\left(\frac{\partial W}{\partial\theta}\right)^{r-1}\right]\ \ldots\right\}\Psi \tag{40}$$

Hence on expanding

$$W = W_o + \hbar W_1 + \ldots \tag{41}$$

and substituting

$$iW_1 = \ln A$$

$$(-i\hbar)^r \frac{\partial^r \Psi}{\partial \theta^r} = \left\{ \left(\frac{\partial W_o}{\partial \theta}\right)^r - \frac{i\hbar}{2} \left[\frac{\partial}{\partial \theta} \, r \left(\frac{\partial W_o}{\partial \theta}\right)^{r-1} \right. \right.$$

$$\left. \left. + \; 2ir \left(\frac{\partial W_o}{\partial \theta}\right)^{r-1} \frac{\partial W_1}{\partial \theta} \right] \right\} \Psi$$

$$= \left\{ \left(\frac{\partial W_o}{\partial \theta}\right)^r - \frac{i\hbar}{2A^2} \left[\frac{\partial}{\partial \theta} \, A^2 \, r \left(\frac{\partial W_o}{\partial \theta}\right)^{r-1} \right] - - - \right\} \Psi \qquad (42)$$

Similar arguments apply to derivatives with respect to R. It follows on combining terms $\hbar^o$ in the Schrodinger equation that $W_o(\theta,R)$ must satisfy the Hamilton-Jacobi equation

$$\frac{1}{2m} \left(\frac{\partial W_o}{\partial R}\right)^2 + \tilde{H}_{int} \left(\frac{\partial W_o}{\partial \theta} + \frac{\hbar}{2}\right) + V\left(\frac{\partial W_o}{\partial \theta} + \frac{\hbar}{2}, \theta, R\right) = E \qquad (43)$$

It follows by comparison with the classical identity

$$\frac{1}{2m} P^2 + \tilde{H}_{int}(N) + V(N,\theta,R) = F \qquad (44)$$

that

$$\frac{\partial W}{\partial \theta} = N - \hbar/2 = n(\theta,R)\hbar$$

$$\frac{\partial W_o}{\partial R} = P = k(\theta,R)\hbar \qquad (45)$$

where $n(\theta,R)$ and $k(\theta,R)$ go over to the constant values n_1 and k_1 in the initial asymptotic region. Equation (43) implies that $W_o(\theta,R)$ may be expressed as the line integral

$$W_o(\theta,R)/\hbar = \int_{\theta_1}^{\theta} n(\theta,R)d\theta + n_1 \theta_1 + \int_{R_1}^{R} k(\theta,R)dr + k_1 R_1 \qquad (46)$$

taken along a classical trajectory from (θ_1,R_1) to (θ,R). By this definition

$$W_o(\theta,R)\hbar \sim n_1\theta + k_1R \tag{47}$$

in the initial asymptotic region, as required by (37) .

Given equations (43) for $(\partial W_o/\partial\theta)$ and $(\partial W_o/\partial R)$, the factor $r(\partial W_o/\partial\theta)^{r-1}$ in the second term in (40) may be identified with a constibution to the classical frequency $(\partial H/\partial N)$ arising from a term N^r in the hamiltonian. But by Hamilton's equations

$$(\partial H/\partial N) = \dot{\theta}. \tag{48}$$

Thus on collecting the terms in $\hbar$ analogous to those in (40) the function $A(R,\theta)$ must satisfy

$$\frac{\partial}{\partial\theta}\left[A^2\dot{\theta}\right] + \frac{\partial}{\partial R}\left[A^2\dot{R}\right] = \nabla\cdot\underset{\sim}{j} = 0 \tag{49}$$

where $\underset{\sim}{j}$ is the two dimensional current with components $(A^2\dot{\theta}, A^2\dot{R})$ But if $\underset{\sim}{j}$ has zero divergence there can be no net outflow from any region bounded by neighbouring trajectories as shown in Fig 4. Therefore

$$A_1^2(\theta_1,R_1)\ v_1\ d\theta_1 = A^2(\theta,R)\ v\ d\theta \tag{50}$$

so that

$$A(\theta,R) = A_1(\theta_1,R_1)\left[\frac{v_1}{v}\left(\frac{\partial\theta_1}{\partial\theta}\right)_{n_1}\right]^{\frac{1}{2}} \tag{51}$$

This shows that in more than one dimension the semi-classical normalisation must accommodate not only a change in radial velocity but also any divergence of the trajectories.

The initial value $A_1(\theta_1,R_1) = (2\pi k_1)^{-\frac{1}{2}}$ is determined by (33) and the total wavefunction may be written

$$\Psi(R,\theta) = \left[\frac{1}{2\pi k(r,\theta)}\left(\frac{\partial\theta_1}{\partial\theta}\right)_{n_1}\right]^{\frac{1}{2}} \exp\left[i\ W_o(R,\theta)/\hbar\right] \tag{52}$$

with the classical action $W_o\ (R,\theta)$ given by (42)

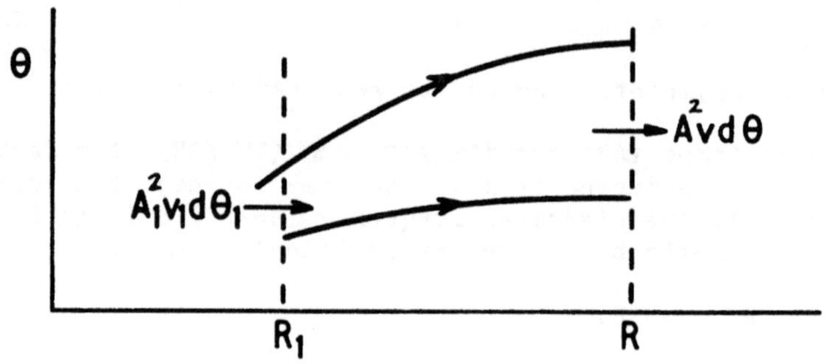

Fig 4. Evolution of adjacent trajectories in R,θ space.

2.2 The Primitive Semi-classical S matrix

An integral representation for the S matrix is readily obtained by comparison between (52) and the standard asymptotically out going part of the wavefunction

$$\Psi(R,\theta) \sim \sum_{n_2} S_{n_1 n_2} k_{n_2}^{-\frac{1}{2}} \Phi_{n_2}(\theta) e^{ik_2 R} \tag{53}$$

so that

$$S_{n_1 n_2} = k_2^{\frac{1}{2}} e^{-ik_2 R} \int_0^{2\pi} \Phi^*_{n_2}(\theta) \Psi(R,\theta) d\theta$$

$$= \frac{1}{2\pi} \int_0^{2\pi} \left[\frac{k_2}{k} \left(\frac{\partial\theta}{\partial\theta_1}\right)_{n_1}\right]^{\frac{1}{2}} \exp[i\Delta_{n_1 n_2}(\theta_1)] \, d\theta_1 \tag{54}$$

where

$$\Delta_{n_1 n_2}(\theta_1) = W_o(R,\theta)/\hbar - n_2\theta - k_2 R$$

$$= \int_{\theta_1}^{\theta} n d\theta - n_2\theta + n_1\theta_1 + \int_{R_1}^{R} k dR - k_2 R + k_1 R_1 \tag{55}$$

Notice that for given R_1 and R, the final angle θ may be treated as a function of θ_1 or vice versa; this flexibility has been used in replacing θ by θ_1 as the integration variable in (54).

Each starting angle defines a particular trajectory starting from n_1 and terminating at a dependent point $n(n_1\theta_1)$, $\theta(n_1,\theta_1)$ as illustrated in Figs 3 and 4; and the integral over θ_1 implies an integral over all such trajectories.

A similar result may be shown [3] to apply with θ and θ_1 replaced by $\bar{\theta}$ and $\bar{\theta}_1$ respectively and with the ratio (k_2/k) replaced by unity.

It is illuminating first to consider the stationary phase approximation to S_{n_1n}. For given n_1 and n_2 the critical initial angles θ_1 are those for which

$$\frac{\partial\Delta}{\partial\theta}{}_{n_1n_2} = \left[n(n_1,\theta_1) - n_2\right] \; (\partial\theta/\partial\theta_1)_{n_1} = 0 \qquad (56)$$

In other words θ_1 must coincide either with $\theta_1^{(a)}$ or $\theta_1^{(b)}$ in fig 3, implying a two term consribution to the integral. The magnitude of these terms depemds on the pre exponent in (50) and on the second derivatives at $\theta_1^{(a)}$ and $\theta_1^{(b)}$

$$\frac{\partial^2\Delta}{\partial\theta}{}_{n_1n_2} = \left(\frac{\partial n}{\partial\theta_1}\right)_{n_1} \left(\frac{\partial\theta}{\partial\theta_1}\right)_{n_1} \qquad (57)$$

the first of which is assumed to be positive and the second negative. Thus on employing the standard integrals of $\exp(\pm i\alpha x^2)$,

$$S_{n_1n_2} = P_a^{\frac{1}{2}} \, e^{i\Delta_a + i\pi/4} + P_b^{\frac{1}{2}} \, e^{i\Delta_b - i\pi/4} \qquad (58)$$

where P_a and P_b are the classical probability's defined by equation (33)

$$P_a = [2\pi(\partial n/\partial\theta_1)_a]^{-1} \quad , \; P_b = [-2\pi(\partial n/\partial\theta_1)_b]^{-1} \qquad (59)$$

This primitive semi-classical approximation shows that interference between the two contributing trajectories leads to a transition probability

$$P_{n_1n_2} = \left|S_{n_1n_2}\right|^2 = P_a + P_b + 2(P_aP_b)^{\frac{1}{2}}\sin\,(\Delta_b - \Delta_a) \qquad (60)$$

which oscillates around the classical value $P_a + P_b$.

The procedure required to implement equation (60) and a geometrical interpretation of the phase difference $(\Delta_b-\Delta_a)$ are illustrated for clarity in a reactive context in Fig 5.

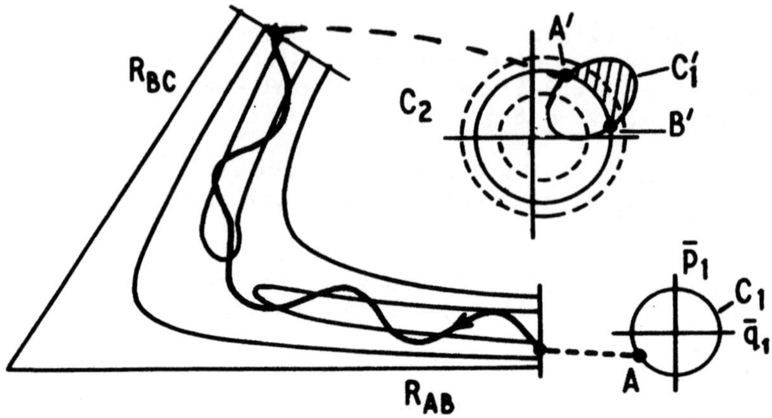

Fig 5. The dynamical transformation from initial to final phase space. C_1 and C_2 correspond to the initial and final quantum states. C_1' is the locus of trajectory end points starting from C_1. The intersections A' and B' identify the stationary phase trajectories. The shaded area determines the semiclassical phase difference $(\Delta_b-\Delta_a)$.

The first necessity is to identify the two (or possible more) stationary phase trajectories with desired initial and final actions, $N_i = (n_i + \frac{1}{2})\hbar$, one of which is shown in Fig 5. Secondly the actions along these trajectories determine the phase Δ_ν, and small variations of θ_1 at fixed n_1 are used to calculate $(\partial n/\partial\theta_1)_{n_1}$. This calculation may also be extended by use of complex trajectories, discussed below, to handle classically inaccessible transitions [15,16].

The geometrical interpretation of the phase difference $(\Delta_b-\Delta_a)$ as the shaded area in Fig 5 follows [17] directly from the canonical nature of the $(N_1,\bar{\theta}_1) \to (N_2,\bar{\theta}_2)$ transformation and from identification of Δ_a and Δ_b with the relevant values of the generator $\Lambda(N_1,N_2)$ defined by equation (28). There is an obvious connection with the phase space interpretation of Franck-Condon amplitudes given in Chapter 4. In both cases the total area enclosed by the initial contour C_1 is conserved during the transition, and the relevant semi-classical phase term is interpreted as the phase overlap between the initial and final states. Just as in the spectroscopic case, it follows in the collisional context that provided sufficient interaction is available [17] the transition probability must follow the form of the initial wavefunction either as a function of final quantum number at given interaction, or as a function of interaction strength at given final quantum number. An example of this behaviour for the exactly soluble forced oscillator model is illustrated in Fig 6.

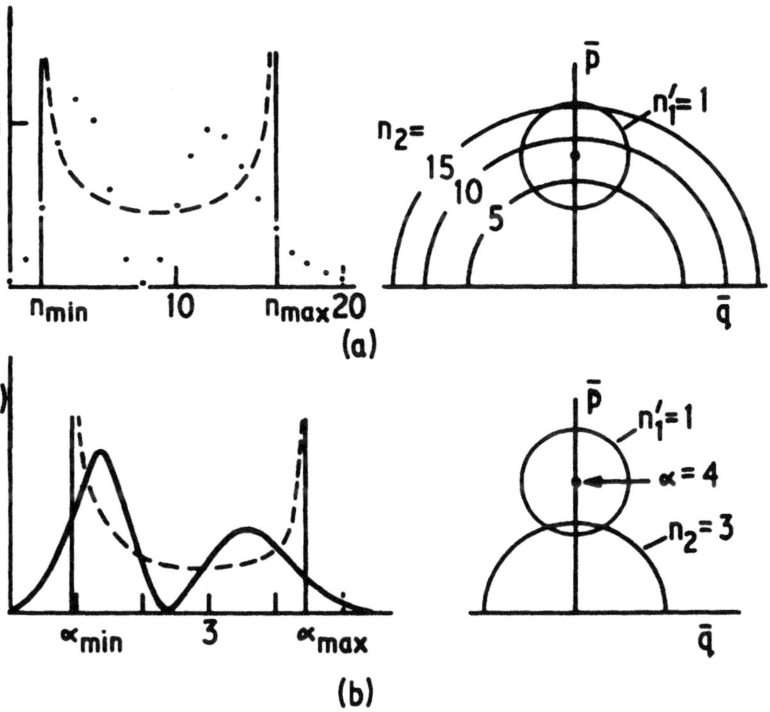

Fig 6. Forced harmonic oscillator transition probabilities from $n = 1$(a) as function of $n_2 = n$ for forcing parameter $\alpha = 4$ (see problem 2); (b) as a function of α for $n_2 = 3$. The dashed lines follow the classical probabilities. Verticals indicate the classical thresholds. The right hand side of each diagram shows the relevant phase space picture.

A more practical aspect of this phase space interpretation is that it may be used to resolve ambiguities of $(2n_i + 1)\pi$ in the magnitude of $(\Delta_b - \Delta_a)$ [18] due to phase discontinuities between neighbouring trajectories which arise from technical aspects of the calculation [19] . Such discontinuities are irrelevant in equation (60), but they become important for the following uniform approximations.

2.3 Uniform Approximations

As usual the primitive expression (60) applies only when the transition $n_1 \rightarrow n_2$ is classically allowed, in other words when the roots of (61) are real. Experience shows however that two complex conjugate roots exist when n_2 lies outside the classical range $n_{min} < n_2 < n_{max}$ in Fig 3, and these may be identified integrating the trajectory in complex time [15,16]. Various uniform expressions have been developed to cover both allowed and forbidden transitions, and to counteract the singularities in P_a and P_b at the classical thresholds, n_{min} and n_{max}. These all rely only on the same classical quantities Δ_ν and P_ν as those required for the primitive expressions (56).

The simplest is the uniform Airy result [7], given in the classically accessible case by

$$P_{n_1n_2} = \pi(P_a^{\frac{1}{2}}+P_b^{\frac{1}{2}})^2\xi^{\frac{1}{2}}\, Ai^2(-\xi) + \pi(P_a^{\frac{1}{2}}-P_b^{\frac{1}{2}})^2\, \xi^{-\frac{1}{2}}\, [Ai'(-\xi)]^2 \tag{62}$$

where

$$\xi = [3(\Delta_b - \Delta_a)/4\,]^{2/3} \tag{63}$$

The analytical continuation for classically inaccessible transitions may be written

$$\begin{aligned} P_{n_1n_2} &= 2\pi\rho(1 + \sin\beta)|\xi|^{\frac{1}{2}}Ai^2(-\xi) \\ &\quad + 2\pi\rho(1 - \sin\beta)|\xi|^{-\frac{1}{2}}\,[Ai'(-\xi)]_1^2 \end{aligned} \tag{64}$$

where, since the two trajectories are complex conjugates

$$\xi = -\,[3|\Delta_b - \Delta_a|/4\,]^{2/3} \tag{65}$$

$$\rho = |P_a| = |P_b|, \quad \beta = \arg P_a = \pi - \arg P_b \tag{66}$$

with the branch chosen so that $\beta \simeq \pi/2$. The derivation [7] assumes a mapping between $\Delta_{n_1n_1}(\theta)$ and a standard cubic form, with the implication, according to (56), that the function $\phi(n_1,\theta_1)$ is roughly quadratic in θ_1. The intrinsic periodicity of $\Delta_{n_1n_2}(\theta_1)$ is therefore ignored, and this is seen in table 1 below to lead to serious inaccuracies for nearly elastic events, such that $n_{min} \simeq n_{max} \simeq n_1$.

This latter deficiency is removed by the following uniform Bessel approximation [20], given in the classically accessible case by

$$P_{n_1 n_2} = (\pi/2)(P_a^{\frac{1}{2}} + P_b^{\frac{1}{2}})^2 \xi \cos\theta \, J_m^{\,2}(\xi) + (\pi/2)(P_a^{\frac{1}{2}} - P_b^{\frac{1}{2}})^2$$

$$\xi [J_m'(\xi)]^2 / \cos\theta, \qquad (67)$$

where

$$m = |n_1 - n_2| \quad , \quad \cos\theta = [1 - (m/\xi)^2]^{\frac{1}{2}} \qquad (68)$$

$$(\xi^2 - m^2)^{\frac{1}{2}} - m \arccos(m/\xi) = (\Delta_b - \Delta_a)/2 \qquad (69)$$

Equations (67) goes over to the following form when the transition is classically forbidden

$$P_{n_1 n_2} = \pi\rho(1 + \sin\beta)\xi \sinh\psi J_m^{\,2}(\xi) + \pi\rho(1 - \sin\beta)$$

$$[J_m'(\xi)]^2 / \sinh\psi, \qquad (70)$$

where ρ and β are given by (66), and

$$m \cosh^{-1}(m/\xi) - (m^2 - \xi^2)^{\frac{1}{2}} = |\Delta_b - \Delta_a| \,/2 \qquad (71)$$

$$\sinh\psi = [(m/\xi)^2 - 1]^{\frac{1}{2}} \qquad (72)$$

This Bessel approximation is undefined when $n_1 = n_2$ in a forbidden transition.

Various other uniform approximations termed Laguerre [21], non-integer Bessel [22] and Harmonic[17] have been suggested. Table 1 gives a short comparison for the Morse oscillator-expotential interaction model of Clark and Dickinson [23] . Similar more extensive tables for this and other systems are available in references [7, 16,21,25]. The most striking feature of all these tables is the remarkable accuracy of the various approximations, with the exceptions of the Harmonic approximation and of the Airy approximation in cases of near elasticity. The choice will therefore normally be made on grounds computational convenience. Here the Airy form will normally be preferred when applicable because the parameter ξ is given directly by (63) or (65), whereas the corresponding quantity in other cases must be derived

Table 1. Comparison between exact and uniform vibrational transition probabilities.

E	n_1	n_2	$P_{n_1 n_2}$ [a]	Airy[b]/Exact	Bessel[c]/Exact	Lag[d]/Exact	Harm[e]/Exact	NIB[f]/Exact
2	0	0	1.000	2.461	1.000	1.000	2.535	-
	1	1	1.000	7.959	1.000	1.000	8.244	-
8	0	0	0.715	1.127	1.078	1.000	1.049	1.114
	0	1	0.247	1.000	0.972	1.000	1.036	0.951,0.980
	0	2†	0.353(-1)	0.997	0.958	1.004	0.929	-
	1	1	0.364	1.000	1.019	0.997	1.023	1.019
	1	2	0.321	1.012	0.997	1.000	1.019	0.991,1.000
	1	3†	0.629(-1)	1.017	1.000	1.011	1.099	-
	1	4†	0.561(-2)	1.004	0.988	1.002	0.581	-
	2	2	0.222	0.964	1.005	0.986	1.009	1.000
	2	3	0.339	1.014	1.000	1.003	1.018	1.000,1.003
	2	4†	0.755(-1)	1.030	1.013	1.019	1.207	-
	2	5†	0.667(-2)	1.015	0.999	1.005	0.565	-
	3	3	0.173	0.963	0.988	0.983	0.925	1.011
	3	4	0.341	1.012	1.000	1.003	1.015	1.000,1.000
	3	5†	0.748(-1)	1.037	1.020	1.023	1.158	-
	4	4	0.175	0.926	0.977	0.971	0.920	0.977

† Classically forbidden

References a[23]; b[35]; c[35]; d[21]; e[36]; f[22].
Two entries are given for $n_1 \neq n_2$ transitions in the final column, because the NIB approximation is not microscopically reversible.

Table 2. Exact and uniform approximations for collinear reactive $H + Cl_2(n_1) \rightarrow HCl(n_2) + Cl$

E(eV)	n_1	n_2	Exact[a]	Airy[b]	Bessel[b]	Laguerre[b]	Harmonic[b]
0.15	0	3	0.320	0.234	0.233	0.236	0.237
	0	4	0.420	0.432	0.429	0.435	0.440
0.17	0	3	0.345	0.318	0.317	0.319	0.321
	0	4	0.529	0.682	0.676	0.691	0.702
0.20	0	3	0.331	0.403	0.401	0.404	0.407
	0	4	0.560	0.505	0.503	0.510	0.514
0.22	0	3	0.357	0.393	0.391	0.394	0.398
	0	4	0.537	0.493	0.492	0.495	0.498
	1	3	0.200	0.159	0.159	0.157	0.157
	1	4	0.059	0.020	0.022	0.019	0.019
	1	5	0.544	0.297	0.294	0.297	0.301
0.25	0	3	0.338	0.344	0.342	0.348	0.351
	0	4	0.527	0.496	0.495	0.491	0.493
	1	2	0.079	0.079	0.078	0.077	0.079
	1	3	0.237	0.235	0.236	0.232	0.231
	1	4	0.061	0.170	0.175	0.165	0.159
	1	5	0.598	0.488	0.484	0.487	0.492
0.30	0	3	0.270	0.256	0.254	0.260	0.263
	0	4	0.467	0.494	0.494	0.472	0.473
	0	5	0.202	0.203	0.201	0.209	0.212
	1	2	0.084	0.108	0.103	0.104	0.107
	1	3	0.294	0.342	0.344	0.340	0.338
	1	4	0.068	0.048	0.049	0.048	0.047
	1	5	0.439	0.405	0.405	0.403	0.402

a ref [37], b ref [38].

from a transcendental equation such as (69) or (71).

Similar calculations have also been performed for collinear reactive models [18,26-29], but the agreement between the semi-classical and exact quantum mechanical transition probabilities shown for example in table 2, is less striking. The reason for this difference between inelastic and reactive systems is not fully understood.

Finally it should be mentioned that extensions of various aspects of the theory to problems involving more than one degree of freedom have been published [30-33]. In particular a generic family of uniform approximations has been developed in outline [32-33]. These are classified in the language of Thom's catastrophe theory according to the dimensionality of the problem and the number of contributing trajectories. Efficient techniques for the evaluation of the canonical integrals remain however to be developed. Work is also required on the root search problem for the trajectories which connect one quantisable state with another.

3. PROBLEMS

1. Given

$$H(p,q) = p^2/2\mu + kq^2/2$$

show that the transformation to angle action variables yields

$$\tilde{H}(N) = N\omega , \omega = (k/\mu)^{\frac{1}{2}}$$

$$q = (2N)^{\frac{1}{2}} (\mu k)^{-\frac{1}{4}} \cos\theta, \; p = -(2N)^{\frac{1}{2}} (\mu k)^{\frac{1}{4}} \sin\theta$$

Note

$$\int_{-a}^{x} [x^2-a^2]^{\frac{1}{2}} dx = [x[x^2-a^2]^{\frac{1}{2}} + a^2(\text{arc sin } (x/a) + \pi/2)]/2$$

2. The scaled forced harmonic oscillator is described by the Hamiltonian

$$H(p,q) = (p^2 + q^2)/2 - f(t)q$$

Show by solving Hamilton's equation for the composite variable

$$a = \bar{p} + i\bar{q} = (p + iq) \exp(-it)$$

or otherwise, that forasymmetric force $(f(t) = f(-t))$

$$\bar{p}_2 = \bar{p}_1 + \alpha, \quad \bar{q}_2 = \bar{q}_1$$

$$\alpha = \int_{-\infty}^{\infty} f(t) \cos t \; dt$$

where subscript 1 and 2 denote times $t = -\infty$ and $t = +\infty$ respectively.

Hence using the notation

$$\bar{q} = (2N)^{\frac{1}{2}} \cos \bar{\theta}, \quad \bar{p} = -(2N)^{\frac{1}{2}} \sin \bar{\theta},$$

show that

$$N_2 = N_1 - \alpha\,(2N_1)^{\frac{1}{2}} \sin \bar{\theta}_1 + \alpha^2/2$$

$$\tan \bar{\theta}_2 = \tan \bar{\theta}_1 - [\alpha/(2N_1)^{\frac{1}{2}}] \sec \bar{\theta}_1$$

3. The equation for the S matrix element for the forced oscillator described in Problem 2 may be shown [34] to be

$$S_{n_1 n_2} = \frac{1}{2\pi} \int_0^{2\pi} \left(\frac{\partial \theta_2}{\partial \bar{\theta}_1}\right)_{N_1} \exp\,[i\, N_1 \bar{\theta}_1 + \frac{i\alpha}{2}\,(2N_1)^{\frac{1}{2}} \cos\, \bar{\theta}_1 - i\, N_2\, \bar{\theta}_2(\bar{\theta}_1)]\; d\bar{\theta}_1$$

where, as shown in Problem 2,

$$\tan \bar{\theta}_2 = \tan \bar{\theta}_1 - \alpha(2N_1)^{-\frac{1}{2}} \sec \bar{\theta}_1 \qquad \text{(a)}$$

and N_2 is a constant. Show that the phase term is stationary with respect to $\bar{\theta}_1$, at points $\bar{\theta}_1^{(a)}$, $\theta_1^{(b)}$, such that

$$N_1 - \alpha(2N_1)^{\frac{1}{2}} \sin \bar{\theta}_1 + \alpha^2/2 = N_2 \qquad \text{(b)}$$

Prove that in the stationary phase approximation

$$S_{n_1 n_2}(\alpha) = 2\, P^{\frac{1}{2}}(\alpha) \sin\; [A_{n_1 n_2}(\alpha) + \pi/4]$$

where

$$P(\alpha) = \frac{1}{2\pi\,[(N_1+N_2)\alpha^2 - (N_2-N_1)^2 - \frac{1}{4}\alpha^4]^{\frac{1}{2}}}$$

$$A_{n_1n_2}(\alpha) = N_1\,\overline{\theta}_1 + \tfrac{1}{2}\alpha(2N_1)^{\frac{1}{2}}\cos\overline{\theta}_1 - N_2\,\overline{\theta}_2$$

where $\overline{\theta}_1$ and $\overline{\theta}_2$ are related by equation (a).

Note that $A_{n_1n_2}(\alpha)$ may be visualised as half the shaded area shown in Fig 6. Hint: The manipulation may be simplified by use of the identity

$$(2N_2)^{\frac{1}{2}}\cos\overline{\theta}_2 = (2N_1)^{\frac{1}{2}}\cos\overline{\theta}_1$$

implied by (a) and (b).

REFERENCES

1. Goldstein, H.: 1950, *Classical Mechanics*. (Addison Wesley)
2. Born, M.: 1960, *Mechanics of the Atom*, (G. Bell, London).
3. Child, M.S.: 1976, in *Dynamics of Molecular Collisions Part B*. ed Miller W.H. (Plenum Press) Chap. 4.
4. Miller, W.H.: 1970, J.Chem.Phys. 53, p.3578
5. Marcus. R.A.: 1970, Chem.Phys. Lett., 7, p.525
6. Marcus, R.A.: 1971, J.Chem.Phys., 54, p.3965
7. Connor, J.N.L. and Marcus, R.A.:1971. J.Chem. Phys. 55, p.5636
8. Pechukas, P.: 1969, Phys.Rev. 181, pp 166, 173.
9. Miller, W.H.: 1970, J.Chem.Phys., 53, p.1949.
10. Feynman, R.E. and Hibbs, A.R.: 1965, *Quantum Mechanics and Path Integrals* (McGraw-Hill).
11. Miller, W.H.: 1974, Adv.Chem.Phys. 25, p.69.
12. Miller, W.H.: 1976, Adv.Chem.Phys. 30, p.77.
13. Child, M.S.: 1974, *Molecular Collision Theory*.(Academic Press)
14. Child, M.S.: 1978, Adv. in At.Mol.Phys., 14, p.225
15. Miller, W.H. and George, R.F.1972 J.Chem.Phys. 36, p.5668
16. Stine, J.R. and Marcus R.A.: 1972, Chem.Phys.Lett. 5, p.536
17. Child, M.S.: 1978, Mol.Phys. 35, p.759.
18. Child, M.S. and Whaley, K.B.: 1979, Far.Disc. 67, (in Press).
19. Fraser S.J., Grottdiener, L., and Murrell, J.N.: 1975, Mol.Phys., 29, 415.
20. Stine, J.R. and Marcus R.A.: 1973, J.Chem.Phys. 59, p5145.
21. Child, M.S. and Hunt. P.M.: 1977, Mol.Phys. 34, p.261
22. Connor, J.N.L. and Mayne, H.R.: 1979, Mol.Phys. 37, pp1,15
23. Clark, A.P. and Dickinson A.S.: 1973, J.Phys.B 6, p.164.
24. Duff, J.W. and Truhlar, D.G.: 1975,Chem.Phys. 9, p.243
25. Connor, J.N.L. and Mayne, H.R.: 1979, Mol.Phys. 37, p.15
26. George, T.F. and Miller, W.H.: 1972, J.Chem.Phys. 57, p.2458,5722.
27. Rankin, C.C. and Miller, W.H.: 1971, J.Chem.Phys. 55, p.3150
28. Bowman, J.M. and Kuppermann, A.: 1973, J.Chem.Phys. 59, p.6524
29. Truhlar, D.G., Merrick, J.A., and Duff, J.W.: 1976, J.Am Chem.Soc. 98, p.6771
30. Kreek, H., Ellis, R.L., and Marcus, R.A.: 1974, J.Chem.Phys. 61, p.4540.
31. Kreek, H., Ellis, R.L., and Marcus, R.A.: 1975, J.Chem Phys 62, p.913
32. Berry, M.V.: 1976, Ann.Phys. (Paris) 25, p.1
33. Connor, J.N.L.: 1976, Mol.Phys. 31, p.33
34. Pechukas P., and Child, M.S.: 1976, Mol.Phys. 31, p.973
35. Connor, J.N.L.: 1974, Mol.Phys. 28 p.1569.
36. Hunt, P.M.H.: 1978, D.Phil Thesis, Oxford University
37. Baer M.: 1973, J.Chem.Phys. 60, p.1057
38. Whaley, K.H.: PtII B.A. Thesis, Oxford University (unpublished).

SEMICLASSICAL THEORY OF ELECTRONICALLY NONADIABATIC TRANSITIONS IN MOLECULAR COLLISION PROCESSES*

Kai S. Lam and Thomas F. George†

Department of Chemistry
University of Rochester
Rochester, New York, U.S.A.

An introductory account of the semiclassical theory of the S-matrix for molecular collision processes is presented, with special emphasis on electronically nonadiabatic transitions. This theory is based on the incorporation of classical mechanics with quantum superposition, and in practice makes use of the analytic continuation of classical mechanics into the complex space or time domain. After a brief discussion of the relevant concepts of molecular scattering theory and related dynamical models, the formalism is developed and illustrated with simple examples - collinear collision of the A+BC type. The theory is then extended to include the effects of laser-induced nonadiabatic transitions. The lecture concludes with a discussion of two bound continuum processes - collisional ionization and collision-induced emission - also amenable to the same general semiclassical treatment.

* Project supported by the U.S. Air Force Office of Scientific Research (Contract No. F49620-78-C-0005), the National Science Foundation (Grant No. CHE77-27826) and the National Aeronautics and Space Administration (Grant No. NSG-2198).

† Camille and Henry Dreyfus Teacher-Scholar (1975-80) and Alfred P. Sloan Research Fellow (1976-80).

M. S. Child (ed.), Semiclassical Methods in Molecular Scattering and Spectroscopy, 179-261.

I. INTRODUCTORY CONCEPTS AND TOOLS

Ia. Scattering Matrices and Scattering Cross-Sections

In scattering experiments of any kind, the fundamental observable quantity is the scattering cross section; and the theoretical object of prime importance associated with it is the scattering matrix (or S-matrix). We begin by giving formal definitions of the S-matrix applicable to general scattering processes and in later sections adapt them to molecular collision problems. The S-matrix will be shown to be directly related to the scattering cross section through an auxillary object, the transition matrix (or T-matrix); and a key objective of the theoretical program of collision processes is the calculation of S- or T-matrices.

S- and T-matrices are quantum mechanical objects and they will be introduced as such in the present subsection. However, under certain conditions, such as in the treatment of heavy-particle dynamics in molecular collisions they can be adapted to semiclassical or even classical evaluations. In the present lecture series we shall focus on the semiclassical evaluation of the S-matrix (called the "semiclassical S-matrix"), with special applications to the case of electronically non-adiabatic transitions. Before discussing this explicitly, we shall review some basic ideas associated with the S-matrix in general.

A collision system can be thought of as being in the states ψ^{in} (incoming) and Ψ^{out} (outgoing) before and after the collision process respectively; and the collision can be visualized as a "blackbox" mathematically described by a transformer operator S (the scattering operator) such that

$$\Psi^{out} = S\Psi^{in} . \qquad \text{(Ia.1)}$$

Suppose Ψ^{in} is specified by a given free state $|i\rangle$ (this state is often chosen to be an eigenstate of the unperturbed Hamiltonian). Ψ^{out} can be expanded in terms of the complete set of which $|i\rangle$ is a member

$$\Psi^{out} = \sum_j c_j |j\rangle . \qquad \text{(Ia.2)}$$

The transition probability amplitude from $|i\rangle$ to a particular $|f\rangle$ is then given by

$$c_f = \langle f|\Psi^{out}\rangle = \langle f|S|\Psi^{in}\rangle = \langle f|S|i\rangle \equiv S_{fi} . \qquad \text{(Ia.3)}$$

Hence, it is seen that S_{fi}, which is a matrix element of the scattering operator S in the $|i\rangle$ representation, gives directly the transition probability amplitude from the state $|i\rangle$ to the state $|f\rangle$. We note that the conservation of probability for Ψ^{in} and Ψ^{out} requires that

$$\langle\Psi^{out}|\Psi^{out}\rangle = \langle\Psi^{in}|\Psi^{in}\rangle \ , \tag{Ia.4}$$

implying that

$$S^{\dagger}S = SS^{\dagger} = 1 \ , \tag{Ia.5}$$

which is a statement of the unitarity of the S-matrix.

The definition of S can also be formulated in the time-dependent formalism, if we set

$$\Psi^{in} = \Psi(t=-\infty)$$

$$\Psi^{out} = \Psi(t=+\infty) \ . \tag{Ia.6}$$

Then S can be defined by

$$\Psi(+\infty) = S\Psi(-\infty) \ , \tag{Ia.7}$$

and is also expressible as

$$S = \lim_{\substack{t\to+\infty \\ t_0\to-\infty}} U(t,t_0) \ , \tag{Ia.8}$$

where U is the time evolution-operator for the interaction picture wave function.

In order to calculate the scattering cross section from the S-matrix, we first evaluate the transition probability per unit time $W_{i\to f}$, since the cross section $\sigma_{i\to f}$ is defined as

$$\sigma_{i\to f} = \frac{W_{i\to f}}{F_i} \ , \tag{Ia.9}$$

where F_i is the incident flux of particles for the state $|i\rangle$. Now S, as defined earlier, is the transition probability amplitude between infinite past and infinite future times. To find the probability per unit time, we must express it as the limit of a time-dependent quantity $\tilde{S}_{fi}(t)$ for $i\neq f$ (transition between different states) such that

$$S_{fi} = \lim_{t\to+\infty} \tilde{S}_{fi}(t) \ . \tag{Ia.10}$$

The transition probability per unit time $W_{i\to f}$ can then be defined as

$$W_{i\to f} = \lim_{t\to\infty} \frac{|\tilde{S}_{fi}(t)|^2}{t} . \qquad \text{(Ia.11)}$$

Since we are interested in transitions between different states $|i\rangle$ and $|f\rangle$, it is reasonable to extract from S a term which is zero when i and f are indeed different, namely δ_{fi}. Also, since energy conservation holds between infinite past and future times, we can further separate out an energy conservation factor $\delta(E_i-E_f)$ in the remainder term. Hence we write

$$S_{fi} \equiv \delta_{fi} - 2\pi i\delta(E_i-E_f)T_{fi} . \qquad \text{(Ia.12)}$$

The factor $2\pi i$ is introduced for convenience; and T, whose matrix elements are $T_{fi} = \langle f|T|i\rangle$, is called the transition or T-matrix. Written as in (Ia.12) with an explicit energy dependence, S_{fi} (for $i\neq f$) can now be expressed as

$$S_{fi} = -2\pi i\delta(E_i-E_f)T_{fi} \equiv \lim_{t\to\infty} \tilde{S}_{fi}(t) . \qquad \text{(Ia.13)}$$

This identifies $\tilde{S}_{fi}(t)$ as

$$\tilde{S}_{fi}(t) = -\frac{i}{\hbar} T_{fi}\int_{-t/2}^{t/2} dt' \, e^{i(E_i-E_f)t'/\hbar}$$

$$= -\frac{i}{\hbar} t \, T_{fi} \frac{\sin\{(E_i-E_f)t/2\hbar\}}{(E_i-E_f)t/2\hbar} \qquad \text{(Ia.14)}$$

which represents a transition probability amplitude for $i\to f$ over a finite period t of observation. The transition probability per unit time, from (Ia.11), is then given by

$$W_{i\to f} = \frac{|T_{fi}|^2}{\hbar^2} \lim_{t\to\infty} \frac{\sin^2\{(E_i-E_f)t/2\hbar\}}{\left(\frac{E_i-E_f}{2\hbar}\right)^2 t} \qquad \text{(Ia.15)}$$

which is equivalent to

$$W_{i\to f} = \frac{2\pi}{\hbar} \delta(E_i-E_f)|T_{fi}|^2 , \qquad f\neq i . \qquad \text{(Ia.16)}$$

In general, in any experimental situation, there will be a group of final states $|f\rangle$ satisfying energy conservation, and the scattering cross-section observed will correspond to the sum over these final states. From (Ia.9),

$$\sigma_{i\to f} = \frac{1}{F_i} \sum_f W_{i\to f} . \qquad \text{(Ia.17)}$$

The sum will go over to $\int dE_f \rho(E_f)$ if we are dealing with continuum states in the product channels; $\rho(E_f)$ is the density of states.

Before leaving our formal discussion of the S-matrix we will point out a formal relation between it and the Green's function to set the stage for the introduction of the semiclassical S-matrix in Section Id.

The operator $(E-H+i\eta)^{-1}$ (η being infinitesimal) is called the retarded Green's function operator and is directly related to the T-matrix, and hence the S-matrix:

$$G^+(E) = G_0^+(E) + G_0^+(E)TG_0^+(E) , \qquad \text{(Ia.18)}$$

where $G_0^+(E) = (E-H_0+i\eta)^{-1}$ is the unperturbed retarded Green's function operator. The Fourier transform $G^+(t)$ of $G^+(E)$ is given by

$$G^+(t) = \frac{1}{2\pi\hbar} \lim_{\eta\to 0} \int_{-\infty}^{\infty} dE\; e^{iEt/\hbar}(E-H+i\eta)^{-1} . \qquad \text{(Ia.19)}$$

By actually evaluating the integral in (Ia.19), $G^+(E)$ can be written as the inverse Fourier transform of $G^+(t)$:

$$G^+(E) = (i/\hbar)\int_0^{\infty} dt\; \exp(iEt/\hbar)\exp(-iHt/\hbar) . \qquad \text{(Ia.20)}$$

(Ia.18) and (Ia.20) will be the starting point for our formulation of the semiclassical S-matrix in Section Id.

lb. Partitioning of the Molecular Collision Hamiltonian

Beginning with this subsection we will restrict our attention to molecular collision problems, and the general formulation of the S-matrix given earlier will be adapted accordingly. To prepare ourselves for these special formulations of the S-matrix we will start with a discussion of the molecular collision Hamiltonian.

Separation of Electronic and Nuclear Motion and the Born-Oppenheimer Approximation. Consider a molecular collision system with n electrons and N nuclei, with electronic coordinates denoted by $\vec{r}_i$ and nuclear coordinates by $\vec{R}_i$. The total Hamiltonian can be written as

$$H = T_R + T_r + V(\vec{R},\vec{r})$$
$$\equiv T_R + H_{el}(\vec{R},\vec{r}) \qquad \text{(Ib.1)}$$

where $H_{el}(\vec{R},\vec{r})$ stands for that part of H besides the nuclear kinetic energy operator. The first-step to simplify the Hamiltonian is to separate out the center of mass motion of the whole system. The dynamics of the system is then described by an arbitrary set of (3N+3n-3) independent internal coordinates. Several choices of these can be made to suit particular problems, but here we will not be concerned with the specific choices. In general one attempts to make coordinate transformations which would leave the total kinetic energy operator in the form

$$T = -\frac{\hbar^2}{2M}\nabla_S^2 - \frac{\hbar^2}{2}\sum_i \frac{\nabla_{R_i}^2}{\mu_i} - \frac{\hbar^2}{2\mu_e}\sum_i \nabla_{r_i}^2 + \text{(m.p.)} \qquad \text{(Ib.2)}$$

where (m.p.) represents mass polarization terms (proportional to $\nabla_{R_i}\cdot\nabla_{R_j}$ and $\nabla_{r_i}\cdot\nabla_{r_j}$) which can be ignored. In (Ib.2), M stands for the 'total mass' corresponding to some C.M. coordinate $\vec{S}$; μ_i and μ_e are the reduced nuclear and electronic masses corresponding to the nuclear internal coordinates $\vec{R}_i$ and electronic internal coordinates $\vec{r}_i$ respectively. The total Hamiltonian can be written in terms of the transformed coordinates as

$$H = -\frac{\hbar^2}{2M}\nabla_S^2 - \frac{\hbar^2}{2}\sum_{i=1}^{N-1}\frac{1}{\mu_i}\nabla_{R_i}^2 - \frac{\hbar^2}{2\mu_e}\sum_{i=1}^{n}\nabla_{r_i}^2 + V(\vec{R},\vec{r}) \,. \qquad \text{(Ib.3)}$$

$V(\vec{R},\vec{r})$ may be taken to include spin-orbit interactions. The Hamiltonian in terms of the 3N+3n-3 internal coordinates becomes

$$H = -\frac{\hbar^2}{2}\sum_i \frac{1}{\mu_i}\nabla_{R_i}^2 - \frac{\hbar^2}{2\mu_e}\sum_i \nabla_{r_i}^2 + V(\vec{R},\vec{r})$$
$$\equiv T_R + H_{el}(\vec{R},\vec{r}) \,, \qquad \text{(Ib.4)}$$

where

$$T_R \equiv -\frac{\hbar^2}{2}\sum_i \frac{1}{\mu_i}\nabla_{R_i}^2 \,.$$

(Ib.4) looks identical to (Ib.1), except now $\vec{R}$ and $\vec{r}$ are understood to be the internal coordinates with the C.M. motion separated out. Hamiltonians will be written in this sense for all our later discussions.

Separation of electronic and nuclear motions ultimately

depends on the great disparity between electronic and nuclear masses. Since the forces acting on them are of comparable orders of magnitude, nuclear motions tend to be much more sluggish than electronic motions and electrons can be assumed to adjust to the nuclear motions adiabatically. This consideration is the basis of the Born-Oppenheimer approximation, which in general means the assumption of the existence of some basis set of electronic wave functions $\phi_k(\vec{r};\vec{R})$ which depend parametrically on the nuclear positions $\vec{R}$, and for any fixed $\vec{R}$ is orthonormal and complete:

$$\int d^3r\phi_k^*(\vec{r};\vec{R})\phi_\ell(\vec{r};\vec{R}) \equiv \langle\phi_k|\phi_\ell\rangle = \delta_{k\ell} \qquad \text{fixed } \vec{R} , \tag{Ib.5}$$

$$1 = \sum_\ell |\phi_\ell\rangle\langle\phi_\ell| \qquad \text{fixed } \vec{R} , \tag{Ib.6}$$

such that the total wave function $\psi(\vec{R},\vec{r})$ of the collision system can be expanded as

$$\psi(\vec{r},\vec{R}) = \sum_\ell \chi_\ell(\vec{R})\phi_\ell(\vec{r};\vec{R}) . \tag{Ib.7}$$

Adiabatic vs Diabatic Representation - Nonadiabatic Couplings and Transitions. The nuclear wave function $\chi_\ell(\vec{R})$ describes the motion of the nuclei on the potential surface associated with electronic state ℓ. We will see below what we mean by these surfaces when different basis sets ϕ_ℓ are chosen. Specifically, we will see how the different choices will lead to the adiabatic and diabatic representations. The function $\chi_\ell(\vec{R})$ will asymptotically ($\vec{R}\to\infty$) give the amplitude for the probability that the system ends up in the electronic state ϕ_ℓ and thus directly gives the S-matrix for this transition (whose initial state is specified by the problem). Hence, our objective of obtaining the S-matrix reduces, quantum mechanically, to the solution of the Schrödinger equation for $\chi_\ell(\vec{R})$.

The time-independent Schrödinger equation to be solved is

$$H\psi(\vec{r},\vec{R}) = E\psi(\vec{r},\vec{R}) \tag{Ib.8}$$

where H is given by (Ib.4). Substitution of (Ib.7) into (Ib.8) results in the coupled set of equations for $\chi_\ell(\vec{R})$:

$$(T_R + T''_{ii} + U_{ii} - E)\chi_i = -\sum_{j\neq i} (T'_{ij} + T''_{ij} + U_{ij})\chi_j \tag{Ib.9}$$

where

$$U_{ij} = \langle\phi_i|H_{el}|\phi_j\rangle \tag{Ib.10}$$

$$T'_{ij} = \sum_{k=1}^{N-1}\left(\frac{-\hbar^2}{\mu_k}\right)\langle\phi_i|\nabla_{R_k}\phi_j\rangle\cdot\nabla_{R_k} \tag{Ib.11}$$

$$T''_{ij} = \sum_{k=1}^{N-1}\left(\frac{-\hbar^2}{2\mu_k}\right)\langle\phi_i|\nabla^2_{R_k}\phi_j\rangle \ . \tag{Ib.12}$$

Brackets denote integration over electronic coordinates $\vec{r}$ only. The diagonal elements $U_{ij}(\vec{R})$ are the effective potential energy surfaces mentioned earlier that govern nuclear motion. The diagonal terms $T''_{ii}(\vec{R})$, which are nonadiabatic corrections to the potential energy surfaces, are usually small and can be ignored. From (Ib.9) it is seen that motion on the different surfaces are coupled by the off-diagonal elements T'_{ij}, T''_{ij} and U_{ij}. T'_{ij} and T''_{ij} are usually called nonadiabatic couplings, of which T'_{ij} are the dominant ones (dependent only on the first derivative of ϕ with respect to nuclear coordinates) and are also nuclear velocity dependent (since it is proportional to ∇_R operating on nuclear wave functions).

The adiabatic representation is defined by the requirement that the potential matrix $\underset{\sim}{U}$ be diagonal for all $\vec{R}$. This representation employs for the electronic basis functions the eigenfunctions of H_{el}:

$$H_{el}\phi_k^a = W_k(\vec{R})\phi_k^a \ ; \tag{Ib.13}$$

the superscript a on the wave functions denoting adiabaticity. The effective potential energy surfaces governing nuclear motion in this representation are

$$U^a_{kk}(\vec{R}) = W_k(\vec{R}) \ . \tag{Ib.14}$$

The only nonzero couplings in (Ib.9) are the nonadiabatic interactions T'_{ij} and T''_{ij}, hence the designation of adiabatic representation. Figure 1 shows a perspective view of the electronically adiabatic surfaces of the H^+D_2 system with a translational degree (R) and a vibrational degree (r) of freedom [$\vec{R}$=(R,r)].

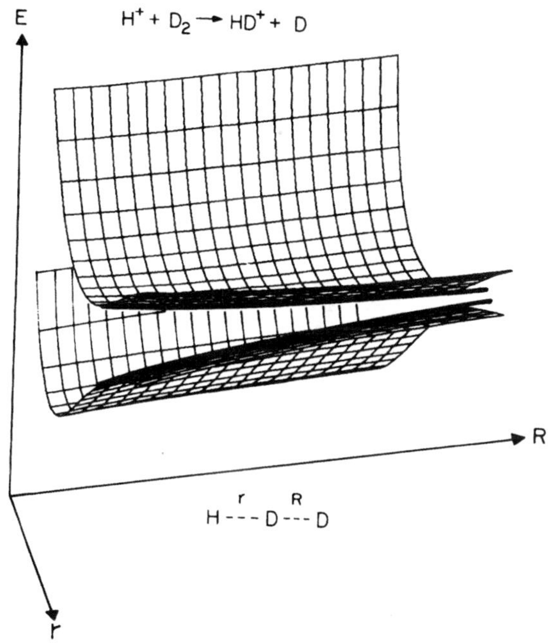

Fig. 1. A perspective view of the electronically adiabatic potential surfaces for H^+D_2.

Writing $\frac{\hbar}{i}\nabla_{R_k}$ as $\vec{p}^{(k)}$, the matrix element $\frac{\hbar}{i} \langle\phi_i|\nabla_{R_k}\phi_j\rangle$ in (Ib.11), can be written as

$$\frac{\hbar}{i} \langle\phi_i|\nabla_{R_k}\phi_j\rangle = \langle\phi_i|\vec{p}^{(k)}|\phi_j\rangle \equiv \vec{P}^{(k)}_{ij} \quad . \quad \text{(Ib.15)}$$

(Ib.11) can then be rewritten as

$$T'_{ij} = \sum_k \frac{1}{\mu_k} \vec{P}^{(k)}_{ij} \cdot \vec{p}^{(k)} \quad . \quad \text{(Ib.16)}$$

Hence in matrix notation

$$\underset{\sim}{T}' = \sum_k \frac{1}{\mu_k} \underset{\sim}{\vec{P}}^{(k)} \cdot \vec{p}^{(k)} . \qquad \text{(Ib.17)}$$

Similarly $\underset{\sim}{T}''$ can be written as

$$\underset{\sim}{T}'' = \sum_k \frac{1}{2\mu_k} (\underset{\sim}{\vec{P}}^{(k)} \cdot \underset{\sim}{\vec{P}}^{(k)} + \vec{p}^{(k)} \cdot \underset{\sim}{\vec{P}}^{(k)}) . \qquad \text{(Ib.18)}$$

Comparison with (Ib.12) shows that (Ib.18) is indeed equivalent to it. Denoting the column matrix χ_i by $\underset{\sim}{\chi}$, (Ib.9) can be rewritten in matrix form as

$$[\sum_k \{\frac{1}{2\mu_k} \underset{\sim}{1}\vec{p}^{(k)} \cdot \vec{p}^{(k)} + \frac{1}{2\mu_k}(\underset{\sim}{\vec{P}}^{(k)} \cdot \underset{\sim}{\vec{P}}^{(k)} + \vec{p}^{(k)} \cdot \underset{\sim}{\vec{P}}^{(k)}) + \frac{1}{\mu_k} \underset{\sim}{\vec{P}}^{(k)} \cdot \vec{p}^{(k)}\}$$

$$+ \underset{\sim}{U}(\vec{R}) - \underset{\sim}{1}E]\underset{\sim}{\chi}(\vec{R}) \equiv (\underset{\sim}{\mathcal{T}} + \underset{\sim}{U}(\vec{R}) - \underset{\sim}{1}E)\underset{\sim}{\chi}(\vec{R}) = 0 ,$$

$$\text{(Ib.19)}$$

where $\underset{\sim}{1}$ is the unit matrix. We see that $\mathcal{T}$ can be identified as the generalized kinetic energy operator of the system. It is useful to write $\underset{\sim}{\mathcal{T}}$ as

$$\underset{\sim}{\mathcal{T}} = \sum_k \frac{1}{2\mu_k} \underset{\sim}{\vec{\mathcal{P}}}^{(k)} \cdot \underset{\sim}{\vec{\mathcal{P}}}^{(k)} , \qquad \text{(Ib.20)}$$

where

$$\underset{\sim}{\vec{\mathcal{P}}}^{(k)} \equiv \underset{\sim}{\vec{P}}^{(k)} + \underset{\sim}{1}\vec{p}^{(k)} \qquad \text{(Ib.21)}$$

is the generalized momentum matrix of the k nuclear coordinate. Thus the matrices $\underset{\sim}{\vec{P}}^{(k)}$ and $\underset{\sim}{U}(R)$ give complete dynamical information of the collision system. The diagonal elements of $\vec{P}_{\ell m}{}^{(k)}$ vanish identically regardless of representation. This can be established by applying the operator $\vec{p}^{(k)} = \frac{\hbar}{i} \nabla_{R_k}$ to the orthonormality condition (Ib.5). This procedure also shows that $\underset{\sim}{\vec{P}}^{(k)}$ is hermitian.

Nonadiabatic couplings tend to peak near avoided crossings, where $U^a_{\ell\ell}-U^a_{mm}$ attains minimum values. Very often these regions arise because the adiabatic states are predominantly mixtures of two simple (molecular orbital or valence bond) structures (described by ϕ_i and ϕ_j) whose corresponding potential energy surface (U_{ii} and U_{jj}) cross. In one-dimensional (atom-atom) systems, if the crossing states are of the same symmetry, then the "non-crossing" rule requires that the adiabatic curves formed from them will exhibit an avoided crossing. The generalization of the non-crossing rule to systems with N internal nuclear degrees of freedom can be stated as follows: the locus of points defined by the intersection of two N-dimensional exact adiabatic potential energy hypersurfaces corresponding to states of the same symmetry forms a hypersurface of at most N-2 dimensions.

During a molecular collision process, if the relative motion of the nuclei is sufficiently slow, they will tend to follow a single adiabatic potential energy surface, even near an avoided crossing. This is so because electronic motion always has time to adjust to the changing nuclear configuration. On the other hand, if the nuclei move very rapidly in the vicinity of an avoided crossing, the probability of nonadiabatic transition (moving from one adiabatic surface to another) will approach unity. This is due to insufficient time for the electrons to adjust their motion adiabatically to rapidly changing nuclear configurations. In the fast motion limit, therefore, the adiabatic representation may not be the most appropriate; and it may be more suitable to choose a representation in which the nonadiabatic couplings are minimal.

Representations which make use of electronic wave functions of desired characteristics (such as ionic or covalent) or those which minimize nonadiabatic couplings are in general known as diabatic representations - a term introduced originally by Lichten. Very often diabatic representations are chosen such that the matrices $\vec{\underset{\sim}{P}}^{(k)}$ vanish. In these representations transitions between surfaces are induced primarily by the off-diagonal elements U_{ij} of the Hamiltonian H_{el}. In the cases where these are small, the corresponding diabatic representation may be preferred over the adiabatic representation, even at low collision energies.

We will now illustrate the foregoing discussion with the simplest possible example of collision dynamics: the one-dimensional two-state problem. The matrix $\underset{\sim}{T}$ (generalized kinetic energy) in (Ib.20) becomes

$$\underset{\sim}{T} = \frac{1}{2\mu}\,\underset{\sim}{1}p^2 + \frac{1}{2\mu}\,(\underset{\sim}{P}\cdot\underset{\sim}{P} + p\underset{\sim}{P}) + \frac{1}{\mu}\,\underset{\sim}{P}p \qquad \text{(Ib.22)}$$

where all matrices are two-by-two and $p = \frac{\hbar}{i}\frac{\partial}{\partial R}$. We start with a diabatic representation in which $\underset{\sim}{P}^d=0$ (all nonadiabatic couplings vanish), and

$$\underset{\sim}{U}^d(R) = \begin{pmatrix} U^d_{11}(R) & U^d_{12}(R) \\ U^d_{12}(R) & U^d_{22}(R) \end{pmatrix}. \qquad \text{(Ib.23)}$$

To find the adiabatic representation, we have to transform $\underset{\sim}{U}^d$ such that

$$\underset{\sim}{U}^a(R) = \underset{\sim}{C}^{-1}(R)\,\underset{\sim}{U}^d(R)\,\underset{\sim}{C}(R) \qquad \text{(Ib.24)}$$

becomes diagonal. The matrix that diagonalizes $U^d(R)$ can be written

$$\underset{\sim}{C}(R) = \begin{pmatrix} \cos\alpha & \sin\alpha \\ -\sin\alpha & \cos\alpha \end{pmatrix} , \qquad \text{(Ib.25)}$$

where

$$\tan 2\alpha(R) = \frac{2U^d_{12}(R)}{[U^d_{22}(R)-U^d_{11}(R)]} . \qquad \text{(Ib.26)}$$

The adiabatic potential matrix $\underset{\sim}{U}^a(R)$ is then

$$\underset{\sim}{U}^a(R) = \begin{pmatrix} 1 & 0 \\ 0 & 1 \end{pmatrix} \bar{U}(R) + \begin{pmatrix} -1 & 0 \\ 0 & 1 \end{pmatrix} U(R) \qquad \text{(Ib.27)}$$

where

$$\bar{U}(R) = \frac{1}{2} [U^d_{11}(R) + U^d_{22}(R)] , \qquad \text{(Ib.28)}$$

$$U^2(R) = U^2_d(R) + [U^d_{12}(R)]^2 , \qquad \text{(Ib.29)}$$

$$U_d(R) = \frac{1}{2} [U^d_{22}(R)-U^d_{11}(R)] . \qquad \text{(Ib.30)}$$

Of course the price we have to pay in using the adiabatic representation is that $\underset{\sim}{P}^a \neq 0$ in general. If we represent the diabatic basis set by ϕ^d_i, the adiabatic set will be given by

$$\phi^a_m = \sum C^{-1}_{mj} \phi^d_j . \qquad \text{(Ib.31)}$$

Hence

$$P^a_{m\ell} = \langle \phi^a_m | \frac{\hbar}{i} \frac{\partial}{\partial R} \phi^a_\ell \rangle = \frac{\hbar}{i} \sum_j C^{-1}_{mj} \frac{\partial}{\partial R} C^{-1}_{\ell j} ; \qquad \text{(Ib.32)}$$

e.g.

$$P^a_{12} = - \frac{\hbar}{i} \frac{\partial\alpha}{\partial R} .$$

We saw earlier that $P_{ii}=0$ in any representation and $\underset{\sim}{P}$ is always hermitian. Therefore, $\underset{\sim}{P}^a$ in our example can be written

$$\underset{\sim}{P}^a(R) = \begin{pmatrix} 0 & i \\ -i & 0 \end{pmatrix} \hbar \frac{\partial\alpha(R)}{\partial R} \quad . \qquad \text{(Ib.33)}$$

$\partial\alpha/\partial R$ can be obtained from differentiating (Ib.26):

$$\alpha' = \frac{\partial\alpha}{\partial R} = \frac{1}{2U^2} [U_d(U_{12}^d)' - U_{12}^d U_d'] \quad . \qquad \text{(Ib.34)}$$

Recalling (Ib.22), we require the operators $\underset{\sim}{P}^a \cdot \underset{\sim}{P}^a$, $p\underset{\sim}{P}^a$ and $\underset{\sim}{P}^a p$ operating on the nuclear wave functions $\underset{\sim}{\chi}(R)$. We will look at the effects of these in turn. First, $\underset{\sim}{P}^a \cdot \underset{\sim}{P}^a$ is diagonal. Therefore, it does not couple the two adiabatic electronic states. Second,

$$\underset{\sim}{P}^a p = \begin{pmatrix} 0 & 1 \\ -1 & 0 \end{pmatrix} \hbar^2 \alpha'(R) \frac{\partial}{\partial R} \qquad \text{(Ib.35)}$$

is the dominant nonadiabatic coupling matrix [which is also hermitian, since it involves $\partial/\partial R$ acting on $\chi_j(R)$]. Finally,

$$p\underset{\sim}{P}^a = \begin{pmatrix} 0 & 1 \\ -1 & 0 \end{pmatrix} \hbar^2 \alpha''(R) \qquad \text{(Ib.36)}$$

is anti-hermitian. We will now consider the physical significance of the quantities α' and α'' by examining the case of the Landau-Zener model. This model assumes straight-line potentials $U_{11}^d(R)$ and $U_{22}^d(R)$ and constant U_{12}^d near the crossing region. From (Ib.30) we see that U_d' is also constant near the crossing region. At the crossing point R_0 itself, $U_d(R)$ vanishes by definition and in general (regardless of the choice of dynamic models)

$$\alpha'(R_0) = -U_d'(R_0)/[2U_{12}^d(R_0)] \quad . \qquad \text{(Ib.37)}$$

Hence in the Landau-Zener model α' is constant throughout the vicinity of the crossing point, and the quantity ΔR_0 defined by

$$\Delta R_0 = 1/\alpha' \qquad \text{(Ib.38)}$$

is a length characterizing the effective width of the crossing region. $\alpha''(R)$, which governs the anti-hermitian nonadiabatic coupling (Ib.36), can be evaluated as [using (Ib.34)]:

$$\alpha'' = \frac{1}{2U^2} [U_d(U_{12}^d)'' - U_{12}^d U_d''] - \frac{2}{U^2} \alpha' [U_d U_d' + U_{12}^d (U_{12}^d)'] \ , \qquad \text{(Ib.39)}$$

which gives

$$\alpha''(R_0) = \frac{-U_d''(R_0)}{2U_{12}^d(R_0)} - \frac{U_d'(R_0)[U_{12}^d(R_0)]'}{[U_{12}^d(R_0)]^2} . \quad \text{(Ib.40)}$$

In the Landau-Zener model, this leads to $\alpha''(R_0)=0$. Thus we can just concentrate on the hermitian nonadiabatic coupling term. Since the off-diagonal elements U_{12}^d are diagonalized away in the adiabatic representation at the expense of the introduction of $\tilde{P}^a p$, the advantage of the adiabatic vs diabatic representations depends on the relative magnitudes of U_{12}^d and $\tilde{P}^a p$. We thus introduce a dimensionless constant reflecting this ratio:

$$\lambda_\ell(R) = \frac{\hbar\alpha'(R)v_\ell(R)}{2U_{12}^d(R)} , \quad \text{(Ib.41)}$$

where $v_\ell(R)$ is a state-dependent nuclear velocity given by

$$\frac{1}{2}\mu v_\ell^2(R) = E - U_{\ell\ell}^d(R) , \quad \text{(Ib.42)}$$

and E is the total energy (kinetic + potential) of the collision system. At R_0 (when v becomes state independent since $U_{11}^d=U_{22}^d$) we then have

$$\lambda(R_0) = \frac{\hbar v(R_0)}{2U_{12}^d(R_0)\Delta R_0} = \frac{v(R_0)}{v_0} \quad \text{(Ib.43)}$$

where

$$v_0 \equiv \frac{2U_{12}^d(R_0)\Delta R_0}{\hbar} \quad \text{(Ib.44)}$$

is a characteristic velocity which also plays a prominent role in Landau-Zener transition probabilities (see Section IIa). v_0 is then the critical velocity determining the suitability of adiabatic vs diabatic representations. If $v(R) \gg v_0$, $\lambda \gg 1$, and (I.41) implies that U_{12}^d is insignificant compared to the strength of nonadiabatic couplings. Hence for high nuclear velocities, the diabatic representation would be more favorable. On the other hand, if $v \ll v_0$, U_{12}^d exceeds the nonadiabatic couplings even at the crossing point, where they tend to be the strongest, and the diabatic representation would lose its advantage compared to the adiabatic. Far from the crossing region ($R\to\infty$), we can assume that

$$U_{12}^d(\infty) = 0 ,$$

$$\frac{\partial}{\partial R} U_{12}^d(\infty) = 0 ,$$

$$U_d(\infty) = \text{const.}$$

and
$$U_d'(\infty) = 0 \; , \tag{Ib.45}$$

such that the logarithmic derivative of U_{12}^d tends to a finite limit

$$\lim_{R\to\infty} \frac{[U_{12}^d]'}{U_{12}^d} = k_{12} \; . \tag{Ib.46}$$

Then, from (Ib.34)

$$\lim_{R\to\infty} \frac{\alpha'}{U_{12}^d} = \lim_{R\to\infty} \frac{(U_{12}^d)'}{2U_d U_{12}^d} = \frac{k_{12}}{2U_d(\infty)} \tag{Ib.47}$$

and, using (Ib.41),

$$\lambda(\infty) = \frac{\hbar v(\infty)}{4U_d(\infty)} \; . \tag{Ib.48}$$

Since $\lambda(\infty)>>1$ for high velocities, (Ib.41) again shows that under these conditions, the diabatic representation is more favorable for large R. In view of the discussion above we can conclude that for large velocities, the diabatic representation is preferable for all R.

Ic. Quantum Mechanical Treatment of Nuclear Motion - Close Coupling

In the last subsection we have separated electronic from nuclear motions by appropriately choosing Born-Oppenheimer electronic basis functions. To deal with nuclear motion we must further separate the coordinates describing relative nuclear motion from those describing internal motion (vibration and rotation). Quantum mechanically this is usually done by expanding $\chi_\ell(\vec{R})$ as [in analogy to (Ib.7)]

$$\chi_\ell(\vec{\rho},\vec{R}') = \sum_\alpha u_{\ell\alpha}(\vec{\rho})\,\xi_{\ell\alpha}(\vec{R}';\vec{\rho}) \tag{Ic.1}$$

where $\vec{\rho}$ stands for the relative motion coordinates and $\vec{R}'$ the internal motion (vibrational-rotational) coordinates. $\xi_{\ell\alpha}$ is the nuclear motion wave function for the α internal vibrational-rotational state of the ℓ electronic state and, as written in (Ic.1), is taken to depend parametrically on $\vec{\rho}$. Similar to the case of purely electronic transitions, the nuclear wave functions $u_{\ell\alpha}$ directly give S-matrix elements for transitions between electronic and internal nuclear states. To solve for $u_{\ell\alpha}$, we have to substitute (Ic.1) into (Ib.9), or, equivalently, (Ib.19). First we write (assuming there is only one internal coordinate $\vec{R}'$)

$$T_R = -\frac{\hbar^2}{2\mu}\nabla_\rho^2 - \frac{\hbar^2}{2\mu'}\nabla_{R'}^2, \qquad \text{(Ic.2)}$$

where μ and μ' are reduced masses for the relative and internal motions respectively. (The case of many internal degrees of freedom will not affect the formal development of the coupled equations). Analogous to (Ib.11) and (Ib.12) we have

$$T'_{ij} = -\frac{\hbar^2}{2\mu}\langle\phi_i|\nabla_\rho\phi_j\rangle\cdot\nabla_\rho - \frac{\hbar^2}{2\mu'}\langle\phi_i|\nabla_{R'}\phi_j\rangle\cdot\nabla_{R'} \qquad \text{(Ic.3)}$$

$$T''_{ij} = -\frac{\hbar^2}{2\mu}\langle\phi|\nabla_\rho^2\phi_j\rangle - \frac{\hbar^2}{2\mu'}\langle\phi_i|\nabla_{R'}^2\phi_j\rangle \ . \qquad \text{(Ic.4)}$$

Multiplying (Ib.9) by $\xi_{i\alpha}$ and integrating over $\vec{R}'$, making use of the property (orthonormality of $\xi_{i\alpha}$ for the same electronic state)

$$\int d^3R'\xi_{i\alpha}\xi_{i\alpha'} = \delta_{\alpha\alpha}\ , \qquad \text{(Ic.5)}$$

we arrive at the coupled equations

$$\left(-\frac{\hbar^2}{2\mu}\nabla_\rho^2 - E\right)u_{i\alpha}(\vec{\rho}) = \sum_{j\alpha'}\Omega_{i\alpha,j\alpha'}(\vec{\rho})u_{j\alpha'}(\vec{\rho}) \qquad \text{(Ic.6)}$$

where

$$\begin{aligned}\Omega_{i\alpha,j\alpha'} &= \int d^3R'\xi_{i\alpha}(U_{ij} + T'_{ij} + T''_{ij})\xi_{j\alpha'} \\ &- \frac{\hbar^2}{\mu}\left\{\int d^3R'\xi_{i\alpha}(\langle\phi_i|\nabla_\rho\phi_j\rangle + \delta_{ij}\nabla_\rho)\xi_{j\alpha'}\right\}\cdot\nabla_\rho \\ &- \delta_{ij}\frac{\hbar^2}{2}\int d^3R'\xi_{i\alpha}\left(\frac{\nabla_\rho^2}{\mu} + \frac{\nabla_{R'}^2}{\mu'}\right)\xi_{j\alpha'}\ .\end{aligned} \qquad \text{(Ic.7)}$$

We note that $\xi_{i\alpha'}$ and $\xi_{j\alpha}$ need not be orthogonal if $i\neq j$. The integral

$$\int d^3R\xi_{i\alpha}\xi_{j\alpha'} \equiv \gamma_{i\alpha,j\alpha'} \qquad \text{(Ic.8)}$$

gives the overlap between internal nuclear wave functions on different electronic potential energy surfaces. The program of close coupling quantum mechanical calculations is the solution of (Ic.6), for which several powerful algorithms have been developed. In principle (Ic.6) is capable of describing rotational, vibrational as well as electronic transitions. However, the problem may become prohibitively difficult as the number of open channels increases.

Id. Semiclassical S-Matrix Elements and the Stationary Phase Approximation

In this subsection we will present a formulation of the classical limit ($\hbar \to 0$) of the S-matrix due to Miller (A complementary formulation has been carried out by Marcus.) This form for the S-matrix for molecular collisions takes advantage of the fact that heavy particle dynamics of the molecular collision system can be well treated by classical mechanics. We will first consider the case of electronically elastic (adiabatic) collisions in which the motion of nuclei follows a single electronically adiabatic surface $U^a(\vec{R})$. In Section III we will generalize our formulation to electronically nonadiabatic collisions based on a path integral formulism.

For the description of adiabatic motion we can use the Hamiltonian

$$H = \frac{P^2}{2\mu} + h(R') + V(R',\rho) \qquad \text{(Id.1)}$$

where the coordinates R' and ρ retain the same meaning as in the last subsection and P is the momentum operator canonically conjugate to ρ. $V(R',\rho)$ is an interaction which vanishes as $\rho \to \infty$, and

$$h|n\rangle = \varepsilon_n|n\rangle \qquad \text{(Id.2)}$$

defines the asymptotic ($\rho \to \infty$) internal states $|n\rangle$ with internal energies ε_n. In a given electronically adiabatic collision process, we are interested in obtaining the transition amplitude between two internal states defined by h, i.e., the S-matrix element

$$S_{n_2,n_1}(E) \ , \qquad \text{(Id.3)}$$

where E is the total energy of the collision system. We will first establish a relationship between the S-matrix element and the Green's function (Ia.20), which will then be computed in the classical limit to give an expression for the semiclassical S-matrix. To obtain $G^+(E)$, we first have to calculate $G_0^+(E)$, the unperturbed retarded Green's function corresponding to the Hamiltonian

$$H_0 = H - V \tag{Id.4}$$

in the translational coordinate and internal quantum number representation $|\rho,n\rangle$. This is given by

$$\langle\rho_2,n_2|G_0^+(E)|\rho_1,n_1\rangle = \delta_{n_2n_1}\left(\frac{-2\mu}{\hbar^2}\right)\frac{\sin(k_1\rho_<)}{k_1}e^{ik_1\rho_>} \tag{Id.5}$$

where

$$k_1 = \left[\frac{2\mu(E-\varepsilon_{n_1})}{\hbar^2}\right]^{1/2} \tag{Id.6}$$

and $\rho_<(\rho_>)$ is the smaller (larger) of ρ_1 and ρ_2. Substituting into (Ia.18) we have, for $n_1 \neq n_2$,

$$\langle\rho_2n_2|G^+(E)|\rho_1n_1\rangle = \int_0^\infty d\rho_2'\int_0^\infty d\rho_1'\langle\rho_2n_2|G_0^+(E)|\rho_2'n_2\rangle \times \langle\rho_2'n_2|T(E)|\rho_1'n_1\rangle\langle\rho_1'n_1|G_0^+(E)|\rho_1n_1\rangle\ . \tag{Id.7}$$

Hence

$$\lim_{\rho_1,\rho_2\to\infty}\langle\rho_2n_2|G^+(E)|\rho_1n_1\rangle = \lim_{\rho_1,\rho_2\to\infty}\left(\frac{2\mu}{\hbar^2}\right)^2\exp(ik_1\rho_1 + ik_2\rho_2) \times \langle k_2n_2|T(E)|k_1n_1\rangle \tag{Id.8}$$

where $\langle k_2n_2|T|k_1n_1\rangle = \int_0^\infty d\rho_1'\int_0^\infty d\rho_2'\,\frac{\sin k_2\rho_2'}{k_2}\langle\rho_2'n_2|T(E)|\rho_1'n_1\rangle \times \frac{\sin k_1\rho_1'}{k_1}\ .$ (Id.9)

Since the T- and S-matrices are related by

$$S_{n_2,n_1} = \delta_{n_2n_1} - 2i\left(\frac{2\mu}{\hbar^2}\right)(k_1k_2)^{1/2}\langle k_2n_2|T(E)|k_1n_1\rangle\ , \tag{Id.10}$$

we can identify the S-matrix in terms of $G^+(E)$ as $(n_1 \neq n_2)$

$$S_{n_2,n_1}(E) = -\frac{i\hbar^2}{\mu}(k_1k_2)^{1/2}\lim_{\rho_1,\rho_2\to\infty}\exp(-ik_1\rho_1 - ik_2\rho_2)$$

$$\times \langle\rho_2 n_2|G^+(E)|\rho_1 n_1\rangle = \lim_{\rho_1,\rho_2\to\infty}\left(\frac{\hbar^2 k_1 k_2}{\mu^2}\right)^{1/2}$$

$$\times \exp(-ik_1\rho_1 - ik_2\rho_2)\int_0^\infty dt\ \exp\left(\frac{iEt}{\hbar}\right)\langle\rho_2 n_2|\exp\left(\frac{-iHt}{\hbar}\right)|\rho_1 n_1\rangle \qquad \text{(Id.11)}$$

where we have made use of (Ia.20).

So far the treatment has been exactly quantum mechanical. In order to evaluate S_{n_2,n_1} in the classical limit, we use the classical Hamiltonian in (Id.11), which is given by [compare (Id.1)]

$$H(P,\rho,\vec{n},\vec{q}) = \frac{P^2}{2\mu} + \varepsilon(\vec{n}) + V(\rho,\vec{n},\vec{q}) \ . \qquad \text{(Id.12)}$$

This Hamiltonian has been expressed in terms of the action-angle variables $\{n_i\}$ and $\{q_i\}$, $i=1,\ldots,N-1$, where N is the total number of nuclear degrees of freedom. These variables describe internal motion classically and, in particular, the action variables $\vec{n}$ are the classical counterpart of the quantum numbers for the internal degrees of freedom. WKB quantum conditions require that n_i be integers at asymptotic regions ($\rho\to\infty$), and this forms the basis for the semiclassical description of the initial and final quantum states of the collision system.

In order to proceed further, we have to digress to present a result for the classical limit ($\hbar\to 0$) Green's function in the coordinate representation:

$$\langle q_2|\exp\left[-\frac{iH(t_2-t_1)}{\hbar}\right]|q_1\rangle = \left[-\frac{\partial^2\phi(q_2,q_1)/\partial q_2\partial q_1}{2\pi i\hbar}\right]^{1/2}$$

$$\times \exp\left[\frac{i\phi(q_2,q_1)}{\hbar}\right] \qquad \text{(Id.13)}$$

where ϕ is the classical action, the time integral of the classical Lagrangian,

$$\phi(q_2,q_1) = \int_{t_1}^{t_2} dt\{p(t)\dot{q}(t) - H[p(t),q(t)]\} \qquad \text{(Id.14)}$$

and p,q are the canonically conjugate momentum and coordinate. The only approximation involved in the derivation of (Id.13) is the stationary phase approximation formula:

$$\int dx\ g(x)\exp\left[\frac{if(x)}{\hbar}\right] \simeq \sum_k g(x_k)\left[\frac{2\pi i\hbar}{f''(x_k)}\right]^{1/2}\exp\left[\frac{if(x_k)}{\hbar}\right] \tag{Id.15}$$

where the points of stationary phase are given by

$$f'(x_k) = 0\ . \tag{Id.16}$$

Since (Id.15) becomes exact as $\hbar\to 0$, one may say that classical mechanics is the stationary phase approximation of quantum mechanics.

Making use of (Id.13) in (Id.11) we have

$$S_{n_2,n_1}(E) = \lim_{\rho_1,\rho_2\to\infty}\left(-\frac{P_1P_2}{\mu^2}\right)^{1/2}\exp\left[\frac{i}{\hbar}(P_1\rho_1 - P_2\rho_2)\right]$$

$$\times\int_0^\infty dt\left[\frac{\partial^2\phi(\rho_2\vec{n}_2,\rho_1\vec{n}_1;t)/\partial(\rho_2\vec{n}_2)\partial(\rho_1\vec{n}_1)}{(-2\pi i\hbar)^N}\right]^{1/2}$$

$$\times\exp\{\frac{i}{\hbar}[Et + \phi(\rho_2\vec{n}_2,\rho_1\vec{n}_1;t)]\} \tag{Id.17}$$

where the action integral in the mixed representation $(\rho,\vec{n})$ is

$$\phi(\rho_2\vec{n}_2,\rho_1\vec{n}_1;t) = \int_{t_1}^{t_2}dt'[P(t')\dot{\rho}(t') - \vec{q}(t')\cdot\dot{\vec{n}}(t') - H(t')]\ , \tag{Id.18}$$

$$t \equiv t_2 - t_1\ ,$$

and the momenta P_1 and P_2, with directions accounted for, are given by

$$P_1 = -\hbar k_1\ ;$$

$$P_2 = +\hbar k_2\ .$$

The stationary phase approximation (Id.15) is now invoked again to do the integral in (Id.17). First we note that

$$\frac{\partial\phi}{\partial t}(\rho_2\vec{n}_2,\rho_1\vec{n}_1;t) = -E(\rho_2\vec{n}_2,\rho_1\vec{n}_1;t) \qquad (\text{Id.19})$$

where $E(\rho_2\vec{n}_2,\rho,\vec{n}_1;t)$ is the total energy for the trajectory with the indicated boundary conditions. It is distinct from the constant quantity appearing in the argument of $S_{n_2,n_1}(E)$. In fact, the stationary phase condition

$$-\frac{\partial}{\partial t}\phi(\rho_2\vec{n}_2,\rho_1\vec{n}_1;t) \equiv E(\rho_2\vec{n}_2,\rho_1\vec{n}_1;t) = E \qquad (\text{Id.20})$$

defines the functional relationship between E and t. The stationary phase in the integrand of (Id.17) can now be written

$$\Phi(\vec{n}_2,\vec{n}_1;E) = P_1\rho_1 - P_2\rho_2 + Et - \int_{t_1}^{t_2} dt'[P(t')\dot{\rho}(t') - \vec{q}(t')\cdot\dot{\vec{n}}(t') - H(t')] = \int_{-\infty}^{\infty} dt[-\rho(t)\dot{P}(t) - \vec{q}(t)\cdot\dot{\vec{n}}(t)]\ , \qquad (\text{Id.21})$$

and

$$S_{n_2,n_1}(E) = \left[-\frac{P_1P_2}{\mu^2}\frac{\partial t}{\partial E}\frac{\partial^2\phi(\rho_2\vec{n}_2,\rho_1\vec{n}_1;t)}{\partial(\rho_2\vec{n}_2)\partial(\rho_1\vec{n}_1)}\Big/(-2i\pi\hbar)^{N-1}\right]^{1/2} \times \exp\left[\frac{i\Phi(\vec{n}_2,\vec{n}_1;E)}{\hbar}\right] \qquad (\text{Id.22})$$

The pre-exponential derivative in (Id.22) is actually the determinant

$$\begin{vmatrix} \dfrac{\partial^2\phi}{\partial\vec{n}_1\partial\vec{n}_2} & \dfrac{\partial^2\phi}{\partial\vec{n}_2\partial\rho_1} \\ \dfrac{\partial^2\phi}{\partial\rho_2\partial\vec{n}_1} & \dfrac{\partial^2\phi}{\partial\rho_2\partial\rho_1} \end{vmatrix} \qquad (\text{Id.23})$$

which can be shown to be (with some messy algebra):

$$\frac{\partial^2\phi}{\partial(\rho_2\vec{n}_2)\partial(\rho_1\vec{n}_1)} = \frac{\mu^2}{P_1P_2}\left(\frac{\partial t}{\partial E}\right)^{-1}\frac{\partial^2\Phi(\vec{n}_2,\vec{n}_1;E)}{\partial\vec{n}_2\partial\vec{n}_1}\ . \qquad (\text{Id.24})$$

The expression (Id.22) for the semiclassical S-matrix then simplifies to

$$S_{\vec{n}_2,\vec{n}_1}(E) = i\left[\frac{\partial^2\Phi(\vec{n}_2,\vec{n}_1;E)/\partial\vec{n}_2\partial\vec{n}_1}{(-2\pi i\hbar)^{N-1}}\right]^{1/2}\exp\left[\frac{i\Phi(\vec{n}_2,\vec{n}_1;E)}{\hbar}\right] \tag{Id.25}$$

We can go one more step to simplify the pre-exponential factor. From (Id.21) we have

$$\phi(\vec{n}_2\rho_2,\vec{n}_1\rho_1;t) = \Phi(\vec{n}_2,\vec{n}_1;E) + P_2\rho_2 - P_1\rho_1 - Et \tag{Id.26}$$

and from (Id.18),

$$\frac{\partial\phi}{\partial\vec{n}_1} = q_1 . \tag{Id.27}$$

Differentiating (Id.26) with respect to $\vec{n}_1$, we obtain

$$\frac{\partial\Phi}{\partial\vec{n}_1}(n_2,n_1;E) = \vec{q}_1 + \rho_1\frac{\partial P_1}{\partial\vec{n}_1}$$

$$= \vec{q}_1 - \hbar\rho_1\frac{\partial k_1}{\partial\vec{n}_1}$$

$$= \vec{q}_1 - \frac{\mu\rho_1}{P_1}\frac{\partial\varepsilon(\vec{n}_1)}{\partial\vec{n}_1} \equiv \vec{q}_1' \tag{Id.28}$$

where in the last equality we have used (Id.6). If we consider the time dependence of $\vec{q}_1$ without the interaction potential in (Id.12), i.e., the time dependence in asymptotic regions, we see that (using Hamilton's equations)

$$\dot{\vec{q}}_1 = \frac{\partial H}{\partial\vec{n}_1} = \frac{\partial\varepsilon(\vec{n}_1)}{\partial\vec{n}_1} \tag{Id.29}$$

$$\dot{\rho}_1 = \frac{\partial H}{\partial P} = \frac{P_1}{\mu} . \tag{Id.30}$$

Hence

$$q(t_1) = \text{const} + \frac{\partial\varepsilon(\vec{n}_1)}{\partial\vec{n}_1}t_1 \tag{Id.31}$$

$$\rho(t_1) = \text{const} + \frac{P_1}{\mu} t_1 \, , \qquad \text{(Id.31)}$$

and $\vec{q}_1'$ as defined in (Id.28) does not have any asymptotic time dependence, that is, it has the "free" time dependence of $\vec{q}(t)$ subtracted out. The pre-exponential factor of S_{n_2,n_1} is then given by

$$\frac{\partial^2 \Phi(\vec{n}_2,\vec{n}_1,E)}{\partial\vec{n}_2 \partial\vec{n}_1} = \left(\frac{\partial\vec{q}_1'}{\partial\vec{n}_2}\right)_{\vec{n}_1} = \left[\left(\frac{\partial\vec{n}_2}{\partial q_1'}\right)_{\vec{n}_1}\right]^{-1} . \qquad \text{(Id.32)}$$

Finally the S-matrix can be expressed most conveniently by

$$S_{\vec{n}_2,\vec{n}_1}(E) = i\left[(-2\pi i\hbar)^{N-1}\left(\frac{\partial\vec{n}_2}{\partial\vec{q}_1'}\right)_{\vec{n}_1}\right]^{-1/2} \exp\left[\frac{i\Phi(\vec{n}_2,\vec{n}_1;E)}{\hbar}\right] , \qquad \text{(Id.33)}$$

where Φ is given by (Id.21). In any application of (Id.33), we must determine classical trajectories specified by initial boundary conditions

$$\begin{aligned} \vec{n}(t_1) &\equiv \vec{n}_1 = \text{specified integers} \\ \rho(t_1) &= \text{asymptotically large} \\ P(t_1) &= -\{2\mu[E - \varepsilon(\vec{n}_1)]\}^{1/2} \\ \vec{q}(t_1) &= \vec{q}_1' + \frac{\mu\rho(t_1)}{P(t_1)} \frac{\partial\varepsilon(\vec{n}_1)}{\partial\vec{n}_1} \end{aligned} \qquad \text{(Id.34)}$$

and final conditions

$$\begin{aligned} \vec{n}(t_2) &\equiv \vec{n}_2 = \text{specified integers} \\ \rho(t_2) &= \text{asymptotically large} \\ P(t_2) &= \{2\mu[E - \varepsilon(\vec{n}_2)]\}^{1/2} \\ \vec{q}(t_2) &= \text{anything} \end{aligned} \qquad \text{(Id.35)}$$

If there is more than one trajectory fitting the prescribed bondary conditions, (Id.33) must consist of a sum of terms, one for each such trajectory. To evaluate the derivatives $\partial\vec{n}_2/\partial\vec{q}_1'$, we must integrate Hamilton's equations to obtain a function

$\vec{n}_2(\vec{q}_1', \vec{n}_1; E)$, whose values, of course, need not be integers. We will then have to solve

$$\vec{n}_2 = \vec{n}_2(\vec{q}_1') \qquad \text{(Id.36)}$$

(with the arguments $\vec{n}_1$ and E of $\vec{n}_2$ suppressed) for the initial angle variables $\vec{q}_1'$ which would lead to the prescribed final integral values for $\vec{n}_2$. (Id.36) is a set of N-1 equations in the N-1 unknowns $\vec{q}_1'$, and the pre-exponential factor $\partial\vec{n}_2/\partial\vec{q}_1'$ is in fact a determinant of partial derivatives, which are to be evaluated at the roots of (Id.36). Each set of roots $\vec{q}_1'$ will then correspond to a distinct trajectory and will lead to a different Φ.

II. SPECIFIC THEORIES AND DYNAMICAL MODELS

In this section we will begin by studying simple dynamical models for nonadiabatic electronic transitions. We will consider low energy collisions such that the characteristic extent of nonadiabatic coupling is small. Under such circumstances the potential surfaces $U_{ii}(\vec{R})$ and the nonadiabatic coupling matrix elements can very often by approximated by analytic functions of $\vec{R}$. Moreover, the transition can be considered as due solely to the coupling between two electronic states, thus the general multistate problem defined by (Ib.9) can be reduced to the relatively simple two-state problem. We will state our problem in terms of the semiclassical time-dependent formulation.

The Schrodinger equation to be solved is

$$H\psi(\vec{R},\vec{r};t) = i\hbar \frac{\partial}{\partial t} \psi(\vec{R},\vec{r};t) \qquad \text{(II.1)}$$

where [compare with (Ib.7)]

$$\psi(\vec{R},\vec{r};t) = \sum_{\ell} \chi_{\ell}(\vec{R};t)\phi_{\ell}(\vec{R},\vec{r}) \quad . \qquad \text{(II.2)}$$

If $\{\phi_\ell\}$ is chosen to be an adiabatic representation, (II.1) becomes

$$\{-\frac{\hbar^2}{2\mu}\frac{\partial}{\partial R^2} + U_{ii}^a(R)\}\chi_i + \sum_{j\neq i} T_{ij}\chi_j = i\hbar\frac{\partial}{\partial t}\chi_i \qquad \text{(II.3)}$$

where $$T_{ij} \equiv \langle\phi_i|-\frac{\hbar^2}{2\mu}\frac{\partial^2}{\partial R^2}\phi_j\rangle - \frac{\hbar^2}{\mu}\langle\phi_i|\frac{\partial}{\partial R}\phi_j\rangle\frac{\partial}{\partial R} , \qquad \text{(II.4)}$$

and a one-dimensional problem has been assumed. [It is seen that $T_{ij} = T_{ij}' + T_{ij}''$, on comparison with (Ib.11) and (Ib.12)]. The

T_{ij} are, of course, the nonadiabatic couplings discussed in Section Ib.2.

So far, our time-dependent formulation is rigorously quantum mechanical, and (II.3) is equivalent to (Ib.9). The semiclassical viewpoint is introduced in assuming a classical trajectory

$$R = R(t) .$$

The electronic Hamilton H_{el} then becomes time-dependent through $R(t)$, and the electronic motion can be described by a time-dependent wave function $\Phi(\vec{r};t)$ satisfying

$$H_{el}(\vec{r},R(t))\Phi(\vec{r};t) = i\hbar \frac{\partial}{\partial t} \Phi(\vec{r};t) . \quad \text{(II.5)}$$

Φ can be expanded in terms of the ϕ_i:

$$\Phi(\vec{r};t) = \sum_i a_i(t)\phi_i(R(t),\vec{r})\exp\{-\frac{i}{\hbar}\int^t U^a_{ii}(R(t))dt\} . \quad \text{(II.6)}$$

The nonadiabatic transition amplitudes, or S-matrix elements, are then determined by $|a_n(t)|^2$. Substitution of (II.6) into (II.5) gives

$$i\hbar\dot{a}_i = \sum_{j\neq i} C_{ij}\exp\{-\frac{i}{\hbar}\int^t dt(U^a_{jj} - U^a_{ii})\}a_j \quad \text{(II.7)}$$

where the nonadiabatic coupling matrix elements are

$$C_{ij} \equiv \langle\phi_i|-i\hbar\partial\phi_j/\partial t\rangle . \quad \text{(II.8)}$$

For a two-state problem we have

$$i\hbar\dot{a}_1 = C_{12}\exp\{\frac{i}{\hbar}\int^t dt(U^a_{11} - U^a_{22})\}a_2 , \quad \text{(II.9a)}$$

$$i\hbar\dot{a}_2 = C_{21}\exp\{\frac{i}{\hbar}\int^t dt(U^a_{22} - U^a_{11})\}a_1 . \quad \text{(II.9b)}$$

The S-matrix element $S_{2\leftarrow 1}$ is given by

$$S_{2\leftarrow 1} = a_2(t\rightarrow +\infty) \quad \text{(II.10)}$$

with the initial conditions

$$a_1(-\infty) = 1 , \quad a_2(-\infty) = 0 , \quad \text{(II.11)}$$

and the coupling C_{12} is assumed to be maximum at t=0.

IIa. The Landau-Zener Model

It is often more convenient to formulate collision problems in the diabatic rather than the adiabatic representation (see discussion in Section Ib.2). The Landau-Zener model of nonadiabatic coupling between electronic states of the same symmetry, for instance, is formulated in terms of the diabatic representation of the electronic Hamiltonian, $\underset{\sim}{U}^d$, defined by

$$U^d_{11} = \bar{U}(R_0) - F_1(R-R_0)$$

$$U^d_{22} = \bar{U}(R_0) - F_2(R-R_0) \qquad \text{(IIa.1)}$$

$$U^d_{12} = a = \text{const.} ,$$

where $$\bar{U}(R_0) \equiv U^d_{11}(R_0) = U^d_{22}(R_0) , \qquad \text{(IIa.2)}$$

R_0 being the point of intersection of the diabatic curves, and F_1, F_2 are constants. The equation of the trajectory in the vicinity of R_0 is approximated by a linear function of time:

$$R - R_0 = v_0 t . \qquad \text{(IIa.3)}$$

From (Ib.27), $\underset{\sim}{U}^a$ is then given by

$$U^a_{11} = \bar{U}(R_0) - \frac{1}{2}(F_1+F_2)(R-R_0) - U(R) \qquad \text{(IIa.4a)}$$

$$U^a_{22} = \bar{U}(R_0) - \frac{1}{2}(F_1+F_2)(R-R_0) + U(R) \qquad \text{(IIa.4b)}$$

where $$4U^2(R) = \{\Delta F(R-R_0)\}^2 + 4a^2 , \ \Delta F=|F_1-F_2| .$$

(IIa.5)

Analogous to the derivation of (Ib.32), we obtain

$$C_{12} = i\hbar\partial\alpha/\partial t \qquad \text{(IIa.6)}$$

where α is defined in (Ib.26). Using (Ib.34) in which ' means $\partial/\partial t$), (IIa.6) reduces to

$$C_{12} = \frac{i\hbar a(\Delta F)v_0}{\{\Delta F(v_0 t)\}^2 + 4a^2} . \qquad \text{(IIa.7)}$$

The exact solution of (II.9) using (IIa.4) and (IIa.7) has been carried out by Zener, and it is found that, for a 'once passage' through the nonadiabatic coupling region, the S-matrix is given by

$$|S_{2\leftarrow 1}|^2 = |a_2(\infty)|^2 = \exp\left[\frac{-2\pi a^2}{v_0 \hbar \Delta F}\right] . \qquad \text{(IIa.8)}$$

As discussed in the previous section, it is thus seen that when the velocity v_0 is small, the nonadiabatic transition probability is also small. On the other hand, for v_0 large, $S_{2\leftarrow 1}$ tends to unity, meaning that the colliding particles tend to follow the diabatic path.

IIb. The Demkov Model

In this model $\underset{\sim}{U}^d(R)$ is given by

$$U_{22}^d(R) = \frac{\Delta}{2} = \text{const} , \quad U_{11}^d(R) = -\frac{\Delta}{2} , \qquad \text{(IIb.1)}$$

$$U_{12}^d(R) = A \exp(-\lambda R) , \qquad \text{(IIb.2)}$$

and the straight-line trajectory $R(t)=v_0 t$ is also employed. Transforming to the adiabatic representation, we have

$$U_{11}^a(R) = -\frac{1}{2}[\Delta^2 + 4A^2\exp(-2\lambda R)]^{1/2} \qquad \text{(IIb.3a)}$$

$$U_{22}^a(R) = \frac{1}{2}[\Delta^2 + 4A^2\exp(-2\lambda R)]^{1/2} . \qquad \text{(IIb.3b)}$$

At asymptotic regions $U_{11}^a(R) \rightarrow U_{11}^d(R)$ and $U_{22}^a(R) \rightarrow U_{22}^d(R)$. Hence in this case, opposite to the Landau-Zener model, the non-adiatic transition amplitudes are the same as the diabatic ones. Expanding $\Phi(\vec{r};t)$ in the diabatic basis set $\{\phi_i^d\}$:

$$\Phi(\vec{r};t) = \sum_i b_i(t)\phi_i^d(R(t),\vec{r})\exp\{-\frac{i}{\hbar}\int^t dt U_{ii}^d(R(t))\} , \qquad \text{(IIb.4)}$$

we obtain, analogous to (II.9), the set of coupled equations

$$i\hbar\dot{b}_1 = U_{12}^d \exp\{\frac{i}{\hbar}\int^t dt(U_{11}^d - U_{22}^d)\}b_2 , \qquad \text{(IIb.5a)}$$

$$i\hbar \dot{b}_2 = U_{21}^d \exp\{\frac{i}{\hbar}\int^t dt(U_{22}^d - U_{11}^d)\}b_1 \ . \tag{IIb.5b}$$

If we use the initial conditions

$$b_1(-\infty) = 1 \ , \ b_2(-\infty) = 0 \ , \tag{IIb.6}$$

the 'once passage' S-matrix element or transition probability is given by $|b_2(+\infty)|^2$. The exact solution of (IIb.5) using (IIb.1) and (IIb.2), as worked out by Demkov, is given by

$$|S_{2\leftarrow 1}|^2 = |b_2(\infty)|^2 = \text{sech}^2\left(\frac{\pi\Delta}{2\hbar\lambda v_0}\right)\sin^2\left(\frac{1}{\hbar}\int_{-\infty}^{\infty} dt \ U_{12}^d(R(t))\right)$$

$$= \sin^2\left(\frac{2A}{\hbar\lambda v_0}\right)\text{sech}^2\left(\frac{\pi\Delta}{2\hbar\lambda v_0}\right) \ . \tag{IIb.7}$$

The Demkov model and the Demkov formula (IIb.7) is widely used to describe charge transfer processes of the type

$$A + B^+ \rightarrow A^+ + B \text{ or}$$

$$A + B^- \rightarrow A^- + B \ .$$

IIc. Stuckelberg Theory - Curve-Crossing Problems

In the diabatic representation (in which all nonadiabatic couplings vanish) the coupled equations of motion for the one-dimensional two-state problem are

$$\chi_1'' + \bar{k}_1^2(R)\chi_1 = \alpha(R)\chi_2 \tag{IIc.1a}$$

$$\chi_2'' + \bar{k}_2^2(R)\chi_2 = \alpha(R)\chi_1 \tag{IIc.1b}$$

where

$$\bar{k}_{1,2}^2 = \frac{2\mu}{\hbar^2}(E - U_{11,22}^d) \ , \tag{IIc.2}$$

$$\alpha = \frac{2\mu}{\hbar^2} U_{12}^d \tag{IIc.3}$$

and "denotes second derivative with respect to R. (IIc.1) is often used to describe collision dynamics with curve crossing, i.e., when $U_{11}^d(R)$ and $U_{22}^d(R)$ intersect at some point. The Stuckelberg method to solve these equations is based on semiclassical phase integrals and their analytic continuation in the complex R plane. For unit amplitude entrance on channel 1, (IIc.1) must be solved subject to the boundary conditions

$$\chi_1(R) \underset{R\to\infty}{\sim} e^{i\bar{k}_1(\infty)R} - S_{11}\, e^{i\bar{k}_1(\infty)R} \qquad \text{(IIc.4a)}$$

$$\chi_2(R) \underset{R\to\infty}{\sim} - S_{21}\, e^{i\bar{k}_2(\infty)R} \qquad \text{(IIc.4b)}$$

where S_{11} and S_{12} are the scattering matrix elements.

The procedure of Stuckelberg to solve the equations is as follows. First one eliminates χ_2 from (IIc.1) to obtain a fourth order differential equation for χ_1. A solution of the form

$$\chi_1 = \exp\{\frac{1}{\hbar} S_0 + S_1 + \hbar S_2 + \ldots .\} \qquad \text{(IIc.5)}$$

is then substituted into the fourth order equation to obtain equations for $S_0, S_1, \ldots .$ when coefficients of the various powers of $\hbar$ are separately equated to zero. Keeping only S_0 and S_1 one can arrive at the general WKB solutions for χ_1 and χ_2. The solutions thus found are observed to have singularities at R_* which satisfies

$$(U^d_{22} - U^d_{11})^2 + 4(U^d_{12})^2 = 0 \ . \qquad \text{(IIc.6)}$$

R_* is in general complex and (IIc.6) implies that R_*^* is also a solution. The WKB forms χ_1 and χ_2 then cannot be used in the neighborhood of R_*, including the immediately adjacent real axis. It is, however, desirable to consider the approximate solutions $\chi_{1,2}$ as analytic in sufficiently remote regions from the conjugate pair R_* and R_*^* so that $\chi_{1,2}$ for $R \gg \mathrm{Re}R_*$ can be joined smoothly to $\chi_{1,2}$ for $R_1, R_2 \ll R \ll \mathrm{Re}R_*$ ($R_{1,2}$ being turning points) by tracing the WKB solutions along a contour remote from the neighborhood of R_* and R_*^*. Any contour chosen will invariably pass through the so-called Stokes lines, which are defined by

$$\mathrm{Re}\int_{R_*}^{R} dR(k_1 - k_2) = 0 \qquad \text{(IIc.7)}$$

where

$$k^2_{1,2} = \frac{2\mu}{\hbar^2}(E - W_{1,2}) \qquad \text{(IIc.8)}$$

and $W_{1,2}$ are the adiabatic curves obtained by diagonalizing $\tilde{U}^d$. The WKB solutions consist of dominant and subdominant solutions in the complex R plane. As one crosses a Stokes line the coefficient of the subdominant solution must change discontinuously in order for the solutions to join. This is known as the Stokes phenomenon, and one can say that a nonadiabatic transition is the physical manisfestation of the Stokes phenomenon.

This connection procedure determines the coefficients of the WKB solutions whose asymptotic forms lead directly to the S-matrix. If there is only one region of nonadiabaticity (in which the WKB solutions are not valid), then there will be two crossing encounters due to the fact that this region is passed twice, once on the way in and once on the way out. Stuckelberg's solution for the S-matrix is given in this case by

$$S_{i,j} = (P_{i,j}{}^{I})^{1/2}\exp(i\phi_{i,j}{}^{I}) + (P_{i,j}{}^{II})^{1/2}\exp(i\phi_{i,j}{}^{II}) \quad \text{(IIc.9)}$$

for i,j=1,2, where W_1 is chosen to be the lower adiabatic curve, the ϕ's are the WKB phase integrals (i.e., classical action integrals), the P's are transition probabilities; and I and II refer to the two crossing encounters and thus to the two different classical trajectories that contribute to the $i\to j$ nonadiabatic transition. The various quantities in (IIc.9) are given below:

$$\phi_{1,1}{}^{I} = \lim_{R\to\infty} 2\left(\xi - k_1R + \int_{R_1}^{R} dR'k_1(R')\right) , \quad \text{(IIc.10a)}$$

$$\phi_{1,1}{}^{II} = \lim_{R\to\infty} 2\left(\xi - k_1R + \int_{R_2}^{\mathrm{Re}R_*} dR'k_2(R') + \int_{\mathrm{Re}R_*}^{R} dR'k_1(R')\right) , \quad \text{(IIc.10b)}$$

$$\phi_{1,2}{}^{I} = \phi_{2,1}{}^{I} = \lim_{R\to\infty} \left(\xi - k_1R - k_2R + \int_{\mathrm{Re}R_*}^{R} dR'k_1(R') + \int_{\mathrm{Re}R_*}^{R} dR'k_2(R') + 2\int_{R_2}^{\mathrm{Re}R_*} dR'k_2(R')\right) , \quad \text{(IIc.10c)}$$

$$\phi_{1,2}{}^{II} = \phi_{2,1}{}^{II} = \lim_{R\to\infty} \left(-\xi-k_1R-k_2R + \int_{\mathrm{Re}R_*}^{R} dR'k_1(R') + \int_{\mathrm{Re}R_*}^{R} dR'k_2(R') + 2\int_{R_1}^{\mathrm{Re}R_*} dR'k_1(R')\right) , \quad \text{(IIc.10d)}$$

$$\phi_{2,2}{}^{I} = \lim_{R\to\infty} 2\left(\xi - k_2R + \int_{R_2}^{R} dR'k_2(R')\right) , \quad \text{(IIc.10e)}$$

$$\phi_{2,2}^{II} = \lim_{R\to\infty} 2(\xi - k_2R + \int_{R_1}^{ReR_*} dR'k_1(R') + \int_{ReR_*}^{R} dR'k_2(R')), \quad \text{(IIc.10f)}$$

$$P_{1,1}^{I} = P_{2,2}^{I} = (1-p)^2 , \quad \text{(IIc.11a)}$$

$$P_{1,1}^{II} = P_{2,2}^{II} = p^2 , \quad \text{(IIc.11b)}$$

$$P_{1,2}^{I} = P_{1,2}^{II} = P_{2,1}^{I} = P_{2,1}^{II} = p(1-p) , \quad \text{(IIc.11c)}$$

where

$$p = \exp(-2\delta) \quad \text{(IIc.12)}$$

is the local nonadiabatic transition probability and

$$2\delta = i\int_{R_*^*}^{R_*} dR[k_2(R) - k_1(R)] = 2i\int_{ReR_*}^{R_*} dR[(k_2(R) - k_1(R)] . \quad \text{(IIc.13)}$$

In (IIc.10), $k_i = k_i(\infty)$ and the phase factor ξ is given by (not originally given by Stuckelberg in this form)

$$\xi = \delta\ell n\delta - \delta - \arg\Gamma(i\delta) - \frac{\pi}{4} . \quad \text{(IIc.14)}$$

For $\delta \ll 1$, $\xi \sim \frac{\pi}{4}$ and for $\delta \gg 1$, $\xi \sim -1/12\delta$.

Interference between the two terms in (IIc.9) leads to oscillatory behavior of $|S_{i,j}|^2$ as a function of E. This phenomenon is known as Stuckelberg oscillations. For example, considering the 1→2 transition, we have

$$|S_{1,2}|^2 = 2p(1-p)[1 + \cos 2(\xi+A-B)] , \quad \text{(IIc.15)}$$

where

$$A = \int_{R_2}^{ReR_*} dR'k_2(R') , \quad \text{(IIc.16a)}$$

$$B = \int_{R_1}^{ReR_*} dR'k_1(R') . \quad \text{(IIc.16b)}$$

IId. Tully-Preston Surface-Hopping Trajectory (SHT) Theory

The SHT approach is basically a classical trajectory approach in which nuclear motion is treated completely classically. Classical trajectories are propagated on the entrance (initial) potential surface until an avoided surface crossing or other point of large nonadiabatic interaction is reached. A relation defining these avoided crossing seams have to be found for each problem:

$$S(\vec{R}) = 0 \ . \qquad \text{(IId.1)}$$

A probability of switching to another surface $P(\vec{R},\dot{\vec{R}})$ is then computed and the trajectory branches. One branch, with the weight (1-P), is resumed on the old surface, whereas a new branch, with the weight P, is resumed on the new potential surface, with a slight velocity correction. More branching will occur if the branched trajectories subsequently encounter the avoided crossing seam again, and at each encounter, a surface-hopping probability has to be determined.

The switching probabilities at each hopping encounter are calculated by integrating the coupled equations (II.9). For instance, if a trajectory is initially propagated on the adiabatic surface 1, then $P = |a_2|^2$ is obtained by integrating (II.9) over the entire nonadiabatic coupling region. The solution obtained would, of course, depend on the choice of the classical trajectory $\vec{R}(t)$ used in the R-dependent quantities in (II.9), which in turn depends on the choice of the potential (either U^a_{11} or U^a_{22}) used for its determination. However, it has been found that for many triatomic systems of interest, especially the H_3^+ system, the choice of the effective potential for the determination of $\vec{R}(r)$ in (II.9) has relatively little effect on its solution, and very often the solutions given by the different choices can be approximated by the Landau-Zener formula with appropriate parameters. Furthermore the transition probability depends almost entirely on the component of velocity perpendicular to the avoided crossing seam.

An important requirement of the model is a prescription for changing velocities to conserve angular momentum and energy when a surface hop occurs. Since the perpendicular (to the seam) component of the velocity is most effective in nonadiabatic coupling, velocity corrections are usually applied to this component only. These corrections reflect the energy defect of the adiabatic surfaces on the seam at the site of hop.

Tully and Preston have obtained classical branching trajectories for the linear $H^+ + D_2$ system using the surface hopping model as described and compared them with trajectories obtained from an effective potential given by

$$W_{eff}(\vec{R}) = |a_1|^2 W_1(\vec{R}) + |a_2|^2 W_2(\vec{R}) \ . \qquad \text{(IId.2)}$$

If the entrance channel is on W_1, they observed that $|a_2|^2$, calculated by (II.9), was essentially zero except in the immediate neighborhood of the avoided crossing seam. In other regions, the trajectories determined from W_{eff} were nearly superimposable on those determined from $W_1(\vec{R})$. Near the seam regions, however, the branched trajectories were significantly different from those determined from W_{eff}. Hence they concluded that, below a certain energy threshold, the probability of nonadiabatic transitions and the magnitude of nonadiabatic corrections to the potential [given by (IId.2)] were both negligible except in regions near the avoided crossing seam.

There is one important difference between the SHT model and the semiclassical S-matrix theory introduced earlier (and which will be generalized to include nonadiabatic transitions in Section III). The SHT theory adds probabilities (rather than probability amplitudes) for different trajectories and hence is incapable of producing interference effects leading to resonances. This defect, however, may be unimportant when a large number of initial vibrational and rotational phases have to be averaged over, since the averaging process would serve to wash out the interference effects.

IIe. The Meyer-Miller Treatment of Classical Electronic Degrees of Freedom

Even though semiclassical methods have been largely successful in the treatment of molecular collision processes, their formulations are plagued by a fundamental shortcoming - electronic and nuclear degrees of freedom are not treated on the same dynamical footing. Since it is never seriously doubted that classical mechanics can be applied safely to nuclear motion, some attempts have been made to redress the situation by advancing schemes to treat electronic motion by classical degrees of freedom also. These attempts by no means aim at reducing electronic motion to the tempting framework of classical particles orbiting around the nucleus. One has rather to search for the right classical dynamical variables which will describe the relevant aspects of electronic motion in molecular collision dynamics. For this purpose we will present a recent formulation due to Meyer and Miller.

To motivate the choice of a suitable set of classical electronic degrees of freedom we will first-formulate the general N electronic-state problem in the diabatic representation. The semiclassical description introduced at the beginning of Section II [assigning a specific classical trajectory R=R(t)] leads to the set coupled equations

$$\dot{\alpha}_i = -\frac{i}{\hbar}\sum_j \alpha_j(t)U^d_{ij}(R(t)), \quad i,j=1,\dots,N \qquad \text{(IIe.1)}$$

for the expansion coefficients of the electronic wave function

$$\Phi(r,R(t)) = \sum_i \alpha_i(t)\phi_i(r,R(t)) \tag{IIe.2}$$

which satisfies the time-dependent electronic Schrödinger equation

$$H_{el}(r,R)\Phi = i\hbar\partial\Phi/\partial t. \tag{IIe.3}$$

[Compare (IIb.5). Knowledge of $\alpha_i(t)$ then gives directly transition amplitudes for electronic transitions between the states ϕ_i. (In this subsection we limit ourselves to one-dimensional nuclear motion.)

One of the ways to assign the trajectory $R(t)$ in (IIe.1) is to assume that nuclei motion follows a force field determined by Ehrenfest's theorem:

$$\dot{p} \equiv \mu\ddot{R}(t) = -\langle\Phi|\frac{\partial H_{el}(r,R)}{\partial R}|\Phi\rangle$$

$$= -\sum_{ij}\alpha_i^*(t)\alpha_j(t)\frac{\partial}{\partial R}U_{ij}^d(R) \ . \tag{IIe.4}$$

Since the α's are unknowns in (IIe.4), this equation has to be adjoined to (IIe.1) to form a set of N+1 equations for the solution of the N+1 unknowns $\alpha_i(t)$ and $R(t)$. Other choices of trajectories can be made (for example, as in the SHT model, in which $R(t)$ is determined by the adiabatic potentials or W_{eff}), but in any case, a single unique classical trajectory is used which is independent of the final electronic state, a clearly dynamically inconsistent situation. (In fact, this is one of the formal weaknesses of the SHT approach).

The Meyer-Miller approach consists in searching for a set of classical electronic degrees of freedom which would lead to Hamilton's equations of motion completely equivalent to (IIe.2) plus the Ehrenfest trajectory. Yet the boundary conditions of these Hamilton's equations, unlike those of the time-dependent semiclassical treatment, would be dynamically consistent, since electronic transitions now correspond to different initial and final values for certain of the classical electronic degrees of freedom, and different transitions entail different classical trajectories. These transitions can be conveniently and unambiguously described by the semiclassical S-matrix formulation of Section Id.

The sought-for classical electronic degrees of freedom are related to the $\alpha_i(t)$ by

$$\alpha_i(t) = \sqrt{n_i(t)}\,\exp[-iq_i(t)] \ . \tag{IIe.5}$$

The n_i and q_i are analogous to those introduced in Section Id for the description of internal nuclear motion and can be correspondingly considered as electronic action and angle variables. The classical Hamiltonian $H_{el}(\vec{n},\vec{q};t)$ is then given by

$$H_{el}(\vec{n},\vec{q};t) = \langle\Phi|H_{el}(\vec{r},R(t))|\Phi\rangle$$

$$= \sum_{ij} \alpha_i^* \alpha_j U_{ij}^d(R(t)) \ , \qquad \text{(II.6)}$$

that is,

$$H_{el}(\vec{n},\vec{q};t) = \sum_{ij} \sqrt{n_i n_j}\ \exp\{i(q_i-q_j)\}U_{ij}^d(R(t)) \ . \qquad \text{(IIe.7)}$$

Hamilton's equations for n_i and q_i are then given by

$$\dot{q}_i = \frac{\partial H_{el}(\vec{n},\vec{q};t)}{\partial n_i} \ , \qquad \text{(IIe.8a)}$$

$$\dot{n}_i = -\frac{\partial H_{el}(\vec{n},\vec{q};t)}{\partial q_i} \ .$$

The full classical Hamiltonian becomes

$$H(p,R,\vec{n},\vec{q}) = \frac{p^2}{2\mu} + \sum_{ij} \sqrt{n_i n_j}\ \exp\{i(q_i-q_j)\}U_{ij}^d(R) \ , \qquad \text{(IIe.9)}$$

and Hamilton's equations for the canonical variables $R,p,\vec{n},\vec{q}$ are

$$\dot{R}(t) = \frac{\partial H}{\partial p} = \frac{p}{\mu} \ , \qquad \text{(IIe.10a)}$$

$$\dot{p}(t) = -\frac{\partial H}{\partial R} = -\sum_{ij} \sqrt{n_i n_j}\ \exp\{i(q_i-q_j)\}\frac{\partial}{\partial R} U_{ij}^d(R) \ , \qquad \text{(IIe.10b)}$$

$$\dot{q}_i(t) = \frac{\partial H}{\partial n_i} = \sum_j \sqrt{\frac{n_j}{n_i}}\ \mathrm{Re}[\exp\{i(q_i-q_j)\}U_{ij}^d] \ , \qquad \text{(IIe.10c)}$$

$$\dot{n}_i(t) = \frac{-\partial H}{\partial q_i} = \sum_j 2\sqrt{n_i n_j}\ \mathrm{Im}[\exp\{i(q_i-q_j)\}U_{ij}^d] \ . \qquad \text{(IIe.10d)}$$

(IIe.10c) and (IIe.10d) can be shown to be equivalent to (IIe.1) if (IIe.5) is used in the latter. Furthermore (IIe.10b) is none other than (IIe.4), the Ehrenfest trajectory equation. Hence the goal of the Meyer-Miller program is achieved by the introduction of the classical electronic degrees of freedom specified by (IIe.5).

As an example we consider the two-electronic-state problem. The classical electronic Hamiltonian (IIe.7) is given by

$$H_{el}(n_1,n_2,q_1,q_2) = n_1U_{11}^d + n_2U_{22}^d + 2\sqrt{n_1n_2}\, U_{12}^d\cos(q_2-q_1) \; . \tag{IIe.11}$$

Making a canonical transformation to the new set of variables

$$\begin{aligned} n_1' &= n_1 + n_2 = 1 \text{ (conservation of electronic probability)} \\ n_2' &= n_2 \; , \\ q_1' &= q_1 \; , \\ q_2' &= q_2 - q_1 \; , \end{aligned} \tag{IIe.12}$$

the two-state Hamiltonian reduces to one with only one degree of freedom $(n_2',q_2') \equiv (n,q)$,

$$H_{el}(n,q) = (1-n)U_{11}^d + nU_{22}^d + 2\sqrt{n(1-n)}\, U_{12}^d\cos q \; . \tag{IIe.13}$$

However, to make the boundary conditions amenable to classical treatment so that $n(t_2) \equiv n_2$ is a function of $q(t_1) \equiv q_1$, we have to introduce the Langer-modified Hamiltonian

$$H_{el}^L(n,q) = (1-n)U_{11}^d + nU_{22}^d + 2\sqrt{n+\frac{1}{2}}\,\sqrt{\frac{3}{2}-n}\; U_{12}^d\cos q \; . \tag{IIe.14}$$

Electronic transitions are then represented by the boundary conditions $n(t_1) \equiv n_1 = 0$, $n_2 = 1$. To compute the S-matrix given by (Id.33), we have to solve the following equation for q_1

$$n_2(q_1) = 1 \tag{IIe.15}$$

and each root will give rise to a separate term in (Id.33).

III. THE MILLER-GEORGE THEORY OF ELECTRONIC TRANSITIONS IN LOW ENERGY MOLECULAR COLLISIONS

In Section I we have introduced the semiclassical S-matrix for electronically adiabatic collisions in which electronic coordinates do not enter into the picture at all. Transitions between internal states of nuclear motion are described within the framework of classical degrees of freedom (action and angle variables) whose boundary values are quantized. In Section

II various ways to deal with nonadiabatic transitions are considered, some based on specific dynamical models, such as the Landau-Zener model, while others, such as the SHT treatment, are built on somewhat dynamically ill-defined boundary conditions, despite successes in many applications. The Meyers-Miller formalism attempts to put everything on firm grounds with the bold step of introducing classical electronic coordinates to complement the classical nuclear ones. This approach, though intellectually challenging and showing great promise, is as yet relatively undeveloped in its applicational aspects. In this section we retreat one step from the Meyers-Miller formalism, as it were, and discuss a treatment of nonadiabatic transitions due to Miller and George, which makes use of quantum mechanical electronic and classical nuclear degrees of freedom. Though suffering from the undesirable feature of having to mix classical and quantum mechanics, this formalism leads to a quite aesthetically appealing interpretation: all dynamics, including nonadiabatic transitions, is essentially described by classical motion of the nuclei on electronically adiabatic potential energy surfaces. Nonadiabatic motion is described by the analytic continuation of classical mechanics into complex coordinate regions. Hence all dynamical information is contained in the analytic structure of the adiabatic potential surfaces, and nonadiabatic coupling matrix elements, while centrally important in quantum mechanical treatments, do not enter explicitly into this formulation. Transitions between internal nuclear states are handled in the same way as described in Section Id.

IIIa. The Feynman Propagator Approach

In our formulation of the S-matrix in Section Id, (Id.13) was the starting point. There we noted that this equation results from the stationary phase approximation of the Green's function propagator $\langle q_2|\exp[-iH(t_2-t_1)/\hbar]/q_1\rangle$, which is equivalent to the choice of a particular trajectory, q(t), for the nuclear coordinate q. This particular trajectory is the one obtained from the solution of Hamilton's equations, and is thus called a classical trajectory. In the treatment of a problem involving electronic coordinates, we will be interested in the more generalized propagator in coordinate representation

$$\langle q_2,x_2|\exp[-iH(t_2-t_1)/\hbar]|q_1,x_1\rangle \qquad \text{(IIIa.1)}$$

where x_2,x_1 are the collective electronic coordinates at t_2 and t_1 respectively. q denotes the collective nuclear coordinates. It was Pechukas who first developed the idea of building a formalism to treat molecular collisions involving electronic transitions based on the propagator (IIIa.1). Instead of starting midway to write down the stationary phase approximation for it, we begin with Feynman's path integral expression:

$$\langle q_2 x_2 | \exp\left(\frac{-iH(t_2-t_1)}{\hbar}\right) | q_1 x_1 \rangle$$

$$= \int_{q_1}^{q_2} Dq \int_{x_1}^{x_2} Dx \, \exp\{\frac{i}{\hbar} \int_{t_1}^{t_2} dt [\frac{1}{2} \mu \dot{q}^2 + T_x - V(x,q)]\} \, , \quad \text{(IIIa.2)}$$

where the path integrals are over all electronic and nuclear trajectories $x(t)$ and $q(t)$ that connect $(x_1 q_1)$ at t_1 and (x_2, q_2) at t_2, and T_x is the electronic kinetic energy. One can imagine doing the electronic path integral first (although in practice its actual evaluation is never carried out) and write

$$\langle q_2 x_2 | \exp\left(\frac{-iH(t_2-t_1)}{\hbar}\right) | q_1 x_1 \rangle = \int_{q_1}^{q_2} Dq \; K[x_2, x_1; q(t)] \exp[\frac{i}{\hbar} \int_{t_1}^{t_2} dt \, \frac{1}{2} \mu \times (\dot{q}(t))^2] \quad \text{(IIIa.3)}$$

where K is the electronic propagator

$$K[x_2, x_1; q(t)] = \int_{x_1}^{x_2} Dx \; \exp[\exp[\frac{i}{\hbar} \int_{t_1}^{t_2} dt \{T_x - V(x, q(t))\}] \, , \quad \text{(IIIa.4)}$$

which is a functional of the nuclear path, i.e., it depends on the particular choice of the nuclear trajectory $q(t)$.

The times t_1 and t_2 are usually taken to be the infinite past and infinite future respectively, when q_1, $q_2 \to \infty$, and the system is in asymptotic electronic state $\phi_1(x_1)$ or $\phi_2(x_2)$, which are eigenfunctions of the Hamiltonian $\lim_{q\to\infty} H_{el}(q,x)$. Instead of the propagator (IIIa.1), Pechukas noted that one actually requires the propagator in 'state' rather than coordinate representation:

$$\langle 2, q_2 | \exp[-iH(t_2-t_1)/\hbar] | 1, q_1 \rangle$$

$$= \int dx_2 \int dx_1 \langle 2 | x_2 \rangle \langle x_2 q_2 | \exp[-iH(t_2-t_1)/\hbar] | q_1 x_1 \rangle \langle x_1 | 1 \rangle$$

$$= \int dx_2 \int dx_1 \phi_2^*(x_2) \langle q_2 x_2 | \exp[-iH(t_2-t_1)/\hbar] | q_1 x_1 \rangle \phi_1(x_1) \quad \text{(IIIa.5)}$$

Using (IIIa.3), the 'state' representation propagator takes the form

$$<2,q_2|\exp\left(\frac{-iH(t_2-t_1)}{\hbar}\right)|1,q_1> = \int_{q_1}^{q_2} Dq\ K_{21}[q(t)]\exp[\frac{i}{\hbar}\int_{t_1}^{t_2} dt\ \frac{1}{2}\mu \times (\dot{q}(t))^2] \qquad \text{(IIIa.6)}$$

where

$$K_{21}[q(t)] = \int dx_1 \int dx_2 \phi_2^*(x_2) \int_{x_1}^{x_2} Dx\ \exp[\frac{i}{\hbar}\int_{t_1}^{t_2} dt\{T_x - V(q(t),x)\}]\phi_1(x_1) \qquad \text{(IIIa.7)}$$

is the electronic transition amplitude. It gives the transition amplitude ($1\to 2$ electronic transition) with the nuclei constrained to follow the trajectory $q(t)$. S-matrix elements for the $1n_1 \to 2n_2$ transition, where n_1 and n_2 are the initial and final values of the quantum numbers for the nuclear degrees of freedom, are then constructed in a way analogous to the development following (Id.11).

Retracing our steps slightly, we see that the calculation of $K_{21}[q(t)]$, which is the determination of the electronic transition amplitude under a fixed nuclear trajectory, is none other than what we set out to do via the semiclassical time-dependent formalism in Section II. K_{21} is none other than $a_2\exp\{\frac{-i}{\hbar}\int^t dt\ U_{22}^a(q(t))\}$ [see (II.6)]; and $q(t)$ was chosen to be the straightline trajectory for both the Landau-Zener and the Demkov models. In the next subsection we will present a semiclassical form for $K[q(t)]$ which is valid for all classical paths. This result is then used in (IIIa.6), in which the path integral over q is done using the stationary phase approximation, which effectively selects out the classical paths, out of infinitely many possible ones. Then everything follows exactly as the development subsequent to (Id.11).

IIIb. The Electronic Propagator and Analytic Continuation of Classical Mechanics

As already remarked the determination of the electronic propagators K_{21} (or K_{11}) is equivalent to the solution of (II.9). The K's are given in terms of the a's by:

$$K_{21}[q(t)] = a_2(t_2)\exp[\frac{-i}{\hbar}\int_{t_1}^{t_2} dt\ W_2(t)] \qquad \text{(IIIb.1a)}$$

$$K_{11}[q(t)] = a_1(t_2)\exp[\frac{-i}{\hbar}\int_{t_1}^{t_2} dt\ W_1(t)] \qquad \text{(IIIb.1b)}$$

where the W's are the adiabatic potential energy surfaces.

We will briefly indicate the semiclassical solution for the two-state case as provided by Miller and George via the first order perturbation solution for a_2 (with the initial conditions $a_1(t_1)=1$, $a_2(t_1)=0$). (II.9) gives

$$a_2(t_2) = -\frac{i}{\hbar}\int_{t_1}^{t_2} dt\ C_{21}\exp\{\frac{i}{\hbar}\int_{t_1}^{t} dt'(W_2-W_1)\} . \qquad \text{(IIIb.2)}$$

Hence, from (IIIb.1a)

$$K_{21}[q(t)] = -\frac{i}{\hbar}\int_{t_1}^{t_2} dt\ C_{21}\exp\{\frac{i}{\hbar}\int_{t_1}^{t} dt'(W_2-W_1)\}\exp\{-\frac{i}{\hbar}\int_{t_1}^{t_2} dt'W_2\}$$

$$= -\frac{i}{\hbar}\int_{t_1}^{t_2} dt\ C_{21}\exp\{-\frac{i}{\hbar}\int_{t_1}^{t} dt'W_1(t') - \frac{i}{\hbar}\int_{t}^{t_2} dt'W_2(t')\} , \qquad \text{(IIIb.3)}$$

which can conveniently be interpreted as an integral over t of transition amplitudes for transitions from W_1 to W_2 at particular times t. One now proceeds to the classical limit by attempting to do the integral using the stationary phase approximation, i.e., one requires that

$$\frac{d}{dt}\left(\int_{t_1}^{t} dt'W_1(t') + \int_{t}^{t_2} dt'W_2(t')\right) = 0$$

or

$$W_1(t) - W_2(t) = 0 . \qquad \text{(IIIb.4)}$$

The solution of (IIIb.4) is in general complex, which means in (IIIb.3) one needs to evaluate the nonadiabatic coupling C_{12} for a complex time. In order to avoid this, one applies the stationary phase approximation to the next order of accuracy by looking for a time at which the phase of the integrand in (IIIb.3) is least rapidly varying, i.e.,

$$\frac{d^2}{dt^2}\left(\int_{t_1}^{t} dt'W_1(t') + \int_{t}^{t_2} dt'W_2(t')\right) = 0$$

or

$$\frac{d}{dt}(W_1(t) - W_2(t)) = -\frac{d}{dt}(\Delta W) = 0 \qquad \text{(IIIb.5)}$$

where $\Delta W = W_2 - W_1$.

Expanding the phase in a Taylor series about t_0, a solution to (IIIb.5), one has

$$\int_{t_1}^{t} dt' W_1(t') + \int_{t}^{t_2} dt' W_2(t') = \tau - (t-t_0)\Delta W_0 + 0 - \frac{1}{6}(t-t_0)^3 \times (\Delta W)''_0 + \ldots\ldots \qquad \text{(IIIb.6)}$$

where

$$\tau = \int_{t_1}^{t_0} dt\, W_1(t) + \int_{t_0}^{t_2} dt\, W_2(t) ,$$

$$\Delta W_0 = \Delta W(t=t_0) ,$$

$$(\Delta W)''_0 = (d^2(\Delta W)/dt^2)_{t=t_0} .$$

(IIIb.3) becomes

$$K_{21}[q(t)] \simeq -\frac{i}{\hbar}(C_{21})_{t_0} \exp\left(\frac{-i\tau}{\hbar}\right)\int_{-\infty}^{\infty} dt \exp\left(\frac{i}{\hbar}\Delta W_0(t-t_0) + \frac{i}{6\hbar}(\Delta W)''_0(t-t_0)^3\right) \qquad \text{(IIIb.7)}$$

where the integration limits have been extended to $\pm\infty$. The integral in (IIIb.5) is recognized as the integral representation of the Airy function, and invoking its asymptotic form gives

$$K_{21}[q(t)] \simeq -\langle\phi_1|\frac{\partial}{\partial t}\phi_2\rangle_{t_0}\left[\frac{2\pi\hbar}{(2\Delta W_0/(\Delta W)''_0)^{1/2}}\right]^{1/2} \times \exp\left[-\frac{2}{3}\frac{\Delta W_0}{\hbar}\left(\frac{2\Delta W_0}{(\Delta W)''_0}\right)^{1/2} - \frac{i}{\hbar}\int_{t_1}^{t_0} dt\, W_1(t) - \frac{i}{\hbar}\int_{t_0}^{t_2} dt\, W_2(t)\right] . \qquad \text{(IIIb.8)}$$

This approximate solution is recognized to resemble the exact Stuckelberg solution (WKB phase integral solution) for (II.9) most closely if the pre-exponential factor is set equal to 1. We thus write

$$K_{21}[q(t)] = e^{-\delta}\exp -\left(\frac{i}{\hbar}\int_{t_1}^{t_0} dt\, W_1(t) - \frac{i}{\hbar}\int_{t_0}^{t_2} dt\, W_2(t)\right) \qquad \text{(IIIb.9)}$$

where
$$\delta \simeq \frac{2}{3}\frac{\Delta W_0}{\hbar}\left(\frac{2\Delta W_0}{(\Delta W)''_0}\right)^{1/2} , \qquad \text{(IIIb.10)}$$

but whose general form is as yet undetermined. This general form can be obtained as follows. We already remarked that (IIIb.4) may not have real solutions, but it will always have complex solutions. For the avoided curve crossing case where t_0 is the solution to (IIIb.5), we may expand ΔW about t_0 to obtain an approximate solution for (IIIb.4):

$$\Delta W(t) \simeq \Delta W_0 + 0 + \frac{1}{2}(t-t_0)^2(\Delta W)''_0 . \qquad \text{(IIIb.11)}$$

Hence t_*, a solution to $\Delta W=0$, is given to the above approximation by

$$t_* = t_0 + i\left(\frac{2\Delta W_0}{(\Delta W)''_0}\right)^{1/2} ; \qquad \text{(IIIb.12)}$$

and it is immediately seen that t_*^* is also a solution. One then obtains

$$\int_{t_*^*}^{t_*} dt\, \Delta W(t) = \int_{t_*^*}^{t_*} dt[\Delta W_0 + \frac{1}{2}(t-t_0)^2(\Delta W)''_0]$$

$$= i\,\frac{4}{3}\,\Delta W_0\left(\frac{2\Delta W_0}{(\Delta W)''_0}\right)^{1/2}$$

$$= 2\int_{t_0}^{t_*} dt\, \Delta W(t) . \qquad \text{(IIIb.13)}$$

Comparing with (IIIb.10) we see that

$$\delta = -\frac{i}{\hbar}\int_{t_0}^{t_*} dt\, \Delta W(t) , \qquad \text{(IIIb.14)}$$

which is the sought for general form, independent of the approximation introduced for $\Delta W(t)$. (IIIb.9) can then be written as

$$K_{21}[q(t)] = \exp\left[-\frac{i}{\hbar}\int_{t_1}^{t_0} dt\, W_1(t) - \frac{i}{\hbar}\int_{t_0}^{t_*} dt\, W_1(t) - \frac{i}{\hbar}\int_{t_*}^{t_0} dt\, W_2(t) - \frac{i}{\hbar}\int_{t_0}^{t_2} dt\, W_2(t)\right]$$

$$= \exp\left[-\frac{i}{\hbar}\int_{t_1}^{t_*} dt\ W_1[q(t)] - \frac{i}{\hbar}\int_{t_*}^{t_2} dt\ W_2[q(t)]\right] . \qquad \text{(IIIb.15)}$$

This last equation gives the general semiclassical solution for the electronic propagator.

With (IIIb.15), the propagator for nonadiabatic transition (IIIa.6) becomes

$$<2,q_2|\exp\left(\frac{-iH(t_2-t_1)}{\hbar}\right)|1,q_1>$$

$$= \int_{q_1}^{q_2} Dq\ \exp\left[\frac{i}{\hbar}\int_{t_1}^{t_2} dt\ \frac{1}{2}\ \mu(\dot{q}(t))^2 - \frac{i}{\hbar}\int_{t_1}^{t_*} dt\ W_1(q(t)) - \frac{i}{\hbar}\int_{t_*}^{t_2} dt\ W_2(q(t))\right].$$

(IIIb.16)

It is instructive to compare this equation with the expression for the electronically adiabatic propagator

$$<q_2|\exp\left(\frac{-iH(t_2-t_1)}{\hbar}\right)|q_1> = \int_{q_1}^{q_2} Dq\ \exp\left[\frac{i}{\hbar}\int_{t_1}^{t_2} dt\ \frac{1}{2}\ \mu(\dot{q}(t))^2 - \frac{i}{\hbar}\int_{t_1}^{t_2} dt\ V(q(t))\right] , \qquad \text{(IIIb.17)}$$

where V(q) is the single potential surface for the adiabatic motion. (IIIb.16) has the following physical interpretation. Instead of moving on the single surface V(q), the nuclei make the nonadiabatic transition at the complex time t_* from one adiabatic surface to another ($W_1 \to W_2$). Since t_* is complex, nonadiabatic transitions are in general classically forbidden processes, and the interesting conclusion emerges that classically forbidden processes can be described by the analytic continuation of classical mechanics into the complex time (or coordinate) domain.

As mentioned earlier, the nuclear path integral in (IIIb.16) (which, like its electronic counterpart, is never evaluated in practice) is carried out again by the stationary phase approximation to give:

$$<2q_2|\exp\left(\frac{-iH(t_2-t_1)}{\hbar}\right)|1q_1> = \left[(-2\pi i\hbar)^N\left(\frac{\partial q_2}{\partial p_1}\right)_{q_1}\right]^{-1/2} \exp\left[\frac{i\phi_{2,1}(q_2,q_1)}{\hbar}\right] ,$$

(IIIb.18)

where N is the number of nuclear degrees of freedom, p_1, is the momentum conjugate to q_1, and $\phi_{2,1}$ is the action integral

$$\phi_{2,1}(q_2,q_1) = \int_{t_1}^{t_2} dt\, \frac{1}{2}\, \mu\dot{q}(t)^2 - \int_{t_1}^{t_*} dt\, W_1 - \int_{t_*}^{t_2} dt\, W_2 \; . \qquad \text{(IIIb.19)}$$

This integral is evaluated along the classical trajectory (obtained by integrating Hamilton's equations) which is determined by double-ended boundary conditions q_1 and q_2 and which change from potential W_1 to W_2 at complex time t_*.

Analogous to (Id.33), the expression for the nonadiatic transition S-matrix element can be written

$$S_{2n_2,1n_1} = i\left[(-2\pi i\hbar)^{N-1}\left(\frac{\partial n_2}{\partial q_1'}\right)_{n_1}\right]^{-1/2}$$

$$\times \exp\left[-\frac{i}{\hbar}\int_{-\infty}^{\infty} dt\{\rho(t)\dot{P}(t) + q(t)\dot{n}(t)\}\right] \qquad \text{(IIIb.20)}$$

where ρ,P are the nuclear translational and q,n the nuclear internal degrees of freedom. (Note that in previous equations of this section q stands for the totality of nuclear degrees of freedom). In the evaluation of the integral in (IIIb.20), the complex trajectories will yield exponential damping factors characteristic of classically forbidden processes.

Finally we point out that in the case of one-dimensional nuclear motion (atom-atom collisions), (IIIb.20) reduces to the Stuckelberg solution given by (IIc.9) to (IIc.13), with $\xi = \pi/4$.

IIIc. Examples of Nonadiabatic Transition Probabilities in the Miller-George Theory

We will consider the Landau-Zener and the Demkov models in this subsection. The Landau-Zener model is given in the diabatic representation by (IIa.1) and (IIa.2). In the adiabatic representation the potential surfaces W_i are given by (IIa.4), (here W_i stands for U^a_{ii}). To calculate the local transition probability $e^{-2\delta}$ [see (IIIb.7)], we need δ as given by (IIIb.12). This can be changed into an integral over R, the translational nuclear coordinate, if we assume the straight-line trajectory (IIa.3). We have, for a 'once passage' through the nonadiabatic coupling region:

$$P_{2,1} = |S_{2\to 1}|^2 = \exp(-2\delta) \qquad \text{(IIIc.1)}$$

where
$$\delta = -\frac{i}{2\hbar}\int_{t_*^*}^{t_*} dt\ \Delta W(t)$$

$$= -\frac{i}{2\hbar v_0}\int_{R_*^*}^{R_*} dR\ \Delta W(R)\ , \qquad \text{(IIIc.2)}$$

and $R_* = R(t=t_*)$ is the complex coordinate at which $\Delta W=0$. From (IIa.4),

$$\Delta W = [(\Delta F)^2 (R-R_0)^2 + 4a^2]^{1/2}\ . \qquad \text{(IIIc.3)}$$

Hence
$$R_* = R_0 + 2i\ a/(\Delta F)\ . \qquad \text{(IIIc.4)}$$

To do the integral in (IIIc.2) it is convenient to change the variable of integration to x, where

$$R = R_0 + 2i\ x\ a/(\Delta F)\ .$$

(IIIc.2) then becomes

$$\delta = \frac{2a^2}{\hbar v_0 \Delta F}\int_{-1}^{1} dx\sqrt{1-x^2}\ . \qquad \text{(IIIc.5)}$$

Elementary integration yields

$$\delta = \frac{\pi a^2}{\hbar v_0 \Delta F}\ . \qquad \text{(IIIc.6)}$$

Finally
$$P_{2,1} = \exp(-2\delta) = \exp\left[\frac{-2\pi a^2}{\hbar v_0 \Delta F}\right] \qquad \text{(IIIc.7)}$$

which is the same result as obtained in (IIa.8).

We now consider the Demkov model given in the adiabatic representation by (IIb.3), which gives

$$\Delta W = [\Delta^2 + 4A^2 \exp(-2\lambda R)]^{1/2}\ . \qquad \text{(IIIc.8)}$$

The solution to $\Delta W=0$ gives a sequence a complex roots

$$R_* = -\frac{1}{\lambda}\ln\left(\frac{\Delta}{2A}\right) + \frac{i\pi}{2\lambda}(2n+1),\ n=0,1,2,\ldots. \qquad \text{(IIIc.9)}$$

To do the integral (IIIc.2), we use the principle root, i.e., the one which is closest to the real axis and leads to the largest $P_{2,1}$. We pick

$$R_* = -\frac{1}{\lambda}\ln\left(\frac{\Delta}{2A}\right) + \frac{i\pi}{2\lambda} . \qquad \text{(IIIc.10)}$$

Effecting the change of variables $R \to x$, where

$$R = -\frac{1}{\lambda}\ln\left(\frac{\Delta}{2A}\right) + \frac{i\pi x}{2\lambda} , \qquad \text{(IIIc.11)}$$

the R-integral in (IIIc.2) becomes

$$\delta = \frac{\pi\Delta}{2\lambda\hbar v_0}\int_{-1}^{1} dx\,(1+e^{-i\pi x})^{1/2} . \qquad \text{(IIIc.12)}$$

This gives

$$\delta = \frac{\pi\Delta}{2\lambda\hbar v_0}$$

and

$$P_{21} = \exp[-\pi\Delta/(\lambda\hbar v_0)] . \qquad \text{(IIIc.13)}$$

For large δ, this is essentially the same result as obtained in (IIb.7).

IIId. Analytic Structure of Adiabatic Potential Surfaces

The most appealing nature of the Miller-George theory lies in its description of nonadiabatic transitions: a nonadiabatic transition can be described by a classical trajectory moving continuously from one adiabatic surface to another, which become degenerate at some complex time (coordinate) point. In fact the two surfaces can be pictured as two pieces of the same surface, so that what appears to be a transition in real time (or coordinate) space is really no transition at all when the trajectory is allowed to be complex. The dynamical information embedded in the nonadiabatic couplings, which are not required in the present formalism at all, is completely contained in the analytic structure of the adiabatic potential surface, i.e., when W_i is regarded as a complex function of either R or t. This is indeed an intellectually pleasing and physically cogent way to look at nonadiabatic transitions.

The theory of analytic functions provides the natural framework for the description of the adiabatic potential surfaces. For the two state case

$$W_{1,2} = \frac{U_{11}^d + U_{22}^d}{2} \mp \frac{1}{2}[(U_{22}^d - U_{11}^d)^2 + 4(U_{12}^d)^2]^{1/2} . \qquad \text{(IIId.1)}$$

The surfaces $W_{1,2}$ are the roots of the quadratic equation

$$(U_{11}^d - W)(U_{22}^d - W) - (U_{12}^d)^2 = 0 \, , \tag{IIId.2}$$

and as such are the two branches of the same analytic function W(R), with the branch point at R_* where

$$\Delta W(R_*) = [(U_{22}^d - U_{11}^d)^2 + 4(U_{12}^d)^2]^{1/2} = 0 \tag{IIId.3}$$

W_1 and W_2 are also known as the two Riemann sheets of the same analytic function W(R), which is double-valued. W can, of course, also be considered as a function of t. Let the complex variable z be either R or t. Starting from an arbitrary point z_0 and following any contour which winds around z_* (the branch point) once and ends up back on z_0, or traverses a branch cut (emanating from z_0) once, one would go from one Riemann sheet of W to the next. Any contour, however, which does not encircle any branch point or which does not traverse any branch cut will force W(z) to stay on the sheet it begins with. It is at the point where the contour traverses the branch cut that W_1 goes over to W_2 or vice versa. Consider the first equality in (IIIb.15). The sum of integrals can be written as one complex integral

$$\int_{C(t_1 \to t_2)} dt \; W(t) \, , \tag{IIId.4}$$

where the contour, from t_1 to t_2, is represented in Figure 2 (solid line). Because of the analytic nature of W, the contour does not have to go exactly through t_*. In fact, any distorted contour from t_1 to t_2 traversing the cut at some point will suffice (such as the dashed contour in Figure 2).

In much the same way we can discuss nonadiabatic transitions in multistate problems. Analogous to (IIId.2) the N adiabatic surfaces of an N-state problem are obtained by the solution of the secular equation

$$\det|\tilde{U}^d - W\tilde{1}| = 0 \, . \tag{IIId.5}$$

The NxN determinant leads to an Nth-order polynomial in W. The N different roots of this polynomial are essentially the N different branches, or Reimann sheets, of the Nth root analytic function W. As in the 2-state case, trajectories that make transitions from one branch to another can be really considered to be propagated on a single surface, although they would be required to go around appropriate branch points.

For a transition from state α to β in an N-state problem, the nuclear path integral can in general be written [compare with (IIIb.18) and (IIIb.19)]:

$$\langle\beta q_2|\exp\left(\frac{-iH(t_2-t_1)}{\hbar}\right)|\alpha q_1\rangle = \left[(-2\pi i\hbar)^M\left(\frac{\partial q_2}{\partial p_1}\right)_{q_1}\right]^{-1/2}$$

$$\times \exp\{\frac{i}{\hbar}\int_{C_{\beta\alpha(t_1\to t_2)}} dt \quad [\frac{1}{2}\,\mu(\dot{q}(t))^2 - W(q(t))]\} \qquad \text{(IIId.6)}$$

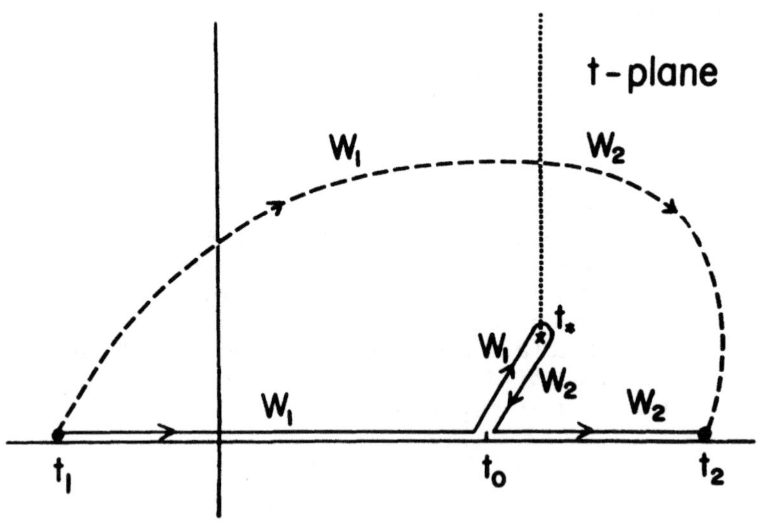

Fig. 2. Contour of integration (——) for the electronic propagator in (IIIb.15). In general $t_0 \sim \mathrm{Re}(t_*)$. The displacement is accentuated here for clarity. The dotted line represents the branch cut. The distorted contour (dashed) is also acceptable for the calculation of the electronic propagator. The W's represent the sheets used for various parts of the contours.

where M is the number of nuclear degrees of freedom. In (IIId.6) W is the N-valued analytic adiabatic surface. $C_{\beta\alpha}$ is a contour in the complex time plane connecting t_1 and t_2 and winds around the appropriate branch points of W in such a way that the trajectory beginning on Reimann sheet α (the α electronic state) at t_1 finishes up on Reimann sheet β (the β electronic state) at t_2. q(t) is the classical trajectory obtained by integrating Hamilton's equations along the complex contour $C_{\beta\alpha}(t_1 \rightarrow t_2)$.

Before leaving our discussion of the analytic structure of adiabatic potential surfaces, we would like to mention a practical problem connected with the calculation of δ in (IIIb.12). In most cases other than the Landau-Zener model of straight-line potentials (for instance, the Demkov model), there will be more than one branch point for W(t). For large δ, that is, small transition probability, it will be usually sufficient to take the branch point closest to the real axis and compute δ as though there were only one branch point (as we did for the Demkov model). For small δ, however, this procedure may not always be sufficient. For this case, the formula for the local transition probability

$$p \simeq \exp(-2\delta_0) ,$$

where δ_0 is computed using the branch point closest to the real axis, may require some correction.

The correction procedure usually employed is to replace $\exp(-2\delta_0)$ by an expression which is the exact solution of a two-state exponential model due to Nikitin. The model is defined by

$$U^d_{22} = B\,\exp(-\beta R) + \frac{\Delta\varepsilon}{2} - \frac{A}{2}\cos\theta\exp(-\beta R) \tag{IIId.7a}$$

$$U^d_{11} = B\,\exp(-\beta R) - \frac{\Delta\varepsilon}{2} + \frac{A}{2}\cos\theta\exp(-\beta R) \tag{IIId.7b}$$

$$U^d_{12} = U^d_{21} = -\frac{A}{2}\sin\theta\exp(-\beta R) \tag{IIId.7c}$$

where $\Delta\varepsilon, \beta, A, B$ and θ are all constants. $\underset{\sim}{U}^d$ can be diagonalized to obtain the adiabatic potentials

$$W_{1,2} = B\,\exp(-\beta R) \mp \frac{\Delta\varepsilon}{2}[1-2\cos\theta\exp(-\beta(R-R_p)) + \exp(-2\beta(R-R_p))]^{1/2} \tag{IIId.8}$$

where R_p is the 'center' of the nonadiabaticity region defined by

$$A\exp(-\beta R_p) = \Delta\varepsilon . \tag{IIId.9}$$

In this model $\Delta\varepsilon$ is the asymptotic splitting of W_1 and W_2, and $1/\beta$ is the characteristic length of the nonadiabatic coupling. If the matrix that diagonalizes $\underset{\sim}{U}^d$ is written as in (Ib.25),

(Ib.26) implies that

$$2\alpha = \tan^{-1}\left(\frac{-A\ \sin\theta\exp(-\beta R)}{\Delta\varepsilon - A\ \cos\theta\exp(-\beta R)}\right) . \qquad \text{(IIId.10)}$$

Hence $$\alpha \to 0 \text{ at } R \to \infty ; \qquad \text{(IIId.11)}$$

and $$2\alpha \to \theta \text{ at } R \ll R_p . \qquad \text{(IIId.12)}$$

Assuming a straight line-trajectory $\Delta R = v_0 t$, the exact solution of (II.9) or (IIb.5) gives the Nikitin result for p, which incorporates the effect of the multiplicity of branch points:

$$p = \exp[-\frac{\pi}{4}\ \delta_0(1-\cos\theta)]\left[\frac{\sinh[(\pi/4)\delta_0(1+\cos\theta)]}{\sinh(\pi\delta_0/2)}\right] . \qquad \text{(IIId.13)}$$

This is the correction formula (sometimes known as the Nikitin formula) used in most applications with exponential-like diabatic curves. δ_0 is calculated by using (IIIb.12), with t_* being the branch point of W(t) closest to the real axis. θ is obtained from (IIId.12).

IV. APPLICATIONS OF THE SEMICLASSICAL S-MATRIX

We will discuss two classes of problems, electronically adiabatic and electronically nonadiabatic problems. The former makes use of the formalism presented in Section Id, and provides the perspective and motivation for its generalization to the latter, which relies on Section III.

IVa. Electronically Adiabatic Collisions

The example we will discuss is vibrational excitation in collinear A+BC, both classically allowed and classically forbidden. Since electronic degrees of freedom are not explicitly considered, all dynamics is treated classically. The system has only two degrees of freedom, one translational R, and one vibrational (internal) r. The completely classical Hamiltonian can be written as

$$H(P,R,p,r) = \frac{P^2}{2\mu} + \frac{p^2}{2m} + v(r) + V(R,r) , \qquad \text{(IVa.1)}$$

where μ and m are the reduced masses for the two degrees of freedom, v(r) is the vibrational potential for BC, and V(R,r) is the interaction potential coupling translation and vibration.

Following (Id.12) we have to assign (n,q), the action-angle variables suitable for this problem. The action variable n is the classical counterpart of the vibrational quantum number for the free BC. The WKB quantum conditions require that

$$(n + \frac{1}{2})\pi = (\frac{2m}{\hbar^2})^{1/2} \int_{r_<}^{r_>} dr[\varepsilon - v(r)]^{1/2} \qquad (IVa.2)$$

where
$$\varepsilon = \frac{p^2}{2m} + v(r) \qquad (IVa.3)$$

and $r_<$, $r_>$ are the classical turning points. The angle variable q is given by

$$q = \frac{(m/2\hbar^2)^{1/2}}{n'(\varepsilon)} \int_{r_<}^{r} dr'[\varepsilon - v(r')]^{-1/2}, \qquad (IVa.4)$$

where $n(\varepsilon)$ is the function defined by (IVa.2). The Hamiltonian (Id.12) becomes

$$H(P,R,n,q) = \frac{P^2}{2\mu} + \varepsilon(n) + V(r(n,q),R) \qquad (IVa.5)$$

$$\equiv H_0 + V$$

where $\varepsilon(n)$ is the inverse function of $n(\varepsilon)$. For a simple harmonic oscillator

$$v(r) = \frac{1}{2} m\omega^2 (r - r_0)^2 . \qquad (IVa.6)$$

Using (IVa.2) and (IVa.4), one obtains

$$\varepsilon(n) = (n + \frac{1}{2})\hbar\omega \qquad (IVa.7)$$

and
$$r(n,q) = r_0 + \left(\frac{(2n+1)\hbar}{\omega m}\right)^{1/2} \sin(q-\pi/2). \qquad (IVa.8)$$

For the general case one can expand the potential:

$$v(r) = \frac{1}{2} m\omega^2 (r-r_0)^2 [1 + a_1(r-r_0)/r_0 + a_2(r-r_0)^2/r_0^2 + \ldots\ldots] ,$$
$$(IVa.9)$$

and $\varepsilon(n)$ and $r(n,q)$ can then be constructed as powers series in $\lambda = (2m\omega r_0^2)^{-1}$. For most diatomic $\lambda \ll 1$ and it is usually sufficient to retain terms up to first or second order in λ.

For a transition between the vibrational states n_1 and n_2, the S-matrix, following (Id.33), is given by

$$S_{\vec{n}_2,\vec{n}_1}(E) = i[(-2\pi i\hbar)\left(\frac{\partial n_2}{\partial \overline{q}_1}\right)_{n_1}]^{-1/2} \exp[i\frac{\Phi}{\hbar}(n_2,n_1)] \ , \tag{IVa.10}$$

$$\text{where } \Phi(n_2,n_1) = -\int_{-\infty}^{\infty} dt[R(t)\dot{P}(t) + q(t)\dot{n}(t)] \ , \tag{IVa.11}$$

and $\overline{q}_1$ is the one-dimensional analog of $\vec{\overline{q}}_1'$ in (Id.34). The classical trajectories q(t) and R(t) are determined by the integration of Hamilton's equations

$$\dot{q} = \partial H(P,R,n,q)/\partial n \ , \tag{IVa.12a}$$

$$\dot{R} = \partial H(P,R,n,q)\partial P \ , \tag{IVa.12b}$$

$$\dot{n} = -\partial H(P,R,n,q)/\partial q \ ,$$

$$\dot{p} = -\ \partial H(P,R,n,q)/\partial R \ .$$

We must now specify the boundary conditions for this problem. First E, the total energy, is an input parameter, as are n_1 and n_2, the initial and final vibrational quantum numbers, both required to be integers. The boundary conditions are then

$$n_1 = \text{specified integer} \tag{IVa.13a}$$

$$n_2 = \text{specified integer} \tag{IVa.13b}$$

$$P_1 = -\{2\mu[E-\varepsilon(n_1)]^{1/2} \tag{IVa.13c}$$

$$P_2 = +\{2\mu[E-\varepsilon(n_2)\}^{1/2} \tag{IVa.13d}$$

Note that once n_1 and n_2 are fixed, P_1 and P_2 will also be fixed by (IVa.13c) and (IVa.13d). One thus starts a trajectory with the initial conditions

$$n_1 = \text{specified integer} \tag{IVa.14a}$$

$$R_1 = \text{asymptotically large} \tag{IVa.14b}$$

$$P_1 = -\{2\mu[E-\varepsilon(n_1)]\}^{1/2} \tag{IVa.14c}$$

$$q_1 = \overline{q}_1 + \varepsilon'(n_1)\ \frac{\mu R_1}{P_1} \tag{IVa.14d}$$

[which is actually the same set of equations as (Id.34)]. In (IVa.14d) $\varepsilon'(n_1)\mu R_1/P_1$ is fixed but $\overline{q}_1$ is as yet unspecified. The idea is to choose representative values of $\overline{q}_1$ in the range $0 \leqslant \overline{q}_1 \leqslant 2\pi$, and integrate Hamilton's equations (IVa.12) forward in time until $R_2[R(t=t_2)]$ is again asymptotically large and $\dot{P}_2[(\dot{P}(t))_{t=t_2}] \to 0$; thus generating many trajectories with many final values for n_2 (which may not be integral). In this way a functional relationship $n_2(\overline{q}_1,n_1)$ is established. One then solves the equation

$$n_2(\overline{q}_1,n_1) = n_2 \tag{IVa.15}$$

for $\overline{q}_1$, where n_2 on the right is the specified final vibrational quantum number of the problem. Associated with each root of (IVa.15) for $\overline{q}_1$ is an action integral Φ and a pre-exponential derivative $(\partial n_2/\partial\overline{q}_1)_{n_1}$; and the S-matrix $S_{n_2n_1}$ is a sum of terms like (IVa.10), one for each root $\overline{q}_1$.

Since for any solution $\overline{q}_1$ to (IVa.15) $\overline{q}_1+2\pi$ is also a solution, there must be an even number of roots to (IVa.15). Moreover, since $n_2(\overline{q}_1)$ is periodic with period 2π, any two solutions $\overline{q}_I$ and $\overline{q}_{II}$ within an interval of 2π must be such that $(\partial n_2/\partial\overline{q}_1)_{\overline{q}_I}$ and $(\partial n_2/\partial\overline{q}_1)_{\overline{q}_{II}}$ have opposite signs. We consider the case where there are two solutions q_I and q_{II} in the interval $0\leq q_1\leq 2\pi$. Let the two action integrals (IVa.11) associated with these two solutions be Φ_I and Φ_{II} respectively. Then the S-matrix is given by

$$S_{n_2n_1} = [-2\pi i\hbar|n_2'(q_I)|]^{-1/2}\exp(\tfrac{i}{\hbar}\Phi_I) + [+2\pi i\hbar|n_2'(q_{II})|]^{-1/2} \times \exp(\tfrac{i}{\hbar}\Phi_{II}) , \tag{IVa.16}$$

where the i outside the pre-exponential factor in (IVa.10) has been suppressed because it does not affect the value of $|S_{n_2n_1}|^2$. Keeping proper track of the signs and the i's in (IVa.16), it can be rewritten as

$$S_{n_2n_1} = p_I^{1/2}\exp(\tfrac{i\pi}{4} + \tfrac{i}{\hbar}\Phi_I) + p_{II}^{1/2}\exp(-\tfrac{i\pi}{4} + \tfrac{i}{\hbar}\Phi_{II}) \tag{IVa.17}$$

where

$$p_I = [2\pi\hbar|n_2'(q_I)|]^{-1} , \tag{IVa.18}$$

$$p_{II} = [2\pi\hbar|n_2'(q_{II})|]^{-1} \tag{IVa.19}$$

are the pre-exponential probability factors. (IVa.17) clearly reveals the quantum superposition effect inherent in the semiclassical formulation. The transition probability is

$$P_{n_2n_1} = |S_{n_2n_1}|^2 = p_I + p_{II} + 2(p_Ip_{II})^{1/2}\sin(\Phi_{II}-\Phi_I) . \tag{IVa.20}$$

It is worthwhile to note here that the oscillatory result in (IVa.20) is also characteristic of the nonadiabatic Stuckelberg type transitions discussed in the next subsection (Stuckelberg oscillations), both being the result of quantum superposition. The completely classical result, which adds probabilities instead of probability amplitudes, is

$$P^{CL}_{n_2n_1} = p_I + p_{II} \; . \tag{IVa.21}$$

So far we have only described the situation when there are real solutions to (IVa.15), i.e., when the transition is classically allowed. However, transitions may be classically forbidden in the sense that no real solutions exist for (IVa.15) even though the state n_2 is energetically accessible $[E>\varepsilon(n_2)]$. In such cases one is forced to consider the complex roots of (IVa.15). To see this explicitly one may expand $n_2(q_1)$ about q_{max}, the real value of q_1 that would lead to the maximum real value of n_2, which we denote n_2^{max}:

$$n_2(\bar{q}_1) = n_2^{max} + \frac{1}{2}\, n_2''(\bar{q}_{max})\,(\bar{q}_1-\bar{q}_{max})^2 + \ldots \tag{IVa.23}$$

Suppose the designated n_2 is larger than n_2^{max}, so that there are no real solutions to (IVa.15). Complex solutions, however, exist, and are approximately given by

$$\bar{q}_1 \simeq \bar{q}_{max} \pm i\left[\frac{2(n_2-n_2^{max})}{|n_2''(\bar{q}_{max})|}\right]^{1/2} \; . \tag{IVa.24}$$

The set of initial conditions (IVa.14), with complex $\bar{q}_1$, is of course still good for the integration of Hamilton's equations (IVa.12). All that is physically required is that n_1, P_1, R_1, n_2, R_2, P_2 are all real at asymptotic regions. q_2 is not physically observable and can acquire any real or complex value. That the initial variables n_1, P_1, R_1 are all real is specified by the problem. n_2 is required to be real, which automatically implies that P_2 is also real [(IVa.13d)]. R_2, however, will in general be complex for large real t if $\bar{q}_1$ is complex, since the Hamiltonian (IVa.5) couples all the variables during the propagation of a trajectory.

Hence we seem to be running into a dilemma with the possible complexity of R_2. But the difficulty only arises if we require t_2 to be real. If t_2 can be made complex in an appropriate way, R_2 can be made real. Actually the constraint of real t_2 need not be applied on any physical grounds: since in this formulation, E has been made exact, and by the uncertainty principle its conjugate variable, t, can be anything, and thus possibly complex.

Suppose at an asymptotic real time $\tilde{t}_2$, $n(t)=n_2$ and $P(t)=P_2$. Propagating further in time will not change either $n(t)$ or $P(t)$. However, Hamilton's equations imply that

$$R(t) = R(\tilde{t}_2) + \frac{P_2}{\mu}(t-\tilde{t}_2) \quad . \tag{IVa.25}$$

Now $R(\tilde{t}_2)$ may in general be complex, but $R(t_2)$ can be made real if the final time t_2 is chosen to be

$$t_2 = \tilde{t}_2 - i\,\frac{\mu}{P_2}\,\mathrm{Im}R(\tilde{t}_2) \quad . \tag{IVa.26}$$

This explicitly shows that in order to have final dynamical variables all real, the final time t_2 may have to be complex when classically forbidden processes are considered.

If q_I is a complex root of (IVa.15), it is seen from (IVa.24) that the other root is q_I^*. Hence, as in the classically allowed case, it appears that there will also be two terms in the S-matrix. Now both Φ_I and Φ_{II} will be complex; but it can be shown that if $\overline{q}_I$ is the root leading to an exponentially damping term $\exp[-\mathrm{Im}\Phi_I]$, the other root will lead to an exponentially growing term $\exp[+\mathrm{Im}\Phi_I]$. Thus in the S-matrix, only one term should be retained, the one leading to damping:

$$S_{n_2,n_1} = [-2\pi i\hbar n_2'(q_I)]^{-1/2}\exp(i\Phi_I/\hbar) \tag{IVa.27}$$

and the transition probability is

$$P_{n_2,n_1} = [2\pi\hbar|n_2'(q_I)|]^{-1}\exp\left(\frac{-2|\mathrm{Im}\Phi_I|}{\hbar}\right) \tag{IVa.28}$$

which is reminiscent of the WKB result for tunnelling in one-dimensional systems.

(IVa.17) and (IVa.27) are called 'primitive' semiclassical results because they only work well for $n_2<n_2^{max}$ and $n_2>n_2^{max}$ independently, i.e., the separate solutions cannot be connected smoothly as n_2 traverses n_2^{max}. In particular, when n_2 approaches n_2^{max} arbitrarily closely, both (IVa.17) and (IVa.27) break down since $|n_2'(q_{I,II})|^{-1}\to\infty$. This situation recalls the classical rainbow effect. To remedy it, one has to introduce uniform semiclassical formulas which, in addition to effecting a smooth connection between $n_2<n_2^{max}$ and $n_2>n_2^{max}$, will also approach the 'primitive' semiclassical results in regions where they work well.

IVb. Electronically Nonadiabatic Collisions

We consider here the collinear A+BC nonreactive collision process again, but with electronic transitions. A typical process of this type can be written

$$X(^2P_{1/2}) + H_2(n_1) \rightarrow X(^2P_{3/2}) + H_2(n_2) \tag{IVb.1}$$

where $n_1(n_2)$ is the initial (final) vibrational quantum number of H_2, and the halogen X undergoes a spin-orbit electronic transition. The total Hamiltonian for this system can be written

$$H = -\frac{\hbar^2}{2\mu}\frac{\partial^2}{\partial R^2} - \frac{\hbar^2}{2\eta}\frac{\partial^2}{\partial r^2} + H_{el} \tag{IVb.2}$$

where R is the translational X-H_2 coordinate and r is the vibrational H_2 coordinate. μ and η are the reduced masses pertaining to R and r respectively.

In the previous subsection, both R and r are treated classically; and the nuclei are considered to be moving on a single electronic potential surface $v(r)+V(R,r)$ [see (IVa.1)]. The transformation $(p,r)\rightarrow(n,q)$ is then made, with the action-angle variables (n,q) still treated classically. Since an electronic transition is involved here, we have to consider at least two electronic surfaces (we will limit ourselves to exactly two). There are two equivalent ways to proceed using the Miller-George theory. One is using the electronic representation, i.e., finding an electronic basis set $\phi(R,r,\vec{x})$ (where $\vec{x}$ stands for electronic coordinates) which will generate the potential surfaces and couplings $U^d_{ij} = \langle\phi_i|H_{el}|\phi_j\rangle_{\vec{x}}$ (the subscript $\vec{x}$ means that brackets denote integration over electronic coordinates only). We can assume that the starting basis set $\{\phi_i\}$ will in general be diabatic. $\underset{\sim}{U}^d$ can then be diagonalized to yield the adiabatic surfaces W_i required for the Miller-George theory. In this approach the hypersurface $W_i(R,r)$ is two-dimensional and one is required to know the analytic structure of a function of two complex variables, very often not an easy task. The second way is to use what is known as the vibronic representation. This approach makes use of (to begin with) a diabatic vibronic basis set

$$\Phi_{in} = \phi_i(\vec{x},r,R)\psi_n(r) \tag{IVb.3}$$

where $\psi_n(r)$ is a vibrational wave function of BC independent of the electronic configuration of A. This set generates diabatic surfaces and couplings given by

$$U^d_{in,jm} = \langle\Phi_{in}|-\frac{\hbar^2}{2\eta}\frac{\partial^2}{\partial r^2} + H_{el}|\Phi_{jm}\rangle_{\vec{x},r}\ , \tag{IVb.4}$$

which can again be diagonalized to produce adiabatic surfaces as required by the Miller-George treatment. The advantage here is that the adiabatic 'surfaces' W(R) here will only be functions of a single variable R, and their analytic structure is much easier to handle than those of $W_i(R,r)$. Hence the classical nuclear problem is reduced to a one-dimensional one, and the result of the application of the Miller-George theory (IIIb.18) reduces to the Stuckelberg solution given by (IIc.9) to (IIc.14). In this subsection we will use the vibronic representation.

The DIM formalism applied to the collinear model (IVb.1) yields a two-by-two form for the electronic Hamiltonian:

$$U^d_{ii}(R,r) = v(r) + D + A_i(R-r/2) + A_i(R+r/2) + i\lambda/3, \quad i=1,2 \tag{IVb.5a}$$

$$U^d_{12} = U^d_{21} = \sqrt{2}\,\lambda/3 \tag{IVb.5b}$$

where D is the depth of the Morse well $v(r)$, the A_i's are the anti-Morse functions, and λ is the spin-orbit coupling of X. $\underset{\sim}{U}^d$ can be diagonalized to yield the adiabatic electronic surfaces $W_i(R,r)$ with the accompanying adiabatic basis set $\{\phi^a_i\}$ such that

$$H_{el}\phi^a_i(r,R,\vec{x}) = W_i(R,r)\phi^a_i(r,R,\vec{x}) \ . \tag{IVb.6}$$

We prefer to use the representation [compare with (IVb.3)]

$$\Phi_{in} = \phi^a_i(\vec{x},r,R)\psi_n(r) \ , \tag{IVb.7}$$

where the vibrational wavefunctions $\psi_n(r)$ are Morse functions satisfying the equation

$$\{-\frac{\hbar^2}{2\eta}\frac{\partial^2}{\partial r^2} + v(r)\}\psi_n(r) = \varepsilon_n\psi_n(r) \ . \tag{IVb.8}$$

where ε_n is the asymptotic vibrational energy for the n vibrational state of H_2. Using (IVb.4) and first integrating over the electronic coordinates $\vec{x}$, we can write

$$U^d_{in,jm} = \delta_{ij}\langle\psi_n|-\frac{\hbar^2}{2\eta}\frac{\partial^2}{\partial r^2} + W_i(R,r)|\psi_m\rangle_r$$

$$-\frac{\hbar^2}{2\eta}\langle\psi_n|\zeta_{ij}(r,R)|\psi_m\rangle_r - \frac{\hbar^2}{\eta}\langle\psi_n|\chi_{ij}(r,R)\frac{\partial}{\partial r}|\psi_m\rangle_r \ , \tag{IVb.9}$$

where

$$\zeta_{ij}(r,R) = \langle\phi^a_i(\vec{x},r,R)|\frac{\partial^2}{\partial r^2}\phi^a_j(\vec{x},r,R)\rangle_{\vec{x}} \ , \tag{IVb.10}$$

$$\chi_{ij}(r,R) = \langle\phi_i^a(\vec{x},r,R)|\frac{\partial}{\partial r}\phi_j^a(\vec{x},r,R)\rangle_{\vec{x}} \,. \qquad \text{(IVb.11)}$$

These coupling matrix elements are similar to the nonadiabatic couplings T'_{ij} and T''_{ij} [(IB.11) and (IB.12)] discussed in Section Ib.

We can also show that

$$\chi_{ij} = -\chi_{ji} \qquad \text{(IVb.12a)}$$

$$\zeta_{ij} = \sum_{k=1}^{2}\chi_{ik}\chi_{kj} + \partial\chi_{ij}/\partial r \,. \qquad \text{(IVb.12b)}$$

The couplings χ and ζ can furthermore be expressed in terms of $\underset{\sim}{U}^d$:

$$\chi_{12} = \sum_{ij}\Gamma_{i1}\Gamma_{j2}\frac{(\partial U_{ij}^d/\partial r)}{W_2-W_1} \,, \qquad \text{(IVb.13)}$$

where

$$\Gamma_{11} = \Gamma_{22} = \frac{U_{12}^d}{[(W_1-U_{11}^d)^2+(U_{12}^d)^2]^{1/2}} \qquad \text{(IVb.14a)}$$

$$\Gamma_{12} = -\Gamma_{21} = \frac{U_{12}^d}{[(W_2-U_{11}^d)^2+(U_{12}^d)^2]^{1/2}} \,. \qquad \text{(IVb.14b)}$$

The 'diabatic' vibronic potential surfaces can then be written, from (IVb.9):

$$U_{in,im}^d(R) = \langle\psi_n|-\frac{\hbar^2}{2\eta}\frac{\partial^2}{\partial r^2} + W_i + \frac{\hbar^2}{2\eta}\chi_{12}^2|\psi_m\rangle_r \,. \qquad \text{(IVb.15a)}$$

The off-diagonal matrix elements become:

$$U_{1n,2m}^d(R) = \frac{\hbar^2}{2\eta}\langle\psi_n|2\chi_{12}\frac{\partial}{\partial r} + \frac{\partial\chi_{12}}{\partial r}|\psi_m\rangle_r \,. \qquad \text{(IVb.15b)}$$

For a particular vibrational transition $n\to m$, we have a one-dimensional two-state problem, i.e., we just have to deal with the 2×2 matrix (IVb.15) with fixed n and m. This matrix can be diagonalized to yield the final adiabatic vibronic surfaces $W_i^a(R)$ and the S-matrix is given directly in terms of the Stuckelberg solution [(IIc.9) to (IIc.14)]. Figure 3 illustrates the diabatic and adiabatic vibronic curves and the coupling χ_{12} for the BrH_2 system. The potential curves correlate asymptotically to the states 11 and 20.

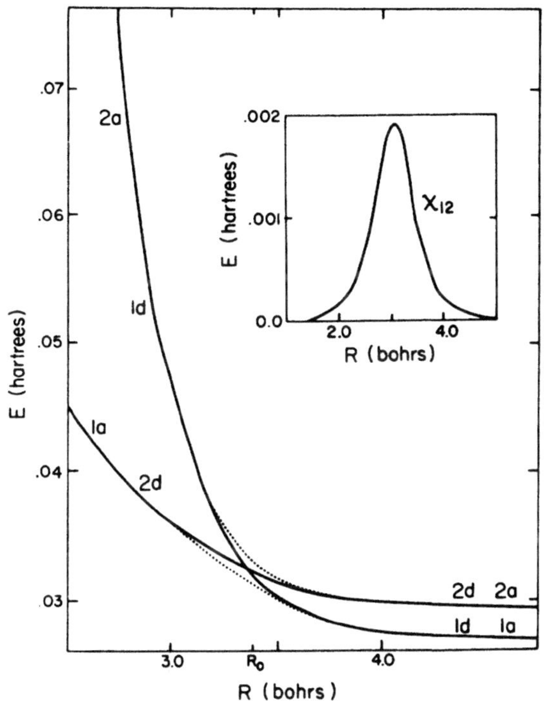

Fig. 3. The diabatic vibronic curves (the solid lines labelled 1d and 2d) are split by the vibrational nonadiabatic coupling χ_{12} (see insert) into the adiabatic vibronic curves, labelled 1a and 2a, which are represented by the dotted lines. Note that the labelling is reversed inside the avoided crossing.

As mentioned before, the formula $p=\exp(-2\delta)$ very often does not yield good results for $P=|S|^2$ when compared with quantum mechanical calculations; and the Nikitin correction formula (IIId.13) is usually employed. There is, however, no cut and fast rule regarding the applicability of the Nikitin formula, it being exact only for exponential diabatic curves and couplings. In the present example, this formula has been applied to Morse-like potentials (IVb.5), and found to help appreciably in bringing the results into good agreement with quantum calculations.

Another slightly disconcerting factor in the application of the semiclassical approach is the difficulty with the phase factor ξ in (IIc.14). The value $\pi/4$ is mot often quoted and used in the literature. In the present applications, since the Nikitin correction formula is sued, the form as given by (IIc.14) would be rendered less rigorous, and the justification for its use in conjunction with the Nikitin formula requires more work. (In any case the use of this form is not very frequently reported in the literature). In our work we have varied ξ between $-\pi/4$ and $\pi/4$ to compare with quantum mechanical results. Reasonable agreements are achieved at either extreme, but it is found that $\xi \sim \pi/10$ would shift the results into phase most closely with the quantum results.

In Figures 4 and 5 we present some results for $|S_{1n,2m}|^2$ (written $n_1 \to m_2$) for X=Br. In each case the semiclassical results are compared with exact quantum mechanical coupled-channel calculations. The reasonable agreement is apparent, pointing to the reliability of semiclassical calculations. Similar results are obtained for the Br-H_2 2_1-1_2 and F-H_2 0_1-0_2 transitions. It should be noted here that the distinct oscillatory behavior may be due entirely to the artifact of the collinear model. In three-dimensional calculations where there is strong coupling between translation and the various internal degrees of freedom, and especially when averages over a large number of quantum states are involved, the interference effects may be quenched.

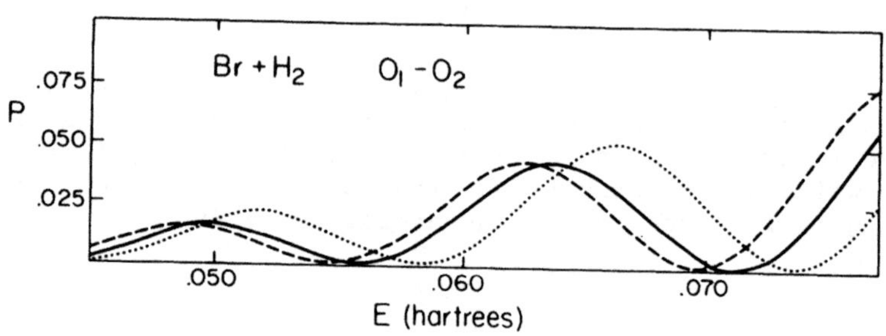

Fig. 4. Transition probability versus energy curves for the 0_1-0_2 transition in the Br + H_2 collision system. The threshold for this transition is 0.027 hartree. The solid curve represents quantum results, and the dashed and dotted curves are the semiclassical results obtained with the phase factor $\xi=0$ and $+\pi/4$, respectively.

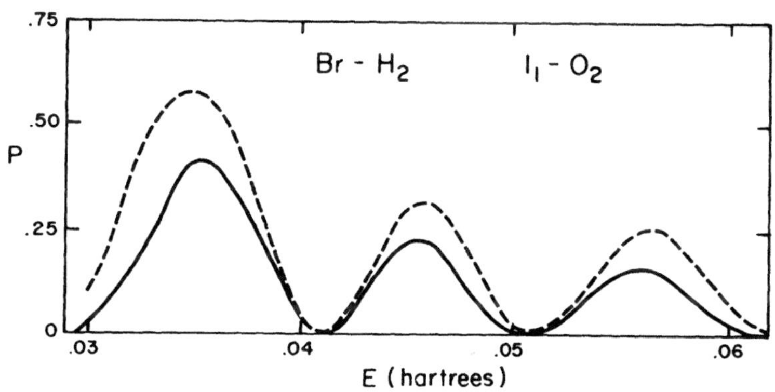

Fig. 5. Transition probability for the 1_1-0_2 transition in the Br + H_2 collision system. The threshold for this transition is 0.029 hartree. The solid curve represents quantum results, and the dashed curve is the semiclassical result obtained using the phase factor $\xi=-\pi/4$.

V. THEORY OF INTERACTION WITH FIELDS - LASER-ENHANCED COLLISIONS

The widespread interest in lasers has generated a quest for a deeper understanding of the effects of intense radiation on molecular collision processes. In this class of phenomena, as distinct from its atomic counterparts, dynamic considerations play a major role, and field-induced molecular events do not have to satisfy strict resonance requirements. Thus the radiation field may be expected to have very prominent effects on the collision dynamics. In particular, such molecular processes as collisional excitation, unimolecular dissociation, radiative collisions, and Penning and associative ionization are expected to be strongly influenced by an intense laser field. In this section we will generalize the Miller-George theory to include radiation interaction.

As soon as a field is put in, many problems suggest themselves. If we are dealing with a laser, the field may approach being monochromatic but always not exactly so. Even a single mode laser has a spectral linewidth owing to broadening effects such as Doppler, collision and radiative lifetime broadening. How many modes of the laser field must one include in the interaction Hamiltonian? This immediately raises the question of what the best representation is for the description of the photon field in laser-molecule interactions. There are a number of candidates: the Fock (photon-number) state representation, the coherent representation, or even the classical field approach, i.e., regarding the field simply as a time-dependent driving force on the system. What part does the polarization of the field play in laser-influenced molecular collisions? Recently there is even some controversy regarding the fundamental validity of using the '$\vec{p}\cdot\vec{A}$' vis a vis the '$\vec{\mu}\cdot\vec{E}$' interaction under the dipole approximation. These and other questions have not as yet been answered definitively. In this section we will restrict ourselves to the simplest possible (although not entirely realistic) situation of a monochromatic field interacting with a two-electronic state molecular system. The dipole approximation is assumed, and the photon-number representation is used, as this lends itself most easily to physical interpretation. Also, complications deriving from the intrinsic angular momentum of the photon will not be considered. Even under these limitations, the radiation field can be demonstrated to have significant effects on the collision system.

Va. The Electronic(Vibronic)-Field Representation

The field-free collision system can be described by the electronically adiabatic potential surfaces $W_i(\vec{R})$ which satisfy:

$$H_{el}\phi_i(\vec{R},\vec{x}) = W_i(\vec{R})\phi_i(\vec{R},\vec{x}) \qquad (Va.1)$$

where $\{\phi_i\}$ is the adiabatic basis set, and $\vec{R}$ ($\vec{x}$) are the nuclear (electronic) coordinates. One can also work with the vibronic surfaces $W_i^{!}(R)$ as discussed in Section IVb, where R represents the translational degree of freedom only. Our discussion in what follows can be phrased equivalently in the electronic or vibronic representations; for the sake of definiteness, we will choose the former. The Hamiltonian due to the presence of the field is given by

$$H_f = \hbar\omega\hat{a}^{\dagger}\hat{a} + \vec{\mu}(\vec{R},\vec{x})\cdot\vec{E}_0^{\prime}(\omega)(\hat{a}^{\dagger}+\hat{a}) \equiv \hbar\omega\hat{a}^{\dagger}\hat{a} + H_1 \ . \qquad \text{(Va.2)}$$

The first term represents the free-field Hamiltonian and the second term, H_1, the interaction Hamiltonian under the dipole approximation. It is understood that the photon creation and annihilation operators, $\hat{a}^{\dagger}$ and $\hat{a}$ respectively, are those for the single frequency ω; and

$$\vec{E}_0^{\prime}(\omega) = (2\pi\hbar\omega/V)^{1/2}\hat{\varepsilon} \qquad \text{(Va.3)}$$

where V is the quantization volume of the collision system and $\hat{\varepsilon}$ is the unit polarization vector of the external field.

A natural basis set to be used for the matter-field system consists of states of the form $\phi_i|n\rangle$, where $|n\rangle$ is the Fock state with n photons (of frequency ω), since they describe the asymptotic ($R\to\infty$) configuration of the matter-field system. For instance one may visualize that before collision, the molecular system is in the ith electronic state, while there are n_i photons floating around in the external field (in the quantization volume V). This state of affairs is described by the asymptotic state $\phi_i|n_i\rangle$. After the collision, again at asymptotic regions, the molecular system may end up in a different electronic state (say the j) whereas the number of photons in the external field may be changed to n_f. The final asymptotic state will then be described by $\phi_j|n_f\rangle$, and the S-matrix element corresponding to this process may be written S_{jn_f,in_i}. If $n_j>n_i$, stimulated emission has taken place; if $n_j>n_i$, absorption has occurred.

Even though $\{\phi_i\}$ is adiabatic in the field-free Hamiltonian H_{el}, the states $\phi_i|n\rangle$ will no longer be adiabatic in the Hamiltonian $H_{el}+H_f$, since the interaction Hamiltonian H_1 mixes them. They are only eigenstates of the Hamiltonian

$$H_R = H_{el} + \hbar\omega\hat{a}^{\dagger}\hat{a} \ , \qquad \text{(Va.4)}$$

$$H_R\phi_i|n\rangle = (W_i+n\hbar\omega)\phi_i|n\rangle \ . \qquad \text{(Va.5)}$$

The potential surfaces $W_i+n\hbar\omega$ can now be considered the diabatic surfaces in the presence of the field. Analogous to the field-free case, at locations (along $\vec{R}$) where resonance occurs ($W_i+n_i\hbar\omega=W_j+n_f\hbar\omega$), avoided crossings may be produced if symmetry conditions permit, i.e., if

$$\int d^3x\phi_j^*\vec{\mu}(\vec{x},\vec{R})\phi_i \neq 0 \ . \qquad \text{(Va.6)}$$

In such cases, one may generate adiabatic curves just as one does in the field-free situation. These adiabatic curves $E_i(R)$ satisfy the equation

$$(H_{el} + H_f)\psi_i(\vec{R},\vec{x}) = E_i(\vec{R})\psi_i(\vec{R},\vec{x}) \qquad \text{(Va.7)}$$

where

$$\psi_i(\vec{R},\vec{x}) = \sum_{\ell,n} \alpha_{i\ell n}(\vec{R})\phi_\ell(\vec{R},\vec{x})|n> \ . \qquad \text{(Va.8)}$$

The representation ψ_i is termed the electronic-field representation while $E_i(\vec{R})$ is known as an electronic-field surface. The E_i's are, of course, eigenvalues of $H=H_{el}+H_f$, where H is expressed in the $\phi_i|n>$ representation. Suppressing i in (Va.8) (i.e., writing it as a general eigenvector) and substituting in (Va.7), we obtain the infinite set of coupled equations for $\alpha_{\ell n}$:

$$E\alpha_{\ell n} = (W_\ell+n\hbar\omega)\alpha_{\ell n} + \sum_i d'_{\ell i}[\sqrt{n+1}\ \alpha_{i,n+1} + \sqrt{n}\ \alpha_{i,n-1}] \ , \qquad \text{(Va.9)}$$

where

$$d'_{\ell i}(\vec{R}) = \vec{E}'_0(\omega)\cdot\int d^3x\phi_\ell^*(\vec{R},\vec{x})\vec{\mu}(\vec{R},\vec{x})\phi_i(\vec{R},\vec{x}) \qquad \text{(Va.10)}$$

is the dipole coupling between ϕ_ℓ and ϕ_i. In obtaining (Va.9) we have used the following properties of the number states $|n>$:

$$<n|\hat{a}|0> = 0 \qquad \text{(Va.11a)}$$

$$<n'|\hat{a}|n''> = \sqrt{n''}\ \delta_{n',n''-1} \qquad \text{(Va.11b)}$$

$$<n'|\hat{a}^\dagger|n''> = \sqrt{n''+1}\ \delta_{n',n''+1} \ . \qquad \text{(Va.11c)}$$

These properties lead to the important fact that only states $\phi_i|n>$ with a difference in photon number equal to one are directly coupled; and hence, to first order in the dipole strength d', only single-photon processes are possible when the interaction term proportional to $\vec{A}^2$ is ignored ($\vec{A}$ is the field vector potential). Even with this simplification, (Va.9) shows that, in order to obtain the adiabatic (electronic-field) surfaces, an infinite-dimensional matrix has to be diagonalized in general.

Vb. The Rotating-Wave Approximation (RWA)

For the present discussion we limit ourselves to the two-state molecular system with $W_2 > W_1$. In this molecular collision problem, the energy detunings $\Delta_\pm$ for a transition $\phi_1|n\rangle \longleftrightarrow \phi_2|n\pm1\rangle$

$$\Delta_\pm(\vec{R}) = W_2 - W_1 \pm \hbar\omega \ , \qquad \text{(Vb.1)}$$

must be examined as a function of all the nuclear degrees of freedom. For $W_2 > W_1$, $\Delta_+ \geq \hbar\omega$ for all $\vec{R}$; hence the transition $\phi_1|n\rangle \rightarrow \phi_2|n+1\rangle$ can be considered anti-resonant (or potential energy non-conserving). Δ_-, however, may vanish for certain nuclear configurations, implying that the transition $\phi_1|n\rangle \rightarrow \phi_2|n-1\rangle$ is resonant at those configurations.

We now assume that $d'_{ii}=0$ (which would be rigorously true for a homonuclear atom-atom collision system). In this case the secular matrix for (Va.9) block-diagonalizes, and it can be written in the general form shown in (Vb.2), where the coupling strength is

$$d_N = \sqrt{N}\, d'_{21} = \vec{E}_0 \cdot \int d^3x \phi_2^* \vec{\mu} \phi_1 \ . \qquad \text{(Vb.3)}$$

N is the number of photons in the external field and $\vec{E}_0$ is the electric field strength. The block diagonalization reflects the fact that, in making the transition $\phi_1 \longleftrightarrow \phi_2$, the system must gain or lose an odd number of photons. The two non-diagonal blocks are completely equivalent; and depending on the initial conditions of the collision environment, one can ignore the appropriate one. For instance, if the collision system is prepared in the state ϕ_1 and the applied radiation field has field strength

$$E_0 = \sqrt{2\pi N\hbar\omega/V} \qquad \text{(IVb.4)}$$

with N odd, then we can work entirely with the upper left block; and with N even, the lower right block. In practice, of course, N is so large that $d_N \sim d_{N+1} \sim d_{N-1}$, and the two blocks would be largely equivalent.

It is in principle possible for a system prepared in ϕ_1 initially to absorb n photons (n odd $\leq$N) or emit n photons ($1 \leq n$ odd $\leq \infty$), and either stay in ϕ_1 or make a transition to ϕ_2. While processes involving transitions $\phi_1|n_1\rangle \longleftrightarrow \phi_2|n_2\rangle$ may or may not be resonant if $n_2 < n_1$, they will always be anti-resonant if $n_2 > n_1$. The RWA in the collision case (similar to the atomic case) consists in throwing away all direct couplings between diabatic states $\phi_1|n\rangle$ of the latter kind ($n_2 > n_1$); for example, between the states $\phi_2|N-1\rangle$ and $\phi_1|N-2\rangle$. It is expected to be

$$
\left(\begin{array}{cccccccc|cccccc}
W_2 & d_1 & & & & & & & & & & & & \\
d_1^* & W_1+\hbar\omega & d_2 & & & \bigcirc & & & & & & \bigcirc & & \\
 & d_2^* & W_2+2\hbar\omega & d_3 & & & & & & & & & & \\
 & & d_3^* & W_1+3\hbar\omega & & & & & & & & & & \\
 & & & & \ddots & & & & & & & & & \\
 \bigcirc & & & & W_1+(N-2)\hbar\omega & d_{N-1} & & & & & & & & \\
 & & & & d_{N-1}^* & W_2+(N-1)\hbar\omega & d_N & & & & & & & \\
 & & & & & d_N^* & W_1+N\hbar\omega & & & & & & & \\
 & & & & & & & \ddots & & & & & & \\
\hline
 & & & & & & & & W_1 & d_1^* & & & & \\
 & & & & & & & & d_1 & W_2+\hbar\omega & d_2^* & & \bigcirc & \\
 & & \bigcirc & & & & & & & d_2 & W_1+2\hbar\omega & d_3^* & & \\
 & & & & & & & & & & d_3 & W_2+3\hbar\omega & & \\
 & & & & & & & & & \bigcirc & & & \ddots & \\
\end{array}\right)
\qquad \text{(Vb.2)}
$$

less accurate when the applied field strength is very large; or in non-resonant situations (as opposed to anti-resonant ones) corresponding to $\phi_1|N\rangle \rightarrow \phi_2|N-1\rangle$ such that the energy detuning,, $W_2-W_1-\hbar\omega$, is not significantly different from that for the anti-resonant transition, $W_2-W_1+\hbar\omega$. In the latter instance, application of the RWA would not be internally consistent, since it would not be justifiable to chop only couplings connecting states with $n_2>n_1$ while keeping those with $n_1>n_2$ if the detunings are roughly equal for the two cases.

In our example of the two-state model, the RWA then leads to complete block diagonalization of (Vb.2) into independent 2×2 matrices. For the initial conditions mentioned previously one has to consider only the diabatic matrix

$$\begin{pmatrix} W_2+(N-1)\hbar\omega & d_N \\ d_N & W_1+N\hbar\omega \end{pmatrix} . \tag{Vb.5}$$

Diagonalization of (Vb.5) will lead to the two adiabatic (electronic-field) surfaces

$$E_{1,2} = N\hbar\omega + \frac{W_1+W_2-\hbar\omega}{2} \mp \frac{1}{2}[(W_2-W_1-\hbar\omega)^2 + 4d_N^2]^{1/2} . \tag{Vb.6}$$

[Compare with (IIId.1)]. In effect the RWA limits the consideration of absorption and emission to single-photon resonant processes. Figure 6 illustrates the construction of the electronic-field surfaces.

Vc. An Example

To apply the Miller-George theory, $E_{1,2}$ are considered as analytic functions of $\vec{R}$. As discussed before they can be regarded as different Riemann sheets of the single analytic function $E(\vec{R})$, with branch points which are roots of the equation

$$E_1(\vec{R}) = E_2(\vec{R}) , \tag{Vc.1}$$

or

$$[W_2(\vec{R})-W_1(\vec{R})-\hbar\omega]^2 + 4d_N^2(\vec{R}) = 0 . \tag{Vc.2}$$

Classical trajectories can be propagated on $E(\vec{R})$, as described before, making transitions by rounding the appropriate branch points.

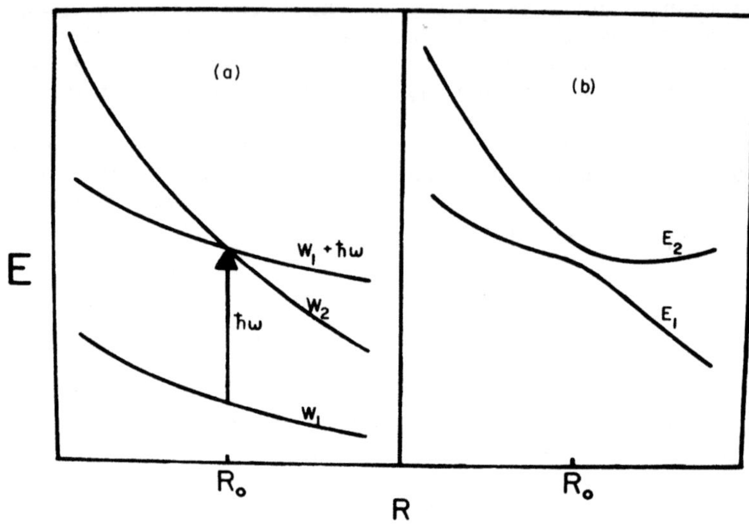

Fig. 6. (a) Schematic drawing of two field-free adiabatic surfaces W_1 and W_2 as functions of the translational coordinate R. R_0 is the point at which the field is in resonance with the two surfaces. (b) Schematic drawing of the electronic-field-surfaces E_1 and E_2. The splitting between them at R_0 is caused by the radiative coupling d_N.

We again consider an example of collinear A+BC collision. The specific process is

$$Br(^2P_{3/2}) + H_2(v=0) + \hbar\omega \rightarrow Br(^2P_{1/2}) + H_2(v=0) , \tag{Vc.3}$$

where v designates the vibrational state of H_2. W_1 and W_2 are taken to be vibronic potential surfaces as before (see section IVb) so that nuclear motion is reduced to one-dimensional. The interaction of the field with the molecular system is approximated by the interaction with the asymptotic collision species alone. In this case the magnetic dipole transition matrix element for the halogen dominates. Hence the magnetic field of the laser radiation appears for $\vec{E}_0$ (formerly considered to be the electric field) in all expressions. The value of $\sqrt{4/3}$ a.u. (atomic units) is used for the magnetic dipole transition moment $|\vec{\mu}_{12}|$. (This value actually holds for all halogens). $\hbar\omega$ is chosen to be 1.001 times the Br spin-orbit splitting (0.4568 eV), such that the laser cannot induce a spin-orbit transition without the aid of the H_2 collision partner. A field strength E_0 of 5.1×10^6 V/cm(10^{-3}a.u.) is used. The vibronic-field surfaces (curves in the present case) can be constructed according to (Vb.6), where $W_{1,2}$ and $E_{1,2}$ resemble the schematic drawings of Figure 6.

Since this is a one-dimensional problem, the Stuckelberg solution (IIc.9) to (IIc.14) again applies, with the slight alteration that W_i in these equations are replaced by E_i. A pair of complex intersection points between E_1 and E_2 is located at $R_*=4.01\pm0.422i$ bohr (1 bohr = 0.529 Å), and classical trajectories were integrated in the complex plane which switch curves smoothly at the intersection point. The Nikitin correction formula (IIId.13) is again employed and ξ is chosen to be $\pi/4$. The semiclassical results are compared with the quantum mechanical as well as the field-free results in Figure 7. The high probabilities for the laser-induced process are in direct contrast to small probabilities for the same process in the absence of the field. The field-free process tends to be electronically adiabatic and the field is the sole agent in inducing electronically nonadiabatic transitions in the present case.

VI. MISCELLANEOUS TOPICS IN THE THEORY OF LASER-ENHANCED COLLISIONS

The majority of collision processes only require a finite number of states (usually two) for their description. These are, however, some in which one or more discrete states of a system can interact with the continuous spectrum. There are in general known as bound-continuum processes. In these processes one is interested in the S-matrix element connecting an initial discrete state and a final continuum channel; and very often there is no straightforward way to reduce the problem to a finite-state one or even one involving infinitely many discrete states. In this section we discuss two examples of field-induced bound-continuum collision processes, spontaneous emission and Penning ionization, in which S-matrix elements may be calculated using discrete-problem techniques by the introduction of certain approximations and discretization procedures. The formalisms for both problems still leave much room for rigorous justification and improvement, and may not be adaptable to the most general collision situations. However, they do provide convenient descriptions, both intuitively and physically, of the respective processes. Our discussion of the bound-continuum

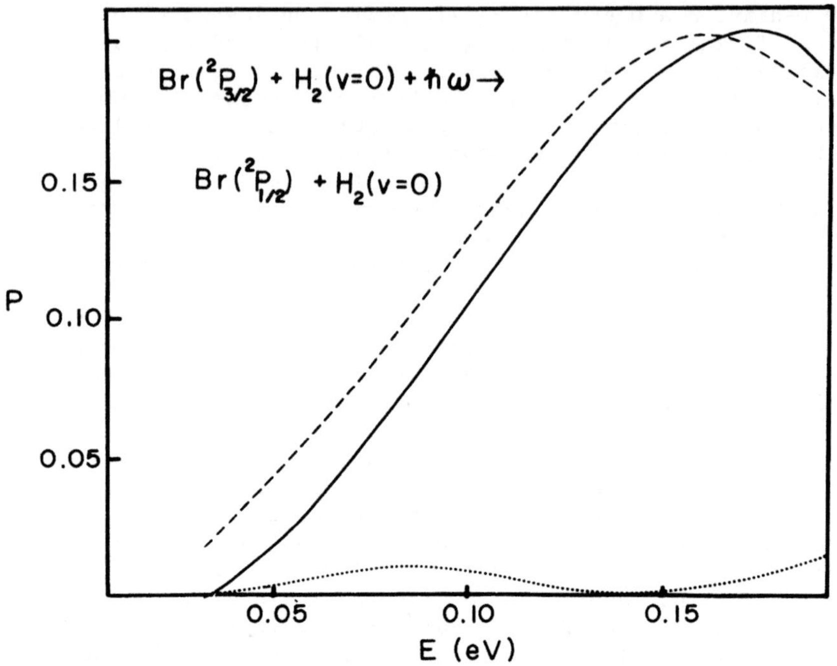

Fig. 7. Probabilities for the collinear reaction Br($^2P_{3/2}$) + H_2(v=0) + $\hbar\omega$ → Br($^2P_{1/2}$) + H_2(v=0) as functions of initial relative translational collision energy. The value of $\hbar\omega$ is 1.001 times the asymptotic spin-orbit splitting. Shown are the results from the quantum calculations using the two-(vibronic) state model (solid line) and results from semiclassical calculations using the two-(vibronic) state model (dashed line). The quantum probabilities for the field-free transition (dotted line) have been multiplied by 100.

problems will be based on the Miller-George semiclassical formalism developed earlier, but we should mention that the discretization procedure introduced below for the ionization problem is also adaptable to a quantum mechanical coupled-channels treatment.

The semiclassical formalism we have developed in the present lecture series for discrete-state field-induced problems can be considered to be exact within the limitations of finite-dimensionality and the RWA. The treatment is non-perturbative, and collisional and radiative interactions are treated on the same footing. Despite these advantages, its application will often raise severe practical problems [e.g., in the analytic continuation of the potential surfaces and the location of branch points (seams)]. For this reason, simpler treatments often retain their appeal and value. In particular, time-dependent perturbation treatments may be most useful when the laser field is of moderate strength and considered as a classical driving force. To conclude the present lecture series, we will give in Section VIc a description of such a theory applied to laser-induced collisional processes as due to Weiner.

VIa. Collision-Induced Spontaneous Emission

We consider non-reactive processes of the type

$$A + B + \hbar\omega \rightarrow A + B + \hbar\omega' \qquad \text{(VIa.1)}$$

(e.g., the Br-H_2 system) where ω is the frequency of the incident laser radiation and ω' that of the emitted radiation. To describe the interaction of the collision system with the incident laser, we can make use of the electronic(vibronic)-field surfaces E_i introduced in Section Va and illustrated in Figure 6.

These electronic-field surfaces [arising from the diagonalization of (Vb.2)] can be regarded as forming a spectrum for spontaneous emission if radiative coupling with the vacuum field is now considered. This interaction involves in principle an infinity of field modes, since a continuum of emission frequencies is possible corresponding to the different nuclear configurations at which emission can take place. The symmetry allowed emission transitions between the electronic-field surfaces are illustrated in Figure 8 (both dotted and solid lines). If we consider the case where the system starting on W_1 asymptotically is de-excited back to W_1 asymptotically after sequential absorption and emission of photons, only the transition marked by the solid line (in Figure 8) need be considered.

The semiclassical picture allows us to think of the photon of frequency ω' as being emitted near the configuration R_ε such that

$$\hbar\omega' = E_2(R_\varepsilon) - E_1(R_\varepsilon) \ . \qquad \text{(VIa.2)}$$

Emission-electronic-field surfaces $E_\pm^{(e)}$ describing the dynamics of this event can then be constructed analogously to the electronic-field surfaces. These are illustrated in Figure 9. It should be noted that the dimensions of the emission avoided crossing are much smaller than those of the absorption one since the former is generated through coupling with the vacuum field. Referring to Figure 9(b), the dynamical picture then emerges that if the system is still propagating on the surface $E_+^{(e)}$ at asymptotic regions, it will have emitted a photon of frequency ω' and become de-excited to the ground state W_1.

The probability of emission at R_ε is determined semiclassically by the trajectories which avoid emission before and which lead to emission when that configuration is reached. If we consider the case of discretized allowable emission configura-

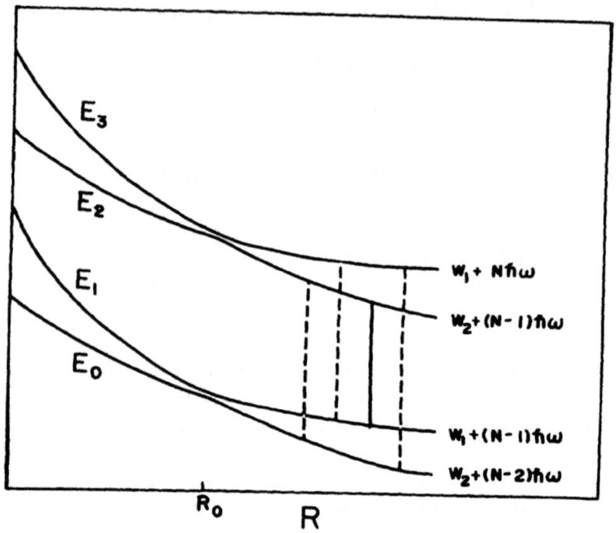

Fig. 8. Symmetry-allowed emission transitions in the two-state electronic field representation. The splitting at R_0 is caused by the absorption radiative coupling. The solid line represents the de-excitation transition from W_2 back to W_1.

tions (in practice, one needs to consider a continuum), a system propagating on the surface E_2 after absorption will encounter a series of configurations at which it can emit. Let one of these be $R_{\varepsilon i}$. At this configuration, the approximation is made that there is emission coupling only at the frequency $\hbar\omega_i=E_2(R_{\varepsilon i})-E_1(R_{\varepsilon i})$. Couplings at frequencies corresponding to more distant configurations can be considered as yet not 'turned on'. Those at frequencies corresponding to configurations prior to $R_{\varepsilon i}$ are assumed to have negligible effects on the dynamics of the system at $R_{\varepsilon i}$, since the system has 'survived' pre-emission loss at these configurations before reaching $R_{\varepsilon i}$. Hence, at each possible emission configuration, emission-electronic-field surfaces can be generated independently of emission couplings at other configurations. A schematic representation for the discrete case

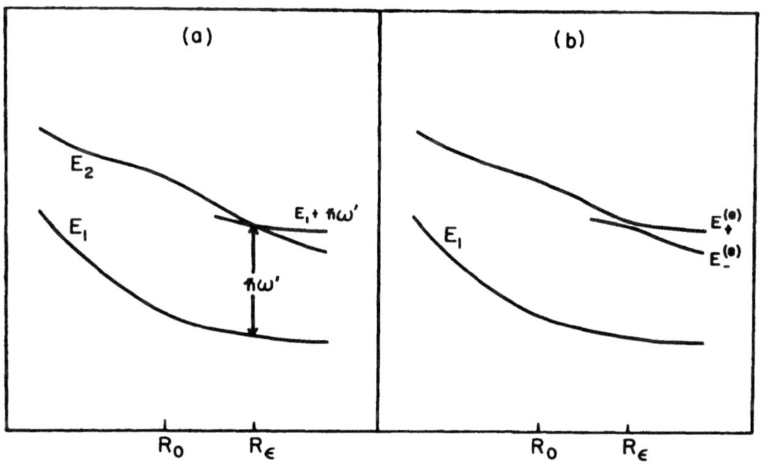

Fig. 9. (a) Schematic drawing of the diabatic surfaces for the description of single-mode emission at R_ε. (b) Schematic drawing of emission-electronic-field surfaces for single-mode emission at R_ε. The avoided crossing is caused by emission radiative coupling.

is illustrated in Figure 10.

When the emission sites become continuously distributed along R, we still retain the approximation of treating each avoided crossing as generated by a localized emission coupling. In this continuous limit the discrete series of branch points will constitute a line of branch points in the complex R-plane. It will be expected to be relatively close to the real axis compared to the location of the absorption branch point because of the relative weakness of the emission couplings. Whereas in the discrete case a trajectory, after rounding each emission branch point, may return to the real axis before going around the next, in the continuum case it will be forced to deviate from the real axis throughout the configurations at which emission coupling is active and be propagated along the line of branch points. The

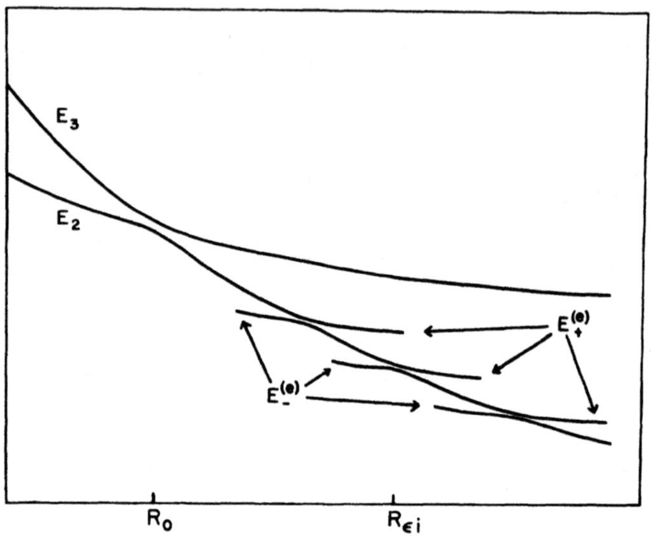

Fig. 10. Schematic drawing of emission electronic-field surfaces for discrete emission configurations. $R_{\varepsilon i}$ is the ith emission site.

imaginary action accumulated along this line will determine the accumulated probability for making local transitions from $E_+^{(e)}$ to $E_-^{(e)}$ continuously, that is, the accumulated probability for not having emitted a photon while the molecular system stays on E_2. At the actual emission site the trajectory reverts back to the real axis and is propagated to asymptotic regions on $E_+^{(e)}$.

The shape of the emission spectrum (as a function of the emission configuration R_ε) is then found to be determined by the shape factor

$$F_S(E,R_\varepsilon) = \exp[-2\chi(E,R_\varepsilon)]\{1 - \exp[-2\xi_\varepsilon(E,R_\varepsilon)]\} \quad \text{(VIa.3)}$$

where E is the total energy of the collision system. The action factor χ accounts for pre-emission effects and $1-\exp(-2\xi_\varepsilon)$ is the local emission probability. Since the emission branch points tend to depart more from the real axis as R_ε increases, 1-exp $(-2\xi_\varepsilon)$ will increase and $\exp(-2\chi)$ decrease as R_ε becomes large. Thus it is seen that pre-emission and localized actual emission effects compete with each other for the total emission probability at a particular configuration. Results of model calculations indicate, however, that this total probability depends critically only on the local emission probability at the actual emission configuration.

VIb. Laser-Induced Collisional Ionization

Let us consider non-reactive processes of the type

$$A(A^*) + B + \hbar\omega \rightarrow A + B^+ + e^-(\varepsilon) \qquad \text{PI} \qquad \text{(VIb.1a)}$$

$$\rightarrow AB^+ + e^-(\varepsilon) \qquad \text{AI} \qquad \text{(VIb.1b)}$$

where ω is the laser photon frequency, ε is the kinetic energy of the emitted electron, and $A(A^*)$ means that A may or may not be in an excited state. Processes (VIb.1a) and (VIb.1b) are the counterparts to field-free (FF) Penning ionization (PI) and associative ionization (AI) respectively. An example of (VIb.1a) is

$$He^*(1s2s,{}^3S) + Ar + \hbar\omega \rightarrow He + Ar^+(3p^5,{}^2P) + e^-(\varepsilon) \ . \qquad \text{(VIb.2)}$$

There are basically two kinds of field-induced ionization processes: field-assisted (FA) and field-modified (FM). FA processes occur in cases where the electronic continuum would not have been energetically accessible in the absence of the field, whereas FM processes occur in cases where field-free ionization is already possible, the field serving to bring new regions of the electronic continuum into accessibility. (VIb.2) is an example of FM ionization. The system can be described by two potential curves: V_d for He*+Ar and V_+ for He+Ar^+. V_d is embedded in the electronic continuum of V_+ and hence field-free ionization is possible. The introduction of the field enables new regions of the continuum to be accessible by collision, as illustrated in Figures 11 and 12. From these figures the qualitative picture emerges that there will be three peaks in the emitted-electron energy spectrum of the cross-section. The middle one is the field-free one, whereas the other two are field-induced, separated from each other by $2\hbar\omega$.

In the field-free case the bound-continuum aspect of the problem consists of the fact that the discrete state $|\phi_d\rangle$, corresponding to the surface V_d, can interact with the continuum of states $|\phi_+,\varepsilon\rangle$ corresponding to the surfaces $V_+ + \varepsilon$, where ε is an arbitrary electronic energy. Any discretization procedure amounts to replacing the infinite set of continuum states $|\phi_+,\varepsilon\rangle$

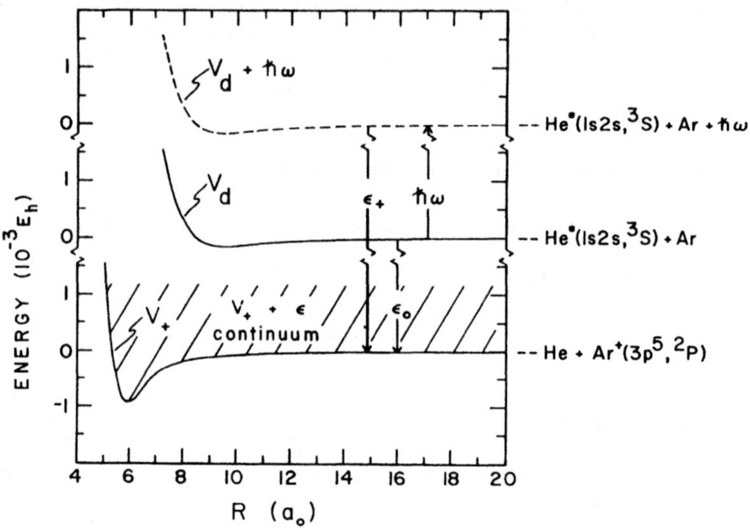

Fig. 11. Relevant potential energy curves for He* + Ar + $\hbar\omega$ collisions. Emission of an electron with energy ε_+ requires absorption of a photon.

by an infinite set of discrete states $|\phi_+,\tilde{\varepsilon}>$, such that the Coulomb couplings, $V_{d\varepsilon}$ and $V_{d\tilde{\varepsilon}}$, leading to ionization are related by

$$V_{d\tilde{\varepsilon}} = \int_0^\infty d\varepsilon \; a_{\tilde{\varepsilon}}(\varepsilon) V_{d\varepsilon} \; , \qquad \text{(VIb.3)}$$

where

$$a_{\tilde{\varepsilon}}(\varepsilon) = <\phi_+,\varepsilon|\phi_+,\tilde{\varepsilon}> \; . \qquad \text{(VIb.4)}$$

$a_{\tilde{\varepsilon}}(\varepsilon)$ must satisfy the orthonormality condition

$$\int_0^\infty d\varepsilon \; a^*_{\tilde{\varepsilon}}(\varepsilon) a_{\tilde{\varepsilon}'}(\varepsilon) = \delta_{\tilde{\varepsilon}\tilde{\varepsilon}'} \; , \qquad \text{(VIb.5)}$$

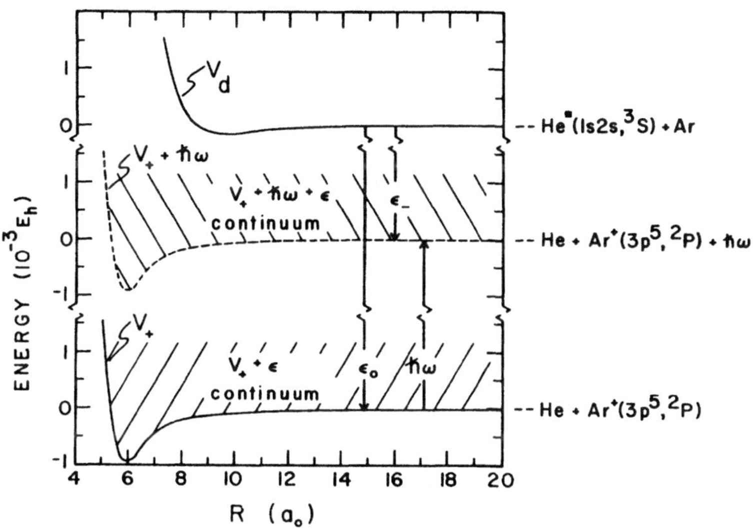

Fig. 12. Relevant potential energy curves for He* + Ar + $\hbar\omega$ collisions. Emission of an electron with energy ε_- leads to emission of a photon.

A particular discretization procedure is then specified by a particular choice for the set $a_{\tilde{\varepsilon}}(\varepsilon)$. Very often the simple choice, satisfying (VIb.5) approximately, is used:

$$a_{\tilde{\varepsilon}}(\varepsilon) = (\Delta\varepsilon)^{-1/2} \quad , \quad \text{for } \tilde{\varepsilon} - \frac{1}{2}\Delta\varepsilon \lesssim \varepsilon \lesssim \tilde{\varepsilon} + \frac{1}{2}\Delta\varepsilon \; ;$$

$$= 0 \quad , \quad \text{otherwise} \; , \qquad \text{(VIb.6)}$$

where the stepsize $\Delta\varepsilon$ can be chosen to be arbitrarily small. If one further introduces the Franck-Condon approximation which implies that only localized bound-discretized state couplings are significant, the diabatic representation of the potential surfaces for nuclear motion with angular momentum L ($\hbar$=1) is simplified to

$$\underset{\sim}{V}^{(L)} = \begin{pmatrix} V_d^{(L)} & V_{d\tilde{\varepsilon}} \\ V_{d\tilde{\varepsilon}}^* & V_+^{(L)} + \tilde{\varepsilon} \end{pmatrix} , \qquad \text{(VIb.7)}$$

where

$$V_d^{(L)} = V_d + L(L+1)/2\mu R^2 , \qquad \text{(VIb.8)}$$

$$V_+^{(L)} = V_+ + L(L+1)/2\mu R^2 , \qquad \text{(VIb.9)}$$

and μ is the reduced mass of the collision system. The coupling $V_{d\tilde{\varepsilon}}$ induces an avoided crossing where the curves $V_d^{(L)}$ and $V_+^{(L)}+\tilde{\varepsilon}$ cross, and the S-matrix describing transition to a discretized state, $S_{\tilde{\varepsilon}}^{(L)}$, can be computed by the semiclassical formalism described earlier. The S-matrix for transition to the true continuum is then given by

$$S_\varepsilon^{(L)} = \sum_{\tilde{\varepsilon}} a_{\tilde{\varepsilon}}(\varepsilon) S_{\tilde{\varepsilon}}^{(L)} . \qquad \text{(VIb.10)}$$

When a field is present, two new regions of the continuum are accessible. The first one is such that $V_d^{(L)}+N\hbar\omega$ cross $V_+^{(L)}+\tilde{\varepsilon}+(N-1)\hbar\omega$, and the second one such that $V^{(L)}+N\hbar\omega$ cross $V_+^{(L)}$ $\tilde{\varepsilon}+(N+1)\hbar\omega$. The new regions are such that both crossings occur where the (field-free) Coulomb coupling is dominated by the longer-ranged radiative coupling. Ionization in the first region requires absorption of a photon while ionization in the second region leads to emission of a photon. Semiclassical treatments can be applied to these field-induced cases entirely analogously to the field-free case.

VIc. Time-Dependent Perturbation Theory

Let us consider the following case of collisional excitation in the presence of a laser,

$$A^* + B + \hbar\omega \rightarrow A + B^{**} , \qquad \text{(VIc.1)}$$

where the asymptotic energy levels of the atoms A and B are illustrated in Figure 13. The formulation of the problem given below is due to Weiner.

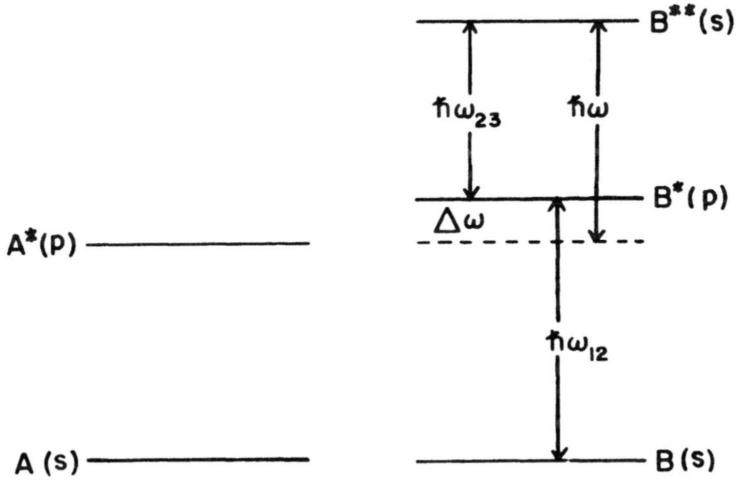

Fig. 13. Asymptotic energy levels of a diatomic collision. The dotted line represents a virtual atomic state of B.

A long-range dipole-dipole interaction between A and B produces a virtual B state at energy equal to the energy difference between A and A*, which is a linear combination of the states corresponding to B and B* (mixing s and p characters). Radiative coupling then carries the system from the virtual state to the final state B**. Direct radiative coupling (dipole coupling) between B and B** is symmetry forbidden (since both are S states). Both the dipole-dipole and the radiative dipole interactions are assumed to be weak so that perturbation theory can be applied. These interactions are respectively given by

$$H_{12} \simeq \frac{\mu_A^{12}\mu_B^{12}}{R^3} \tag{VIc.2}$$

$$H_{23} = E\mu_B^{23}\cos\omega t \tag{VIc.3}$$

where E is the field strength of the laser and the μ are the dipole moments of the respective atoms:

$$\mu_A^{12} = \langle A(s)|\mu|A^*(p)\rangle \ , \tag{VIc.4a}$$

$$\mu_B^{12} = \langle B(s)|\mu|B^*(p)\rangle \ , \tag{VIc.4b}$$

$$\mu_B^{23} = \langle B^*(p)|\mu|B^{**}(s)\rangle \ ; \tag{VIc.4c}$$

and R is the internuclear distance between A and B. The coupled equations (through second order) for the time-dependent expansion coefficients of the atomic wave function for B are ($b_2^{(0)}=0$, $b_3^{(0)}=0$):

$$\dot{b}_1^{(0)} = 0 \tag{VIc.5}$$

$$\dot{b}_2^{(1)} = \frac{1}{i\hbar} b_1^{(0)} H_{12} e^{i\omega_{12}t} \tag{VIc.6}$$

$$\dot{b}_3^{(2)} = \frac{1}{i\hbar} b_2^{(1)} H_{23} e^{i\omega_{23}t} \ . \tag{VIc.7}$$

Putting $b_1^{(0)}=1$ and using the straight line approximation $R(t)=vt$ with impact parameter b, we have

$$b_2^{(1)} = \frac{\mu_A^{12}\mu_B^{12}}{i\hbar}\int_0^\infty dt\, \frac{e^{i\omega_{12}t}}{(b^2+v^2t^2)^{3/2}}$$

$$= \frac{2}{i\hbar}\,\frac{\mu_A^{12}\mu_B^{12}}{v^2 b}\,\omega_{12}K_1\left(\frac{\omega_{12}b}{v}\right)$$

$$\xrightarrow{\text{asymptotically}} \frac{2}{i\hbar}\left(\frac{\mu_A^{12}\mu_B^{12}}{vb^2}\right) \tag{VIc.8}$$

where K_1 is a first-order modified Bessel function and the asymptotic limit holds when $\omega_{12}b/v<<1$. Substituting (VIc.8) into (VIc.7) and using (VIc.3), we have

$$\dot{b}_3^{(2)} = -\frac{1}{\hbar^2}\frac{E\mu_A^{12}\mu_B^{12}\mu_B^{23}}{vb^2}e^{i\Delta\omega t} \qquad \text{(VIc.9)}$$

where the RWA has been applied and $\Delta\omega\equiv\omega_{23}-\omega$. Hence

$$b_3^{(2)} = -\frac{1}{i\hbar^2}\frac{E\mu_A^{12}\mu_B^{12}\mu_B^{23}}{(\Delta\omega)vb^2}(e^{i\Delta\omega t}-1) , \qquad \text{(VIc.10)}$$

and the finite-time transition probability is given by

$$P(t) = |b_3^{(2)}|^2 = 2\left(\frac{1}{\hbar}\frac{\mu_A^{12}\mu_B^{12}}{vb^2}\right)^2\left(\frac{E\mu_B^{23}}{\hbar\Delta\omega}\right)^2(1-\cos\Delta\omega t) . \qquad \text{(VIc.11)}$$

P(t) is proportional to E^2, i.e., linear with respect to the intensity of the radiation (implying a single-photon process).

PROBLEMS

1. Using (Ib.4) to (Ib.8) verify (Ib.9), in which the matrix elements are given by (Ib.10) to (Ib.12).

2. Show that the diagonal elements of $\vec{P}^{(k)}$ [defined in (Ib.15)] vanish identically regardless of representation and that $\vec{P}^{(k)}$ is hermitian.

3. Verify (Id.24).

4. Give an approximate, perturbative derivation of the Landau-Zener formula [(IIa.8)] in the low velocity limit.

5. Following Stuckelberg's procedure write down the fourth-order differential equation for χ_1 [where χ_1 and χ_2 satisfy (IIc.1)]. Then assume a general WKB solution of the form (IIc.5) and obtain equations for S_0 and S_1 when coefficients of the powers of $\hbar$ are separately equated to zero. Finally, retaining only S_0 and S_1 in the expansion (IIc.5) derive the general WKB solutions for χ_1 and χ_2.

6. In the Meyer-Miller treatment establish the equivalence between Hamilton's equations of motion for the classical electronic degrees of freedom [(IIe.10b), (IIe.10c) and (IIe.10d)] and the time-dependent coupled equations (IIe.1) with the trajectory R(t) determined by Ehrenfest's theorem.

7. Show that for the three-state nonadiabatic transition problem the branch point structure of the adiabatic surfaces is still derived from square root functions.

8. Show that $C(R)$ [given by (Ib.25)] with α given by (IIId.10) diagonalizes $U^d(R)$ as given by (IIId.7).

9. For a simple harmonic potential verify (IVa.7) and (IVa.8).

10. Verify (IVb.12), (IVb.13) and (IVb.14).

11. Write down an expression for the nonadiabatic transition S-matrix for the case of a one-dimensional two-state, two-intersection-point system.

12. Derive (Va.9).

13. Using the Miller-George theory in the context of the two-state electronic-field representation, show that in the weak field limit the local transition probability is proportional to the field intensity.

14. Assuming single-mode emission write down the matrix for the diabatic representation of the emission-electronic-field surfaces and simplify.

15. Verify (VIb.3) and (VIb.5).

KEY REFERENCES

1. Y. N. Demkov: 1964, Sov. Phys. JETP 18, pp.138-142.

2. T. F. George, I. H. Zimmerman, P. L. DeVries, J. M. Yuan, K. S. Lam, J. C. Bellum, H. W. Lee, M. S. Slutsky and J. T. Lin, in " Chemical and Biochemical Applications of Lasers ", Vol. IV, edited by C. B. Moore (Academic Press, New York, 1979), pp.253-354.

3. H. -D. Meyer and W. H. Miller: 1979, J. Chem. Phys. 70, pp.3214-3223.

4. W. H. Miller: 1974, Adv. Chem. Phys. 25, pp.69-177.

5. W. H. Miller and T. F. George: 1972, J. Chem. Phys. 56, pp.5637-5652.

6. E. E. Nikitin, " Theory of Elementary Atomic and Molecular Processes in Gases " (Clarendon Press, Oxford, 1974).

7. F. T. Smith: 1969, Phys. Rev. 179, pp. 111-123.

8. J. C. Tully and R. K. Preston: 1971, J. Chem. Phys. 55, pp.562-572.

CORRESPONDENCE-PRINCIPLE METHODS FOR MOLECULAR COLLISIONS

A.S. Dickinson

Department of Atomic Physics, The University, Newcastle upon Tyne, NE1 7RU, U.K.

ABSTRACT: The basic classical and semiclassical results required for the derivation of the Strong-Coupling Correspondence Principle (SCCP) are discussed. The SCCP is derived and related to other approximate methods. Application to collinear vibrational excitation and to excitation of a rigid rotor are examined in some detail. Atom-vibrating rotor and rigid rotor-rigid rotor collisions are discussed more briefly.

1. INTRODUCTION

These lectures are concerned with the application to inelastic scattering processes of approximate methods relying to varying degrees on classical mechanics.

Semiclassical methods were employed in the early days of quantum mechanics but gradually fell into disuse. Current interest in them may be reckoned to date from the work of Grysinski (1) about which Bates (2) has commented: "To revitalize an apparent corpse is no mean feat".

The methods to be described here grew out of work on electron-hydrogen atom collisions where high principal quantum numbers, typically 100, were involved. These cross sections were required for analysis of the observations on radio-recombination lines emitted by highly excited hydrogen atoms in a very tenuous plasma. More recently the methods have been of interest for laboratory work on Rydberg atoms (3). In this approach much of the novelty lay in the use of classical mechanics to describe the motion of the target electron - a classical description of the relative

M. S. Child (ed.), Semiclassical Methods in Molecular Scattering and Spectroscopy, 263–296.

motion being a frequent occurrence. These techniques are generally termed Correspondence-Principle Methods. Applications in atomic and molecular physics have been reviewed (4,5).

Somewhat later, and quite independently, classical S-matrix theory was introduced by Miller and Marcus and their collaborators (6,7). Connections between the two approaches will be discussed.

We begin by considering briefly the quantization of internal states of the target system and the use of classical approximations to simple matrix elements. Next various relevant aspects of collision theory, classical and quantal, are discussed as a preliminary to deriving the Strong-Coupling Correspondence Principle. We then show how this can be related to various other approximations. Applications follow, principally to collinear vibrational excitation and atom rigid-rotor collisions, and more briefly to simultaneous rotational and vibrational excitation and rigid-rotor—rigid-rotor collisions. We end with a final discussion and conclusions.

2. BOUND STATES AND MATRIX ELEMENTS

While these topics will be covered extensively by Dr N. Fröman (8) we summarize the basic, lowest-order results here for later use. Even for highly excited systems, in many cases observations of quantized levels are made, particularly using radiation, so that the use of at least some quantum mechanics is necessary to describe their properties.

In all the applications to be discussed here the target Hamiltonian is separable. We consider initially one-dimensional systems and then examine multidimensional systems. Energy quantization is by the implicit Bohr-Sommerfeld quantization of the action :

$$I_n = (2\pi)^{-1} \oint p(E_n,q)\,dq = (n+\varepsilon)\hbar, \tag{2.1}$$

where $p(E_n,q)$ denotes the momentum conjugate to the coordinate q for energy E_n and the integration is over a complete cycle of the bounded motion. The constant ε, independent of n, depends upon the nature of the motion; for libration $\varepsilon = \frac{1}{2}$ and for rotation $\varepsilon = 0$. In the important cases of the simple harmonic oscillator, the Coulomb potential and the Morse potential, the quantization condition (2.1) is exact, while for the rigid rotor all the semi-classical energy levels are in error by a constant amount. More elaborate techniques are needed for non-separable systems (9,10) but no applications of these in collision problems have yet been reported.

Frequently we shall require differences between neighbouring energy levels

$$\begin{aligned} E_{n+s} - E_n &\simeq s\,\frac{dE_n}{dn} \\ &= s\hbar\omega^c(E_{\bar{n}}), \end{aligned} \qquad (2.2)$$

where $\omega^c(E_{\bar{n}})$ is the classical angular frequency at some energy $E_{\bar{n}}$, where $\bar{n}$ is some mean of n and n+s; often n+s/2 is used. Physically when this result (2.2) is employed in the theory of radiation, it relates quantal transition frequencies between adjacent highly excited levels to harmonics of the fundamental classical frequency of the system.

For matrix elements we use Heisenberg's form of the correspondence principle. For bounded motion in one dimension

$$\langle n+s|F(q)|n\rangle \simeq \frac{1}{T}\int_0^T F\left[q(t,E_{\bar{n}})\right]\,\exp\left[i\,s\,\omega^c(E_{\bar{n}})t\right]dt, \qquad (2.3)$$

when $\bar{n}$ is again some mean of n and n+s, and T is the period of the classical motion. For unbounded motion, matrix elements between states of energy E and E' are approximated by

$$\langle E'|F(x)|E\rangle \simeq \tfrac{1}{2}(vv')^{\frac{1}{2}}\int_{-\infty}^{\infty} F\left[x(t,\bar{E})\right]\,\exp\left[i\Delta Et/\hbar\right]dt, \qquad (2.4)$$

using the quantum normalization

$$\lim_{x\to\infty} \langle E|x\rangle = \sin\left(\frac{\sqrt{2mE}}{\hbar}x + \delta\right). \qquad (2.5)$$

Here v and v' are the speeds when the energies are E and E' respectively, $\Delta E = E' - E$ and $\bar{E}$ is some mean of E and E'.

For multidimensional systems the natural extension of these one-dimensional results is through the use of action-angle variables (11,12). Briefly, let $H(\underline{p},\underline{q})$ be the Hamiltonian of an N-dimensional system for which the motion is bounded in configuration space. We introduce $S(q,P)$ as the solution of the time-independent Hamilton-Jacobi equation

$$H\left(\frac{\partial S}{\partial \underline{q}}, \underline{q}\right) = E, \qquad (2.6)$$

where the $\underline{P}$ are the N constants of integration. For integrable bound systems we can now find a canonical transformation (11) to new conjugate variables $\underline{P},\underline{Q}$ where

$$\underline{p} = \frac{\partial S}{\partial \underline{q}}, \quad \underline{Q} = \frac{\partial S}{\partial \underline{P}}. \qquad (2.7)$$

For separable systems we now define the action variables $\underline{I}$ and the angle variable $\underline{\theta}$ by

$$I_k = (2\pi)^{-1} \oint \partial S/\partial q_k \, dq_k \qquad 1 \leq k \leq N$$
$$\underline{\theta} = \partial S/\partial \underline{I} \, . \tag{2.8}$$

By construction the actions I are constants of the motion and may be shown to be adiabatic invariants. The surfaces in the $(\underline{I},\underline{\theta})$ phase space given by $\underline{I}$ = constant are N-dimensional tori, and the angle variables $\underline{\theta}$ are the angular coordinates on each torus. From Hamilton's equations we have

$$\dot{\underline{\theta}} = \frac{\partial H}{\partial \underline{I}} = \underline{\omega}^c = \text{constant}; \quad \underline{\theta} = \underline{\omega}^c t + \underline{\theta}_o \, , \tag{2.9}$$

where the $\underline{\omega}^c$ are the fundamental frequencies of the system.

Many systems of physical interest, such as the rigid rotor, are degenerate, that is there exist relationships between the fundamental frequencies of the form

$$\sum_{k=1}^{N} n_{ik} \, \omega^c{}_k = 0 \qquad 1 \leq i \leq j < N \tag{2.10}$$

where the n_{ik} are integers. For such cases another, more convenient, set of action-angle variables may be found (12) such that

$$\theta_k = \text{constant} \qquad 1 \leq k \leq j$$
$$\theta_k = \omega_k t + \theta_o^k \qquad j+1 \leq k \leq N \tag{2.11}$$
$$\underline{I} = \text{constant.}$$

We quantize these multidimensional systems using

$$\underline{I} = (\underline{n} + \underline{\varepsilon})\hbar \tag{2.12}$$

where, as in (2.1), the constants $\underline{\varepsilon}$, independent of $\underline{n}$, depend on the nature of the motion. The generalization of (2.2) is

$$E(\underline{n} + \underline{s}) - E(\underline{n}) \simeq \underline{s}.\underline{\omega}^c \hbar \tag{2.13}$$

while for matrix elements we have

$$\langle \underline{n} + \underline{s} | F(\underline{q}) | \underline{n} \rangle \simeq (2\pi)^{-N} \int d\underline{\theta} \; F[\underline{q}(\underline{\theta},\bar{\underline{I}})] \; \exp[i \, \underline{s} \, . \, \underline{\theta}] \tag{2.14}$$

where $\bar{\underline{I}}$ is some mean action.

3. CROSS SECTIONS

3.1 Classical Cross Sections

As most people are more familiar with quantal scattering than with classical we consider briefly here some basic concepts of classical scattering theory. We restrict ourselves to inelastic processes although the generalization to reactive scattering is straightforward and provides one of the most fruitful areas for the application of classical trajectory methods.

We examine initially a very simple case - a collinear collision between a particle of energy E and an oscillator of energy ε_i, where the phase of the angle variable before the collision is specified by θ_o. After the collision the oscillator has energy $\varepsilon'(\theta_o,\varepsilon_i,E)$. Let $P(\varepsilon_i \to \varepsilon_f;E)d\varepsilon_f$ be the probability of the oscillator having final energy between ε_f and $\varepsilon_f + d\varepsilon_f$. Then, averaging over a uniform distribution of initial phases θ_o,

$$P(\varepsilon_i \to \varepsilon_f;E) = (2\pi)^{-1} \int_0^{2\pi} \delta(\varepsilon_f - \varepsilon')d\theta_o \tag{3.1}$$

$$= (2\pi)^{-1} \sum_j |\partial\varepsilon'/\partial\theta_o|_j^{-1}, \tag{3.2}$$

where the summation is over all real values of θ_o satisfying

$$\varepsilon'(\theta_o,\varepsilon_i,E) = \varepsilon_f . \tag{3.3}$$

Since ε' is a periodic function of θ_o generally this equation has an even number of roots. Exceptionally two roots coalesce and then P has a square-root singularity, similar to the familiar classical rainbow singularity in potential scattering.

In three dimensions we consider a structureless particle with energy E and impact parameter vector $\underline{b}$ colliding with an N-dimensional system described by action-angle variables $(\underline{I}_i,\underline{\theta})$, with initial phases $\underline{\theta}_o$. After the collision we have the incident particle moving in a direction $\Omega(\underline{I}_i,\underline{\theta}_o,E,\underline{b})$ and the target left with action $I'(\underline{I}_i,\underline{\theta}_o,E,\underline{b})$. Of interest now is the probability density $P(\underline{I}_i \to \underline{I}_f, \Omega_f;E,\underline{b})$ that a collision results in the incident particle moving off along the direction Ω_f leaving the target in state $\underline{I}_f$. Then

$$P(\underline{I}_i \to \underline{I}_f,\Omega_f;E,\underline{b}) = (2\pi)^{-N} \int d\underline{\theta}_o \; \delta(\Omega_f - \Omega)\; \delta(\underline{I}_f - \underline{I}') \tag{3.4}$$

and the classical differential cross section $\partial^2\sigma/\partial\Omega\partial\underline{I}$ for scattering in the direction Ω_f leaving the target in state $\underline{I}_f$ is

$$\frac{\partial^2\sigma}{\partial\Omega\partial\underline{I}}(\underline{I}_i \to \underline{I}_f, \Omega_f;E) = \int d\underline{b}\; P(\underline{I}_i \to \underline{I}_f, \Omega_f;E,\underline{b}). \tag{3.5}$$

Frequently we are not interested in the angular distribution and need only the cross section integrated over all angles:

$$\frac{\partial\sigma}{\partial \underline{I}}(\underline{I}_i \to \underline{I}_f;E) = (2\pi)^{-N} \int d\underline{b}\, d\underline{\theta}_o\, \delta(\underline{I}_f - \underline{I}'). \quad (3.6)$$

When $\underline{I}'$ and Ω are calculated numerically, the integral (3.4) can be evaluated only by turning the initial-value problem into a boundary value problem, that is we must search the space of initial phases $\underline{\theta}_o$ for all those values satisfying the required final boundary conditions and then determine the density of paths in these neighbourhoods. Except for the one-dimensional case this is a non-trivial problem. The most common approach is the Monte-Carlo technique where we approximate, say $\partial\sigma/\partial\underline{I}$, by

$$\frac{\partial\sigma}{\partial \underline{I}}^{MC}(\underline{I}_i \to \underline{I}_f; E) = C \int_{\underline{I}_f - \underline{x}}^{\underline{I}_f + \underline{x}} \frac{\partial\sigma}{\partial\underline{I}}\, d\underline{I}, \quad (3.7)$$

where $\underline{x}$ is a vector of small constants and C is the appropriate normalization factor. This approximation is readily evaluated numerically since now essentially we count the fraction of orbits falling in each possible interval of final states. This is a direct classical simulation of an experiment.

3.2 Quantization

Next we turn to the problem of estimating cross sections between quantized levels using these classical results. Since classical variables take on continuous values the natural classical cross section (3.5) was differential and must be integrated in some fashion to approximate quantal results. Possibly the most satisfactory conceptually is to use the density of states correspondence principle

$$\sigma^{Cl}(\underline{n} \to \underline{n}';E) = \hbar^N \frac{\partial\sigma}{\partial\underline{I}}(\underline{I}_i(\underline{n}) \to \underline{I}_f(\underline{n}');E), \quad (3.8)$$

where the $\underline{I}(\underline{n})$ are obtained using (2.12). Primitive classical S-matrix theory (6,7) uses this type of quantization but with important additional phase terms arising from interference between different contributing classical trajectories.

If the eigenstate of the target are degenerate with degeneracy g_n in level n then the quantum detailed-balance requirement (13,14) is that

$$E g_n\, \sigma(n \to n';E) = E' g_{n'}\, \sigma(n' \to n;E'), \quad (3.9)$$

where the total energy is the same for both collisions. It can be shown (4) that quantization as in (3.8) satisfies this requirement, which is a consequence of the invariance of the system under time reversal.

However, as noted above, in general there are severe practical difficulties in finding the trajectories required by eq.(3.6). Accordingly it is appealing to quantize by integrating the differential cross sections across a band of final actions of width $\hbar$, that is to employ (3.7) with $x_i = \frac{1}{2}\hbar$. Such a procedure leads to cross sections which do not, in general, satisfy the detailed-balance requirement (3.9). For a fuller discussion of this type of quantization, see (5,15).

3.3 Classical Perturbation Theory

Much of the following will use results based explicitly or implicitly on classical perturbation theory (CPT). Since this is generally less familiar than its quantal equivalent, it is useful to review it here.

We are interested in a Hamiltonian of the form

$$H(\underline{I},\underline{\theta},\underline{R},\underline{P}) = H_o(\underline{I}) + P^2/2\mu + V_o(R) + V(\underline{R},\underline{\theta},\underline{I}), \tag{3.10}$$

where $H_o(\underline{I})$ is the Hamiltonian of the target, $\underline{I}$ and $\underline{\theta}$ are the action-angle variables for the target, $\underline{R}$ is the position vector of the incident particle relative to the centre of mass of the target, $\underline{P}$ is the momentum of the incident particle, μ is the reduced mass and $(V_o + V)$ is the interaction between the incident particle and the target. This interaction is split into a part V_o which can cause only elastic scattering and a part V which can induce transitions in the target. The quantity of most relevance is the change in $\underline{I}$. Hamilton's equations are

$$\begin{aligned}
\dot{\underline{I}} &= -\frac{\partial V}{\partial \underline{\theta}} , \\
\dot{\underline{\theta}} &= \frac{\partial H_o}{\partial \underline{I}} + \frac{\partial V}{\partial \underline{I}} , \\
\dot{\underline{P}} &= -\underline{\nabla}_R(V_o + V) , \\
\dot{\underline{R}} &= \underline{P}/\mu .
\end{aligned} \tag{3.11}$$

If V can be neglected we have the zero-order solution

$$\begin{aligned}
\underline{I} &= \underline{I}_o = \text{constant}, \\
\underline{\theta} &= \underline{\omega}^c t + \underline{\theta}_o , \quad \text{where } \underline{\omega}^c = \frac{\partial H_o}{\partial \underline{I}_o} , \ \underline{\theta}_o \text{ constant}, \\
\underline{R} &= \underline{R}_o(t),
\end{aligned} \tag{3.12}$$

where $\underline{R}_o(t)$ is obtained from the solution for motion in the potential $V_o(R)$. Then to first order in V we have

$$\underline{I}(t) = \underline{I}_o - \int_{-\infty}^{t} \frac{\partial V}{\partial \underline{\theta}} \left[R_o(t'), \underline{\omega}^c t' + \underline{\theta}_o, \underline{I}_o\right] dt' \quad (3.13a)$$

$$\underline{\theta}(t) = \underline{\omega}^c t + \underline{\theta}_o + \int_{-\infty}^{t} \frac{\partial V}{\partial \underline{I}} \left[R_o(t'), \underline{\omega}^c t' + \underline{\theta}_o, \underline{I}_o\right] dt' \quad (3.13b)$$

and the changes $\Delta\underline{I}^{CPT}$, $\Delta\underline{\theta}^{CPT}$ due to the collision can be written

$$\Delta\underline{I}^{CPT} = \underline{I}(\infty) - \underline{I}_o = \partial A(\underline{\theta}_o, \underline{I}_o)/\partial\underline{\theta}_o , \quad (3.14a)$$

$$\Delta\underline{\theta}^{CPT} = - \partial A(\underline{\theta}_o, \underline{I}_o)/\partial\underline{I}_o , \quad (3.14b)$$

where

$$A(\underline{\theta}_o, \underline{I}_o) = - \int_{-\infty}^{\infty} V\left[R_o(t), \underline{\omega}^c t + \underline{\theta}_o, \underline{I}_o\right] dt. \quad (3.15)$$

Thus we can view A as a first-order approximation to the change in action of the system during the collision.

The conditions for the validity of CPT are difficult to state precisely but roughly we require $\Delta I/I_o \ll 1$ and $\Delta\underline{\theta} \ll 1$, that is changes during the collision should be relatively small. This is often a less restrictive condition than quantal perturbation theory. Clearly we can use CPT with or without separating off V_o to include in the zero-order problem. In quantum mechanics a similar choice arises between the use of the Born and the Distorted-Wave approximations (16). For atom-molecule collisions generally the curvature of the orbit is important, particularly for strong, close, collisions so the incorporation of V_o into the zero-order problem is essential.

In calculations on electron-atom collisions using CPT it has been found (17) that averages over CPT results are significantly more accurate than the CPT results for individual trajectories.

3.4 Time-Dependent Collision Theory

While the rigorous formulation of atom-molecule collision theory employs a quantal description of the relative motion, we shall concentrate on approximations where the relative motion is described by classical mechanics. This approximation is generally successful because the potentials change slowly over a de Broglie wavelength of the relative motion.

A fundamental approximation of this method is that the relative motion is determined in a potential which is independent of that part of the potential which causes transitions in the target. This means that back coupling from the target to the trajectory is neglected, energy and angular momentum are not conserved and detailed balance is not satisfied.

While this approximation is sometimes termed a semiclassical method, we prefer the term 'Time-Dependent Close Coupling' (TDCC) and, following Balint-Kurti (18) reserve the term 'semiclassical' for methods which make classical approximations to the motion of the target.

We require the transition amplitude $a(\underline{n}\to\underline{m})$ for the target, coordinates $\underline{r}$, to make a transition from state $|\underline{n}\rangle$ to state $|\underline{m}\rangle$ due to the time-dependent potential $V[\underline{r},\underline{R}(t)]$, where $\underline{R}(t)\equiv\underline{R}_o(t)$, determined in §3.3. Then the time-dependent Schrödinger equation can be written

$$i\hbar\frac{\partial}{\partial t}\,a(\underline{n}',\underline{n};t) = \sum_{m} \langle\underline{n}'|V|\underline{m}\rangle \exp[i(E_{\underline{n}'}-E_{\underline{m}})t/\hbar]\,a(\underline{m},\underline{n};t), \tag{3.16a}$$

$$a(\underline{n}',\underline{n};-\infty) = \delta_{\underline{n}'\underline{n}} \tag{3.16b}$$

and

$$a(\underline{n}\to\underline{m}) = a(\underline{m},\underline{n};\infty).$$

Cross sections are obtained from

$$\sigma(\underline{n}\to\underline{m}) = \int d\underline{b}\,|a(\underline{n}\to\underline{m})|^2 \tag{3.17}$$

The basic conditions for the validity of the TDCC approximation are that the change in energy, ΔE and angular momentum, ΔL of the incident particle be small compared to their initial values. These changes need not be small compared to the corresponding initial values for the target. Except for very close collisions, both conditions are usually readily satisfied in heavy-particle collisions.

In general, an analytic solution of (3.16) cannot be obtained and the equations as they stand are solved numerically only rarely in atom-molecule collisions (15). However, the first-order approximation is often useful when the potential V is weak:

$$a^{FO}(\underline{n}\to\underline{m}) = -\,i/\hbar\int_{-\infty}^{\infty} dt\,\langle\underline{m}|V|\underline{n}\rangle\,\exp[i(E_{\underline{m}}-E_{\underline{n}})t/\hbar]. \tag{3.18}$$

When the collision is sudden, that is the collision time is much shorter than the transition time in the target, then the equations (3.16) can be solved yielding

$$a^S(\underline{n}\to\underline{m}) = \langle\underline{m}|\exp(iA^S/\hbar)|\underline{n}\rangle\,, \tag{3.19}$$

$$A^S(\underline{r}) = -\int_{-\infty}^{\infty} V[\underline{r},\underline{R}(t)]\,dt\,. \tag{3.20}$$

We note that in both (3.18) and (3.19) the amplitude is expressed in terms of a matrix element so that correspondence principle approximations may be derived using (2.14).

3.5 The Regular Spacing Approximation

An interesting solution for correspondence principle approximations has been obtained (19) to the TDCC equations (3.16) under quite different approximations to those discussed above. It is assumed that energy differences and potential matrix elements depend only on quantum number differences, essentially a large quantum number approximation:

$$E_{\underline{n}} - E_{\underline{m}} = \hbar(\underline{n} - \underline{m}).\underline{\omega} \tag{3.21}$$

$$\langle\underline{n}|V|\underline{m}\rangle = V_{\underline{n}-\underline{m}}[\underline{R}(t)], \tag{3.22}$$

where the frequency vector $\underline{\omega}$ is independent of $\underline{n}$ and $\underline{m}$.

We introduce the generating function

$$G(\underline{\theta},t) = \sum_{\underline{m}} a(\underline{m},\underline{n};t) \exp[-i(\underline{n}-\underline{m}).\underline{\theta}] \tag{3.23}$$

which, using (3.16a) can be shown to satisfy

$$i\hbar\frac{\partial G}{\partial t} = G \sum_{\underline{s}} V_{\underline{s}}(t) \exp[-i\,\underline{s}.(\underline{\omega}t + \underline{\theta})]. \tag{3.24}$$

The solution of this is trivial and on inverting the Fourier series (3.23) we obtain

$$a(\underline{n}\rightarrow\underline{m}) = (2\pi)^{-N}\int_{0}^{2\pi} d\underline{\theta}\, \exp\{-i[(\underline{m}-\underline{n}).\underline{\theta} - A^{o}(\underline{\theta})/\hbar]\} \tag{3.25}$$

where

$$A^{o}(\underline{\theta}) = -\int_{-\infty}^{\infty} dt \sum_{\underline{s}} V_{\underline{s}}(t) \exp[-i\,\underline{s}.(\underline{\omega}t + \underline{\theta})]. \tag{3.26}$$

As it stands, the presence of an infinite sum in (3.26) means that, in general, it cannot be used without further approximation.

3.6 The Strong-Coupling Correspondence Principle (SCCP)

We now examine a semiclassical approximation where the motion of both projectile and target are treated classically. It was originally derived (20) from the integral equation satisfied by the quantal evolution operator but a simpler derivation based on (3.25) is possible (21).

The SCCP rests on three basic assumptions:

A1. The equivalent purely classical model of the collision does not deviate far from its unperturbed motion.

A2. All strongly coupled energy levels obey a regular spacing condition such as (3.21).

A3. Heisenberg's form of the correspondence principle (2.14) is valid.

We express the interaction potential as a multiple Fourier series in the action-angle variables $(\underline{I},\underline{\theta})$ of the unperturbed target,

$$V(\underline{I},\underline{\theta};t) = \sum_{\underline{j}} V_{\underline{j}}(\underline{I},t) \exp(-i\,\underline{j}.\underline{\theta}) \qquad (3.27)$$

To use this in §3.5 we note that from (2.14)

$$V_{\underline{s}}(t) = \langle \underline{m}+\underline{s}|V|\underline{m}\rangle \simeq V_{\underline{s}}(\underline{I},t). \qquad (3.28)$$

Thus the assumption that $V_{\underline{s}}$ is independent of $\underline{n}$ is equivalent to assuming that $V_{\underline{s}}(\underline{I},t)$ is independent of $\underline{I}$ for all relevant matrix elements. Then, using (3.28) we have

$$\sum_{\underline{s}} V_{\underline{s}}(t) \exp\left[-i\,\underline{s}.(\underline{\omega}t + \underline{\theta})\right] = V(\underline{I},\, \underline{\omega}t + \underline{\theta};\, t) \qquad (3.29)$$

Using (2.13) we see that the $\underline{\omega}$ defined by (3.21) becomes equal to the classical fundamental frequency vector $\underline{\omega}^c$. Thus we have, using (3.29) in (3.25, 3.26)

$$a(\underline{n}\rightarrow\underline{m}) = (2\pi)^{-N} \int_0^{2\pi} d\underline{\theta} \exp\{-i\left[(\underline{m}-\underline{n}).\underline{\theta} - A(\underline{\theta})/\hbar\right]\} \qquad (3.30)$$

$$A(\underline{\theta}) = -\int_{-\infty}^{\infty} dt\, V(\underline{I},\underline{\omega}^c t + \underline{\theta};\, t) \qquad (3.31)$$

Note that A is identical to the action arising in CPT eq.(3.15). This result, eqs.(3.30, 3.31), constitutes the SCCP approximation.

It is readily seen that when $\underline{\omega}^c$ is negligible, giving a sudden collision, (3.30) reduces to the correspondence principle limit of (3.19, 3.20), while when V is weak so that $A \ll \hbar$, then (3.30) reduces to the correspondence principle limit of (3.18). This illustrates the difference between the condition for a first-order quantal treatment to be valid and the conditions for classical perturbation theory to be valid, §3.3.

3.7 The Classical Limit of the Strong-Coupling Correspondence Principle

To emphasize the connection of the SCCP with CPT we examine first its classical limit. We consider just the one-dimensional form - the argument may readily be generalized to the multi-dimensional case.

In the classical limit, A in (3.31) becomes large compared to $\hbar$ so the significant contributions to the integral in (3.30) come from the M neighbourhoods of the stationary phase points where

$$\partial A(\theta_o^{\,k})/\partial\theta_o = \hbar(m-n) = s\hbar, \quad 1 \leqslant k \leqslant M. \tag{3.32}$$

Then, using the standard stationary-phase result in (3.30)

$$a(n\to m) = (2\pi)^{-1}\sum_{k=1}^{M} \exp\{-i[s\theta_o^{\,k} - A(\theta_o^{\,k})/\hbar]\}\ |2\pi i\hbar/A_k''|^{\frac{1}{2}} \tag{3.33}$$

$$P(n\to m) = (\hbar/2\pi)\sum_k |A_k''|^{-1}, \tag{3.34}$$

since the interference terms can be neglected because the differences between the actions $A(\theta_o^{\,k})$ are assumed large compared to $\hbar$.

Recalling (3.14) and (3.15) we see that $\partial A/\partial\theta_o = \Delta I^{CPT} = s\hbar$

$$\therefore\ P(n\to m) = (\hbar/2\pi)\sum_k |\partial\Delta I^{CPT}/\partial\theta_o|_k^{-1}. \tag{3.35}$$

We see that this is identical with the one-dimensional form of (3.8) and (3.4), where ΔI is calculated using CPT.

3.8 Relation to Classical S-Matrix Theory

We examine now the relation between the SCCP and the classical S-matrix theory, discussed here by Child (22). For simplicity we consider the collinear excitation of an oscillator with actions I_i, I_f before and after the collision respectively and initial and final phases, θ_o, θ_f of the angle variables. If the initial quantum number is n, using (2.12) we have

$$I_i = (n+\varepsilon)\hbar. \tag{3.36}$$

Then, in classical S-matrix theory (23)

$$a(n\to m) = (2\pi)^{-1}\int_0^{2\pi} d\theta_o(\partial\theta_f/\partial\theta_o)^{\frac{1}{2}}\exp\{i[\chi(I_i,\theta_f) - m\theta_f]\}, \tag{3.37}$$

where $\chi(I_i, \theta_f)$ is the F_2 type generator (11) for transformations between the initial and final states. It satisfies

$$I_f/\hbar - \varepsilon = \partial\chi/\partial\theta_f, \qquad \theta_o = \hbar\,\partial\chi/\partial I_i \ . \tag{3.38}$$

We now approximate χ to first order in V using

$$\chi(I_i,\theta_f) = \left[A(I_i,\theta_f) + (I_i - \varepsilon\hbar)\,\theta_f\right]/\hbar \ , \tag{3.39}$$

where A is given by (3.15), except that A is now a function of the final phases of the angle variables. Differentiating and using (3.14),

$$\begin{aligned} \partial\chi/\partial\theta_f &= (\partial A/\partial\theta_f + I_i - \varepsilon\hbar)/\hbar \\ &= (\Delta I^{CPT} + I_i - \varepsilon\ )/\hbar \\ &= I_f/\hbar - \varepsilon \ , \end{aligned} \tag{3.40a}$$

$$\begin{aligned} \partial\chi/\partial I_i &= (\partial A/\partial I_f + \theta_f) \\ &= -\Delta\theta^{CPT} + \theta_f = \theta_o \ , \end{aligned} \tag{3.40b}$$

so this approximation (3.39) yields the CPT values for θ_o and I_f. Retaining only the leading term, $\partial\theta_f/\partial\theta_o = 1$ and converting (3.37) to an integral over θ_f,

$$a^{CP}(n \to m) \simeq (2\pi)^{-1} \int_0^{2\pi} d\theta \, \exp\{-i\left[(m-n)\theta - A(I_i,\theta)/\hbar\right]\} \tag{3.41}$$

in agreement with (3.30).

For an alternative discussion of the relation between classical S-matrix theory and the SCCP, avoiding the use of time-dependent scattering theory, see Miller and Smith (24). The connection between the SCCP and other semiclassical approximations is discussed more fully in (5).

4. COLLINEAR VIBRATIONAL EXCITATION

We begin our discussion of applications of the SCCP by examining collinear vibrational excitation. This model has long served as a 'test-bed' for approximations in atom-molecule collisions. Generally low vibrational quantum numbers are of interest so this is not expected to be a particularly favourable case for applying correspondence principle methods. Nowadays quantal close-coupling solutions are generally available which provide results of better than 1% precision for low quantum number transitions in harmonic and anharmonic oscillators (25).

Physically, collinear collisions are clearly a fairly rare occurrence, however if the widely used Infinite-Order-Sudden approximation (26) is employed for the rotational degrees of freedom then the full atom-vibrating rotor problem reduces essentially to a series of collinear vibrational excitation collisions.

For simplicity we examine the particularly straightforward case of a simple harmonic oscillator perturbed by a time-dependent potential linear in the oscillator coordinate. The Hamiltonian is

$$H = \frac{P^2}{2m} + \tfrac{1}{2}m\omega^2x^2 + x\,f(t), \qquad (4.1)$$

where f(t) specifies the time-dependence of the driving potential. An exact quantal solution for this problem is available (27):

$$P(n\to m) = (n_<!/n_>!)\exp(-\alpha)\alpha^s[L^s_{n_<}(\alpha)]^2, \qquad (4.2)$$

where $s = |n-m|$, $n_<(n_>)$ is the smaller (larger) of n and m, and

$$\alpha = (2m\hbar\omega)^{-1}\left|\int_{-\infty}^{\infty} f(t)\exp(i\omega t)dt\right|^2, \qquad (4.3)$$

and $L^m_n(x)$ is the generalized Laguerre polynomial (28). This parameter α appearing in the quantal solution is essentially a classical quantity, being the energy transferred to an oscillator of zero energy, in units of $\hbar\omega$, or the energy transferred to an oscillator of arbitrary energy, averaged over all initial phases.

To calculate the SCCP result we first evaluate the action A from (3.31). For an oscillator with energy $(\bar{n}+\tfrac{1}{2})\hbar\omega$ the unperturbed motion is given by

$$x(\theta_o) = [2(\bar{n}+\tfrac{1}{2})\hbar/m\omega]^{\frac{1}{2}}\sin(\omega t+\theta_o), \qquad (4.4)$$

since the classical frequency for this system is ω, independent of $\bar{n}$. Then

$$A(\theta_o) = 2\hbar[(\bar{n}+\tfrac{1}{2})\alpha]^{\frac{1}{2}}\sin\theta_o, \qquad (4.5)$$

assuming f is an even function of t, as is generally the case in collision problems. Substituting (4.5) in (3.30) we find, for the SCCP approximation,

$$a^{CP}(n\to m) = (2\pi)^{-1}\int_0^{2\pi} d\theta\,\exp\{-i[(m-n)\theta - X\sin\theta]\} \qquad (4.6)$$

$$= J_s(X), \qquad X = 2[(\bar{n}+\tfrac{1}{2})\alpha]^{\frac{1}{2}}, \qquad (4.7)$$

$$P^{CP}(n\to m) = J_s{}^2(X), \qquad (4.8)$$

using the integral representation of a Bessel function (26). It remains to specify the mean quantum number, $\bar{n}$. The slightly elaborate choice

$$\bar{n} + \tfrac{1}{2} = (n_>!/n_<!)^{1/s} \qquad s \neq 0 \tag{4.9}$$

has been suggested (27).

We note first that the Bessel function in (4.8) automatically provides a uniform approximation valid both in the classically forbidden case, where $X \ll s$, and in the classically allowed case, $X \gg s$. This is a general feature of the SCCP. Complex trajectories are not required to describe classically forbidden processes using this approximation.

It may be shown that (4.8) is the leading term in the asymptotic expansion of (4.2) for large n. The SCCP result has the correct oscillatory behaviour as α increases provided $\alpha < \alpha_M, \simeq \pi^2 n/4$ ($s \ll n$), the largest zero of the exact result (4.2). This limitation arises basically because Heisenberg's correspondence principle for matrix elements (2.14) assumes that the quantity whose matrix element is being approximated varies more slowly than the wavefunctions involved. From (4.6) we see that this requires $\alpha \lesssim \bar{n}$.

A direct comparison may be made on expanding in powers of α:

$$\frac{P^{CP}(n \to n+s)}{P(n \to n+s)} = 1 + \frac{(s-1)\alpha}{6(2n+s+1)} + O\{(\alpha/n)^2\}. \tag{4.10}$$

Clearly, for fixed α, which is a good measure of the strength of the collision, the SCCP result improves with increasing n. By contrast, the quantal first-order time-dependent perturbation theory result (3.18), P^{QFO} is generally much poorer:

$$\frac{P^{QFO}(n \to n \pm 1)}{P(n \to n \pm 1)} = 1 + n_>\alpha + O(\alpha^2) \tag{4.11}$$

so the range of validity of P^{QFO} decreases as n increases. It is interesting that the first-order limit of the SCCP result (4.8), using (4.9), is exact for all n. This is a consequence of the choice (4.9) for the harmonic oscillator (30).

Further insight into the SCCP can be obtained by examining the exact classical energy transfer, ΔE and the CPT approximation to this, ΔE^{CPT} for an oscillator with initial energy E and phase θ_o:

$$\begin{aligned} \Delta E &= \alpha\hbar\omega + \Delta E^{CPT} \\ \Delta E^{CPT} &= -2\sqrt{\alpha\hbar\omega E}\,\sin\theta_o. \end{aligned} \tag{4.12}$$

Provided $\alpha\hbar\omega/E$ is small, then the CPT term is generally dominant but for large α clearly the CPT result breaks down.

In figure 1 we show, as a function of α, the exact value for $P(0\to1)$, eq.(4.2), the SCCP result, eq.(4.8), P^{QFO} and results obtained using N-state quantum mechanical solutions (29). Figure 2 contains a similar comparison for the $5\to6$ transition. Examining the two figures we see that the validity of the SCCP is much greater than that of P^{QFO}, even for the $0\to1$ transition and its validity extends into the strong-coupling region. The large number of coupled channels required to obtain comparable agreement with the SCCP, even for such low quantum numbers, is striking. For larger values of α than those shown in figure 2 the SCCP result slowly gets out of phase with the exact result, though the qualitative behaviour is correct up to $\alpha \simeq 12$. The deterioration is due to a combination of the breakdown of CPT and of Heisenberg's correspondence principle.

Other semiclassical approximations for this Hamiltonian (4.1) have been investigated (31).

Application of the SCCP to the more general problem of collinear vibrational excitation without the additional approximations leading to (4.1) are discussed in (5). Generally similar agreement to that obtained here is found. For this one-dimensional problem the more elaborate classical S-matrix theory has been very successful (32) since it uses exact trajectories.

The success of classical methods for oscillator problems may be related to the existence of the 'Feynman Identity' for systems having a Lagrangian at most quadratic in coordinates and velocities. For a further discussion of this see (5).

5. ROTATIONAL EXCITATION

5.1 Introduction

We now consider the application of the SCCP to rotational excitation of linear molecules by atoms. Excitation by molecules will be discussed in §6.1. For collision energies well below the threshold of vibrational excitation the neglect of vibration is not expected to be serious and a rigid-rotor model of the molecule may be used. The effects of vibration can be included when required, see §6.2 below.

Rotational excitation of moderately heavy systems quickly leads to high quantum numbers and a very large number of coupled channels, for example Li^+-CO collisions involving $j=0\to\simeq50$ transitions (j being the rotor quantum number) have been investigated

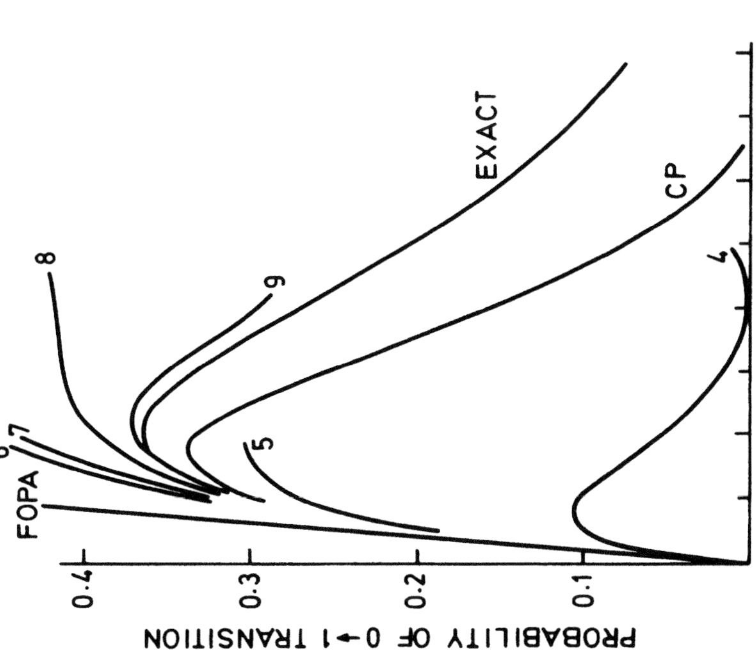

Figure 1. P(0 → 1) as a function of α. EXACT is from (4.2), CP from (4.8), FOPA using (3.18). The curves marked 4,5,...9 result from 4,5,...9 state computer solutions.

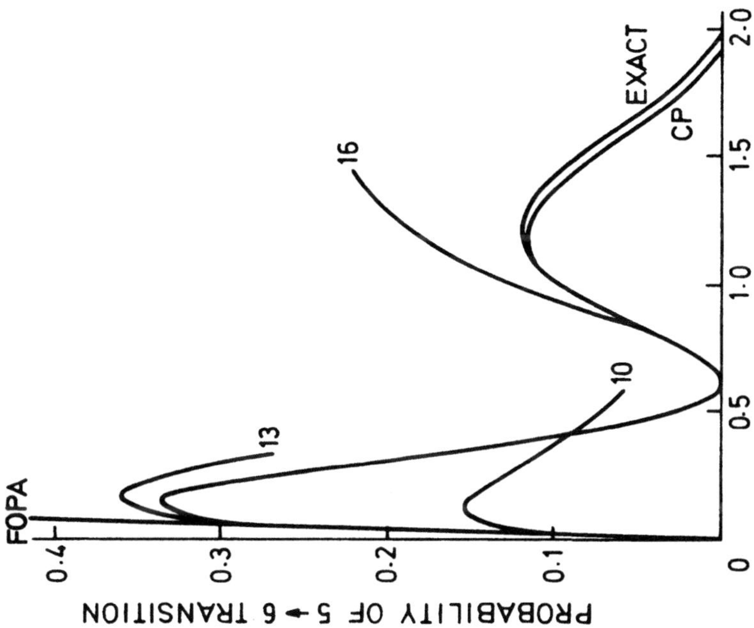

Figure 2. P(5 → 6) as a function of α. Notation as in Figure 1.

(33) for comparison with experiment (34).

As in the quantal close-coupling calculations (35), it is convenient to make a single-centre expansion of the potential about the centre of mass of the molecule:

$$V(R,\Theta) = \sum_{\lambda=0}^{N} F_\lambda(R)\, P_\lambda(\cos\Theta), \tag{5.1}$$

where R is the separation of the atom from the centre of mass of the molecule and Θ is the angle between the molecular axis and the line joining the centre of mass of the molecule to the atom. For homonuclear systems the index λ in eq.(5.1) must be even.

5.2 Classical Perturbation Theory

Before proceeding to use the SCCP we need the CPT approximation to the change in action A, eq.(3.31). To employ CPT we must introduce a trajectory for the relative motion, independent of the target coordinates. We assume that the relative motion takes place in the potential $F_o(R)$. Classical trajectory calculations show that this is generally quite a good approximation for finding the deflection of the atom (36).

To simplify the calculation of the action we choose the orbit of the relative motion to lie in the x-y plane with the x-axis lying along the line of closest approach. Because the rigid rotor is a degenerate system we use action-angle variables defined as in (2.11). The end of the rotor traces out a disc and the angle of rotation, θ_2 in this plane is one of the angle variables: $\theta_2 = \omega^c t + \gamma$. The conjugate action variable is I_j, the magnitude of the angular momentum. The orientation of this disc is specified by the spherical polar angles (β,α) of $\underline{J}$, which is perpendicular to this disc (see figure 3). The angle α is the other, constant, angle variable. It is conjugate to the action $I_m = I_j \cos\beta$. The polar angle, β is not an angle variable. The action variables are quantized:

$$I_j = (j + \tfrac{1}{2})\hbar, \qquad I_m = m\hbar, \tag{5.2}$$

where j and m are the target quantum numbers.

To evaluate A let $(\theta_a;\phi_a)$ and (θ_m,ϕ_m) be the polar coordinates of the atom and the molecular axis respectively. Then $\theta_a = \pi/2$ and ϕ_a is determined from the trajectory in F_o and satisfies $\phi_a(t) = -\phi_a(-t)$, because of our choice of x-axis. The determination of the values of $\theta_m(t),\phi_m(t)$ may be simplified if we note that the rotor axis has polar coordinates $(\theta_m',\phi_m') \equiv (\pi/2,\omega^c t + \gamma)$ in a new frame of reference obtained from the old by the rotation $(\alpha,\beta,0)$ in the notation of Edmonds (37). Then using (3.15), (5.1) and the

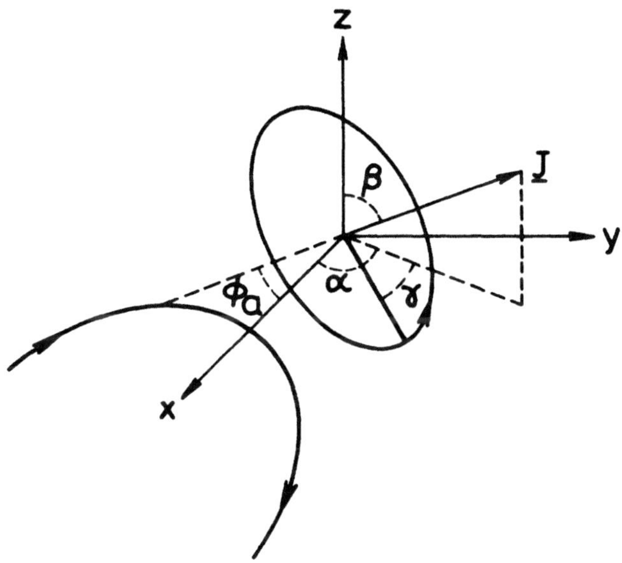

Figure 3. The angles used in the atom-rot... problem.

spherical harmonic addition theorem:

$$A(\alpha,\beta,\gamma;E,b) = -4\pi \sum_{\lambda=1}^{N} (2\lambda+1)^{-1} \int_{-\infty}^{\infty} dt\ F_\lambda[R(t)] \sum_{\mu=-\lambda}^{\lambda} Y^*_{\lambda\mu}(\theta_a,\phi_a)\ Y_{\lambda\mu}(\theta_m,\phi_m) \tag{5.3}$$

$$= -4\pi \sum_{\lambda,\mu,\nu} (2\lambda+1)^{-1}\ Y_{\lambda\mu}(\pi/2,0)\ D^{\lambda}_{\nu\mu}(\alpha,\beta,0) \int_{-\infty}^{\infty} dt\ F_\lambda[R(t)]\ \exp(-i\mu\phi_a)\ Y_{\lambda\nu}(\theta'_m,\phi'_m)$$

$$= -4\pi \sum_{\lambda,\mu,\nu} (2\lambda+1)^{-1}\ N_{\lambda\nu} N_{\lambda\mu} D^{\lambda}_{\nu\mu}(\alpha,\beta,\gamma)\ V^{\lambda}_{\nu\mu}(E,b), \tag{5.4}$$

where $N_{\lambda\mu} = Y_{\lambda\mu}(\pi/2,0)$, $D^{\lambda}_{\nu\mu}$ is the matrix element of finite rotations (37) and

$$V^{\lambda}_{\nu\mu}(E,b) = \int_{-\infty}^{\infty} dt\ F_\lambda(R)\ \exp[i(\nu\omega^c t - \mu\phi_a)]$$

$$= 2\int_{0}^{\infty} dt\ F_\lambda(R)\ \cos(\nu\omega^c t - \mu\phi_a). \tag{5.5}$$

The dependence of A on the action I_j is implicit through the dependence on the frequency ω^c appearing in V^λ. Since $N_{\ell m} = 0$ for $(\ell - m)$ odd, comparatively few of the $V^\lambda_{\gamma\mu}$ integrals are required. A slightly different expression for A is given in (5) but, since all non-vanishing contributions have $(\nu - \mu)$ even, the value of A is identical.

Much of the simplification due to the use of CPT in evaluating A arises because the dependence on the orientation of the rotor, contained in the D functions, has been separated from the dynamical factors in the V^λ integrals. Once these have been calculated for a specified E, b and ω^c the change in action can be obtained trivially for any orientation. This factorization is possible provided a single-centre expansion of the potential, (5.1) is employed.

Using (3.14a) we see that

$$\Delta I_j = \partial A/\partial\gamma; \qquad \Delta I_m = \partial A/\partial\alpha. \tag{5.6}$$

As an example of the use of (5.4) we consider the case of charged particle scattering by a dipolar molecule, where the potential (5.1) may be approximated using $F_\lambda(R) = eD/R^2\delta_{\lambda,1}$. Using the symmetry $V^\lambda_{\nu\mu} = V^\lambda_{-\nu-\mu}$ we have

$$A(\alpha,\beta,\gamma;E,b) = V^1_{1-1}\sin^2\beta/2\,\cos(\alpha-\gamma) - V^1_{1\,1}\cos^2\beta/2\,\cos(\alpha+\gamma), \tag{5.7}$$

where

$$V^1_{1\pm1} = 2eD\int_0^\infty dt\; R^{-2}\left[(b/R)\cos(\omega^c t) \pm (vt/R)\sin(\omega^c t)\right], \qquad R^2 = b^2 + v^2t^2$$

$$= 2\omega^c eD/v^2\;\left[K_1(z) \pm K_0(z)\right], \qquad z = \omega^c b/v. \tag{5.8}$$

Here $K_n(z)$ is the modified Bessel function (28) and v is the speed of relative motion.

Some tests of the accuracy of CPT have been performed (36a) for a potential of the form (5.1) with F_0 and F_2 terms only, these being of the Lennard-Jones (12,6) form. It was found that over a wide range of relevant parameters CPT is accurate to within 10%, tending to be better for large incident energies, large rotational energies and for rotors with large moments of inertia.

5.3 Evaluation of the Scattering Amplitude

Next we consider evaluating the scattering amplitude for $|jm\rangle \rightarrow |j+\Delta j, m+\Delta m\rangle$ transitions. Using (5.4) in (3.30) we have

$$a(jm \to j+\Delta j, m+\Delta m; E,b) = (4\pi^2)^{-1} \int_0^{2\pi}\int_0^{2\pi} d\alpha d\gamma \; \exp\{-i[\Delta j\gamma + \Delta m\alpha - A(\alpha,\beta,\gamma;E,b)/\hbar]\} \quad (5.9)$$

The angle β is determined from some mean of the initial and final polar angles of j:

$$\cos\beta = [m/(j+\tfrac{1}{2}) + (m+\Delta m)/(j+\Delta j+\tfrac{1}{2})]/2. \quad (5.10)$$

As an example we consider the charge-dipole collision examined in the previous section. Introducing new variables $\theta_\pm = \alpha \pm \gamma$ in (5.9) the double integral is separable yielding

$$a(\Delta j,\Delta m) = J_{k+}[(V^1_{1\,1}/\hbar)\cos^2\beta/2]\; J_{k-}[(V^1_{1-1}/\hbar)\sin^2\beta/2]$$
$$k_\pm = (\Delta j \pm \Delta m)/2, \quad (5.11)$$

provided $k_\pm$ is integral. For half-integral $k_\pm$, $a=0$. The origin of this selection rule is clarified if we note that, with our choice of coordinates, classically $\Delta\ell = -\Delta j_z$, where $\Delta\ell$ is the change in the angular momentum of relative motion. Thus, on quantizing, $(\Delta j \pm \Delta\ell)$ must be even, which is the parity condition in the close-coupling solution (35). This result (5.11) has been employed in an extensive Classical S-Matrix study of electron-polar molecule collisions (38).

Again we note that the Bessel functions in (5.11) automatically provide a uniform approximation valid both in the classically forbidden, perturbation, limit where the trajectory integrals, V are small and in the classically allowed, strong-coupling, region where the V integrals are large compared to $k_\pm\hbar$.

Generally we are more interested in the degeneracy-averaged transition probability, $\mathcal{P}(j \to j+\Delta j)$. In this instance the quantal sum over degenerate states is approximated by an integral over a microcanonical ensemble of the target. We require

$$\mathcal{P}(j \to j+\Delta j) = (2j+1)^{-1} \sum_m \sum_{\Delta m} |a(jm \to j+\Delta j,\; m+\Delta m)|^2. \quad (5.12)$$

Using the relations

$$\sum_{\Delta m=-\infty}^{\infty} \exp(i\Delta m\theta) = 2\pi\delta(\theta), \qquad \sum_{m=-j}^{j} \simeq (j+\tfrac{1}{2})\int_{-1}^{1} d(\cos\beta) \quad (5.13)$$

we obtain

$$\mathcal{P}(j \to j+\Delta j) = (4\pi)^{-1}\int_0^{2\pi} d\alpha \int_0^{\pi} \sin\beta d\beta\; |\bar{a}(j \to j+\Delta j;\alpha,\beta)|^2, \quad (5.14)$$

$$\bar{a}(j \to j + \Delta j; \alpha, \beta) = (2\pi)^{-1} \int_0^{2\pi} d\gamma \; \exp[-i(\Delta j \gamma - A/\hbar)] \; . \qquad (5.15)$$

Physically, we average the transition probability for each initial rotor orientation over all orientations.

For a P_1 anisotropy only, $\bar{a}$ again involves a Bessel function of integer order. For a P_2 anisotropy only the change in action, A also takes a simple form. Neglecting the term independent of γ, which cannot produce $\Delta j \neq 0$ transitions,

$$A = R_2 \cos(2\gamma + \psi_2), \qquad (5.16)$$

where R_2 and ψ_2 depend on α and β (39). Then, because A depends on (2γ), only transitions with even Δj have non-zero amplitudes. For Δj odd complete destructive interference occurs between phases γ and $(\pi + \gamma)$. For Δj even

$$|a(j \to j + \Delta j)| = |J_{\Delta j/2}(R_2)|. \qquad (5.17)$$

By noting that the term $N_{\lambda\nu}$ in (5.4) causes only terms in (2γ) to appear in A for λ even, we can easily extend this result to get the quantum mechanical selection rule Δj even for homonuclear systems (35).

To satisfy the detailed-balance condition (3.9) we modify the probability given by (5.14):

$$P(j \to j'; E, b) = (E'/E)^{\frac{1}{2}} \; [(2j'+1)/(2j+1)]^{\frac{1}{2}} \, \mathcal{P}(j \to j'; \bar{E}, b), \qquad (5.18)$$

where E and E' are the initial and final translational energies respectively and $\bar{E}$ is a mean energy, based on the average speed in our work. This correction is required because we have employed CPT but the precise form is arbitrary. For the average rotational frequency, needed in the trajectory integrals, we employ, for a rotor of moment of inertia I,

$$\omega^c = (j + j' + 1)\hbar / 2I. \qquad (5.19)$$

With this choice, quantal energy differences are given exactly using (2.2). Then the cross section

$$\sigma(j \to j'; E) = 2\pi \int_0^\infty b\,db \; P(j \to j'; E, b) \qquad (5.20)$$

satisfies the exact detailed-balance requirement:

$$E(2j+1)\,\sigma(j \to j'; E) = E'(2j'+1)\,\sigma(j' \to j; E'). \qquad (5.21)$$

The equations (5.20, 18, 14, 15, 4, 5 and 19) constitute the

Transition	CC[a]	SCCP[b]	CS[c]	BFCP[d]
		Method		
0→2	(15.7)[e]	18.1	15.2	17.6
2→4	14.4	10.8	14.8	10.9
4→6	12.8	9.6	13.0	9.8
6→8	11.0	9.1	10.7	9.1
8→10	10.8	8.7	9.5	8.7
10→12	9.8	8.6	8.2	8.5
0→4	11.4	13.9	11.9	14.1
2→6	10.5	7.3	11.0	7.5
4→8	7.9	6.0	7.7	6.0
6→10	5.9	5.2	5.5	5.2
8→12	(2.08)	4.0	3.11	3.93
0→6	10.8	10.8	11.2	11.2
2→8	6.6	5.1	6.6	5.3
4→10	3.5	3.6	3.3	3.6
6→12	(0.81)	2.02	1.26	1.96
0→8	5.7	6.5	5.8	6.8
2→10	1.86	2.39	1.87	2.41
4→12	(0.32)	0.96	0.48	0.93
0→10	1.15	2.07	1.17	2.11

a from (42)
b from (41)
c from (44)
d from (40)
e Close-coupling results in parentheses have uncertainties exceeding 4%.

Table I

Comparison of Close-Coupling (CC) and Coupled-States (CS) results with Strong-Coupling Correspondence Principle (SCCP) and Body-Fixed Correspondence Principle (BFCP) results for Ar-N_2 collisions. All cross sections are in $Å^2$, and the initial translational energy is 450 k_B when the molecule is in j=10 level (k_B is Boltzmann's Constant).

working equations for the SCCP evaluation of rotation excitation cross sections. The numerical methods that have been employed are discussed in (39). In contrast to quantal close-coupling methods the computing time is largely independent of the magnitude of the quantum numbers involved, though it does increase slowly with the strength of the collision.

As we have seen the SCCP is expected to be most successful

for large quantum numbers but comparisons with close-coupling and other quantal methods have shown satisfactory results for $\sigma(j \to j')$ in Ar-TℓF (40), Li-N_2, and Li-CO collisions with $j,j' \leq 4$, Ne-N_2 collisions with $j,j' \leq 8$ and Ar-N_2 collisions with $j,j' \leq 12$ (41). We compare in Table I some SCCP results for Ar-N_2 collisions with close-coupling calculations (42). For the $\Delta j = 2$ transition the SCCP error is about 25% for low initial levels j and decreases surprisingly non-uniformly as j increases. For $\Delta j = 4$ the errors are somewhat larger but there is clear evidence of improvement as j increases. For $\Delta j = 6$ or 8 the errors are only slightly larger than for $\Delta j = 2$ or 4 and it is only for $\Delta j = 10$ that the SCCP results are out by a factor of two.

5.4 Decoupling Approximations

There has been considerable interest in the past few years in non-perturbative approximations for atom-molecule collisions which introduce an extra symmetry into the system thus reducing the dimensionality of the problem (15). In one widely used approximation, the J_z Coupled-States (CS) method (43), an approximate Hamiltonian is employed for which j_z is conserved. In the SCCP analogue of this result (40), termed the Body-Fixed (BF) approximation, we neglect the transverse motion of the atom and set $\phi_a(t) = 0$, its value at the mid-point of the collision. Then $V^\lambda_{\nu\mu} = V^\lambda_{\nu o}$ and the orientation dependence of A can be simplified. While still neglecting the transverse motion of the atom, a more direct procedure is to use alternative, primed, coordinates with z' axis along the line of closest approach. Then the approximate change in action, A^{BF} is dependent only on β' and γ' so $\Delta I_m = 0$ and the average (5.14) to obtain transition probabilities involves only a one-dimensional integral.

Some results obtained using the CS method (44) and the BF method for Ar-N_2 collisions are shown in Table I. Clearly the effect of the additional Coupled States or Body-Fixed approximations on the corresponding full results is about the same.

Classical perturbation theory can be employed to investigate the magnitude of the terms omitted in the BF approximation and these are relatively small for close collisions but can be appreciable for distant collisions, especially for slowly decreasing potentials.

An alternative approach using these coordinates, the M-Conserving approximation (45), retains the terms of the full CPT action which are independent of α'. This involves only replacing $V^\lambda_{\nu o}$ by

$$\bar{V}^\lambda_\nu = \int_{-\infty}^{\infty} F_\lambda(R)\, P_\lambda[\cos \phi_a(t)] \cos(\nu\omega^c t)\, dt , \qquad (5.22)$$

which now makes some allowance for the transverse motion of the atom. This generally gives an improvement over the BF approximation, particularly for distant collisions.

6. SIMULTANEOUS ROTATIONAL-VIBRATIONAL EXCITATION AND ROTOR-ROTOR COLLISIONS

6.1 Introduction

Having discussed relatively fully the application of the SCCP to one- and two-dimensional targets, we conclude by examining briefly its application to these more complex systems. No new techniques are required but an additional integration is required for each new degree of freedom. Because of the large number of coupled channels only a few converged close-coupling calculations are available for these collisions (15).

6.2 Atom-Vibrating Rotor Collisions

A full treatment of atom-diatomic molecule collisions requires that we allow for both vibrational and rotational excitation of the molecule. Because of the high density of rotational levels compared to vibrational levels very many states are coupled at energies for which vibrational excitation is possible. Virtually all quantal studies have been for H_2 and it has been suggested (46) that only for Li^+-H_2 (47) and He-H_2 (46) are converged quantal results available.

As for rigid rotors, §5.1, we use a single-centre expansion of the potential

$$V(R,r,\Theta) = \sum_{\lambda=0}^{N} F_\lambda(R,r)\, P_\lambda(\cos\Theta), \tag{6.1}$$

where r is the separation of the atoms forming the diatomic. Since the vibrational and rotational motion of the diatomic are separable the action-angle variables describing the unperturbed target are $(I_v, I_j, I_m;\ \theta_3, \theta_2, \alpha)$. The dependence of the vibrational energy on I_v and of r on θ_3 will, of course, depend on the precise vibrational potential employed. To evaluate A, eq.(3.31) it is convenient first to make a Fourier expansion of F_λ, since it is a periodic function of θ_3.

$$F_\lambda(R,r) = \sum_n C_{n\lambda}(R)\, \exp(i\, n\, \theta_3) \tag{6.2}$$

and using the same notation as §5.2, and setting $\theta_3 = \omega^v t + w$, where ω^v is the classical angular frequency of vibration, we find

$$A(\alpha,\beta,\gamma,w;E,b) = -4\pi \sum_{\lambda,n,\mu,\nu} (2\lambda+1)^{-1} N_{\lambda\nu} N_{\lambda\mu} D^{\lambda}_{\nu\mu}(\alpha,\beta,\gamma) \exp(inw)\, V^{\lambda n}_{\nu\mu}, \tag{6.3}$$

$$V^{\lambda n}_{\nu\mu}(E,b) = 2 \int_0^\infty dt\, C_{n\lambda}(R) \cos(\nu\omega^c t + n\omega^v t - \mu\phi_a). \tag{6.4}$$

The relative motion is calculated in the potential $F_0(R,r_{eq})$, where r_{eq} is the equilibrium separation in the target. For the degeneracy-averaged cross section

$$\mathcal{P}(v\to v+\Delta v,\ j\to j+\Delta j) = (4\pi)^{-1} \int_0^{2\pi} d\alpha \int_0^{\pi} \sin\beta d\beta\, |\bar{a}(v\to v+\Delta v, j\to j+\Delta j;\alpha,\beta)|^2 \tag{6.5}$$

$$\bar{a}(v\to v+\Delta v,\ j\to j+\Delta j;\alpha,\beta) = (4\pi^2)^{-1} \int_0^{2\pi} dw \int_0^{2\pi} d\gamma\, \exp[-i(\Delta v w + \Delta j\gamma - A/\hbar)] \tag{6.6}$$

Some results have been obtained using this method for He-H_2 collisions (48) employing the Gordon-Secrest potential energy surface (49). Vibrationally elastic but rotationally inelastic results were generally in fairly good agreement with the close-coupling calculations. Vibrational excitation cross sections were poorer, though improving significantly as the collision energy increased. Since the energy was close to threshold for vibrational excitation these conditions were unfavourable for any time-dependent method, but at higher energies the SCCP should be more reliable. Typical SCCP computing times were about 15 seconds per cross section on an IBM 370/168.

6.3 Rotor-Rotor Collisions

Finally we examine the application of the SCCP to collisions between two rigid linear molecules. We use suffices 1 and 2 to denote properties of the two molecules and denote the polar coordinates of the rotor axes by $\Omega_i \equiv (\theta_i,\phi_i)$, $i=1,2$. Otherwise we follow the notation of §5.2. As previously, the pattern of the SCCP calculation is set by the expansion of the potential:

$$V(R;\theta_a,\phi_a,\Omega_1,\Omega_2) = \sum_{\lambda_1\lambda_2\lambda} F(\lambda_1,\lambda_2,\lambda;R) \sum_{\mu_1\mu_2\mu} (-1)^{\lambda_1-\lambda_2-\mu} \begin{pmatrix}\lambda_1 & \lambda_2 & \lambda\\ \mu_1 & \mu_2 & \mu\end{pmatrix} Y_{\lambda_1\mu_1}(\Omega_1)\, Y_{\lambda_2\mu_2}(\Omega_2)\, Y^*_{\lambda\mu}(\theta_a,\phi_a), \tag{6.7}$$

where R is now the separation of the centres of mass of the two molecules. The calculation of the action A is straightforward:

$$A = -\sum (-1)^{\lambda_1-\lambda_2-\mu} \begin{pmatrix}\lambda_1 & \lambda_2 & \lambda \\ \mu_1 & \mu_2 & \mu\end{pmatrix} N_{\lambda\mu} N_{\lambda_1\nu_1} N_{\lambda_2\nu_2}$$

$$D^{\lambda_1}_{\nu_1\mu_1}(\alpha_1,\beta_1,\gamma_1)\, D^{\lambda_2}_{\nu_2\mu_2}(\alpha_2,\beta_2,\gamma_2)\, V(\lambda_1,\lambda_2,\lambda;\nu_1,\nu_2,\mu_1,\mu_2), \tag{6.8}$$

$$V(\lambda_1,\lambda_2,\lambda;\nu_1,\nu_2,\mu_1,\mu_2) = 2\int_0^\infty F[\lambda_1,\lambda_2,\lambda;R(t)]$$
$$\cos[\nu_1\omega_1 t + \nu_2\omega_2 t - (\mu_1 + \mu_2)\phi_a(t)] \ . \tag{6.9}$$

The calculation of the degeneracy-averaged transition probability now requires a quadruple integral, averaging $|\bar{a}|^2$ over rotor orientations $\alpha_1,\alpha_2,\beta_1,\beta_2$, where

$$\bar{a}(j_1 \to j_1 + \Delta j_1;\, j_2 \to j_2 + \Delta j_2) = (4\pi^2)^{-1}\int_0^{2\pi} d\gamma_1 \int_0^{2\pi} d\gamma_2$$
$$\exp[-i(\Delta j_1\gamma_1 + \Delta j_2\gamma_2 - A/\hbar)] \ . \tag{6.10}$$

The relative motion is calculated using the spherically symmetric component of the potential (6.5).

This method has been applied to HF-HF collisions (50), using only a modified dipole-dipole anisotropic component in (6.5), to compare with close-coupling calculations (51). Similarly to the ion-dipole example, §5.3, the introduction of coordinates $(\gamma_1 \pm \gamma_2)$ makes the integral in eq.(6.10) separable:

$$\bar{a}(j_1 \to j_1 + \Delta j_1;\, j_2 \to j_2 + \Delta j_2) = \exp(i\delta)\, J_{n_+}(R_+) J_{n_-}(R_-),$$
$$n_\pm = (\Delta j_1 \pm \Delta j_2)/2 \ , \tag{6.11}$$

where δ and the $R_\pm$ are functions of the rotor orientations and the $n_\pm$ must be integral. This selection rule, both Δj even or both odd also holds in the quantum solution. Satisfactory results for some transitions were obtained but the weaker transitions at low energies were overestimated. The convergence of the close-coupling results here was not as well established as in typical atom-rigid rotor studies.

6. CONCLUSIONS

We have shown how the SCCP approximation may be employed to study rotational and vibrational excitation of diatomic molecules. The SCCP is relatively simple to calculate while satisfying detailed-balance and being close to unitary. Perturbation, sudden and decoupling approximations may readily be incorporated. Reliance on classical perturbation theory, which generally ensures that a reasonable approximation to the dynamics is being employed, yields a robust method which is a seldom grossly in error. The computational effort needed for each additional degree of freedom grows moderately and certainly more slowly than in close-coupling methods. Since few molecular potential energy hypersurfaces are known to high accuracy the SCCP can often provide all the accuracy that is warranted. It is particularly appropriate for high quantum numbers and we are currently applying it (52) to rotational excitation in $A + Li_2(A^1\Sigma_u^+)$ collisions, with $j,j' \simeq 20$, where A is He, Ar or Xe, for comparison with collision-induced fluorescence experiments (53).

For certain special physical systems Correspondence Identities exist, that is quantum observables can be obtained exactly for all quantum numbers in terms of classical quantities (54). In general, the use of the SCCP, or Classical S-Matrix Theory, for low quantum numbers, while often successful, has not been theoretically justified and this remains a challenge.

Acknowledgements

Dr. D. Richards (Open University) has shared in the development of correspondence principle methods for molecular collisions and he has made constructive criticisms of these lectures.

PROBLEMS

1. Matrix elements associated with a Morse potential are typical of those arising in the vibrational excitation of anharmonic oscillators. We take the reduced mass to be μ and write the potential as

$$V(q) = D\{1 - \exp[-a(q - q_e)]\}^2,$$

where D, a and q_e are constants.

(a) Using (2.1) show that the relation between the energy and the action is

$$E/D = I(2 - I/\hbar t)/\hbar t, \quad t = (2\mu D)^{\frac{1}{2}}/a\hbar.$$

Hence, using (2.9), show $\omega^c(I) = 2D(1 - I/\hbar t)/\hbar t$.

(b) Prove that

$$q(I,\theta) = q_e + a^{-1} \ln\left[(1 - \sqrt{E/D} \sin\theta)/(1 - E/D)\right],$$

where E(I) is obtained from part (a) and the phase has been chosen so that $\theta = 0$ when $q = q_o = q_e - a^{-1} \ln(1 - E/D)$.

(c) Using (2.14) (or (2.3)) and the result of part (b) show

$$\langle n + s|q|n\rangle \simeq \left[\bar{I}/(2\hbar t - \bar{I})\right]^{s/2} / as, \quad s \neq 0.$$

(Integrate by parts and then convert to a contour integral round the unit circle using $z = \exp(i\theta)$.)

(d) For $t = 50$ compare the values of $a^2|\langle n+1|q|n\rangle|^2$ obtained from (c) for $n = 1$, 7 and 15 with the quantal values 0.020406; 0.086944 and 0.19044 respectively, using $\bar{I} = (n+1)/\hbar$ (the arithmetic mean of the initial and final actions, using $\varepsilon = \frac{1}{2}$ in (2.1)).

Compare also $a^2|\langle 10|q|5\rangle|^2$ with the exact value, 1.83×10^{-7}, using both the arithmetic mean action and that obtained from

$$\bar{I} = \hbar\left[(n+s)!/n!\right]^{1/s}.$$

For further matrix elements of the Morse potential, see ref. (30).

2. As an example of classical perturbation theory we examine the small-angle scattering of a fast, charged, particle by a polar molecule. We assume that the long-range charge-dipole potential is dominant and that the sudden approximation is valid, i.e. the parameter z in eq. (5.8) $\simeq 0$. Show that in this limit the action A of (5.7) is given by

$$A = 2eD\,(\sin\alpha\sin\gamma - \cos\alpha\cos\beta\cos\gamma)/bv \qquad (1)$$

(Note: $\lim z \to 0$ $K_0(z) = -\ln z$, $K_1(z) = 1/z$)

For sudden collisions it is more convenient to work with the coordinates $\hat{\underline{r}} = (\theta,\phi)$ of the rotor axis in the space-fixed frame. Verify that (1) can be rewritten

$$A = -2eD\,\hat{\underline{r}}\cdot\hat{\underline{b}}/bv. \qquad (2)$$

Taking the straight-line path $\underline{R} = b_x\hat{\underline{i}} + b_y\hat{\underline{j}} + vt\,\hat{\underline{k}}$, show that for any anisotropic potential (not just the dipole case considered here)

$$\partial A/\partial b_x = \Delta P_x,$$

where ΔP_x is the x-component of the change of momentum of the incident particle. Using this, and (2), prove that the polar and azimuthal angles of deflection, Ψ and $\chi(\hat{\underline{b}},\hat{\underline{r}})$, respectively are given by

$$\Psi = eD\sin\theta/Eb^2, \qquad \chi = 2\psi - \phi + \pi \qquad (3)$$

where ψ is the azimuthal angle of the impact-parameter vector. (Remember $\Psi = (\Delta P_x{}^2 + \Delta P_y{}^2)^{\frac{1}{2}}/P$; $\chi = \tan^{-1}(\Delta P_y/\Delta P_x)$.)

Now we may calculate the differential cross section $\frac{d\sigma}{d\Omega}(\Theta,\Phi;\hat{\underline{r}})$ for each rotor orientation using

$$\sin\Theta\,\frac{d\sigma}{d\Omega}(\Theta,\Phi;\hat{\underline{r}}) = \int_0^\infty\int_0^{2\pi} b\,db\,d\psi\;\delta(\Theta-\Psi)\,\delta(\Phi-\chi). \qquad (4)$$

Use (3) to verify that the RHS of (4) is given by $\int_0^\infty b\,db\;\delta(\Theta-\Psi)$.

Finally prove that the differential cross section, averaged over all rotor orientations, $\frac{d\bar{\sigma}}{d\Omega}$, is given by

$$\frac{d\bar{\sigma}}{d\Omega} = \frac{\pi e D}{8E\Theta^3}\,.$$

This approach has been employed for both ion (55) and electron (56) collisions with polar molecules. The electron work discusses effects due to quantal interference. The Glauber approximation employs the same first-order classical action but treats the target quantum mechanically.

3. The first-order SCCP approximation to the degeneracy-averaged transition probability for a $j\to j+\Delta j$ transition can be calculated using the small A limit of (5.15) in (5.14). Using (5.4), show that

$$\mathcal{P}(j\to j+\Delta j) = 16\pi^2\sum_{\lambda\mu}(2\lambda+1)^{-3}\,N^2_{\lambda\Delta j}\,N^2_{\lambda\mu}\,(V^\lambda_{\Delta j\mu}/\hbar)^2\,.$$

$$\left(\int_0^{2\pi} d\alpha\int_0^{\pi}\sin\beta\,d\beta\; D^j_{mn}(\alpha,\beta,0)\,D^{j'*}_{mn'}(\alpha,\beta,0)\right.$$

$$\left. = 4\pi\,(2j+1)^{-1}\,\delta_{jj'}\,\delta_{nn'}\right)$$

Compare this result with that obtained using quantal first-order time-dependent perturbation theory. First use (3.18), proceeding as in (5.3) and use (5.5) to approximate $a(jm\to j+\Delta j\; m+\Delta m)$:

$$a(jm\to j+\Delta j\; m+\Delta m) = i\sqrt{4\pi}\,\left[(2j+1)(2j+2\Delta j+1)\right]^{\frac{1}{2}}$$
$$\sum_\lambda (2\lambda+1)^{-\frac{1}{2}}\begin{pmatrix} j & j' & \lambda\\ 0 & 0 & 0\end{pmatrix}^2\begin{pmatrix} j & j' & \lambda\\ m & -m-\Delta m & \Delta m\end{pmatrix} N_{\lambda\Delta m}\,V^\lambda_{\Delta j\Delta m}/\hbar\,.$$

$$\left(\iiint Y^*_{j_1m_1}(\Omega)\,Y_{j_2m_2}(\Omega)\,Y_{j_3m_3}(\Omega)\,d\Omega = \right.$$

$$(-1)^{m_1}\left[(2j_1+1)(2j_2+1)(2j_3+1)/4\pi\right]^{\frac{1}{2}}$$
$$\begin{pmatrix} j_1 & j_2 & j_3 \\ 0 & 0 & 0 \end{pmatrix}\begin{pmatrix} j_1 & j_2 & j_3 \\ -m_1 & m_2 & m_3 \end{pmatrix} ;$$

Remember from (5.19), $E_{j+\Delta j} - E_j = \Delta j\, \omega^c \hbar)$. Now use (5.12) to verify:

$$P^{FO}(j \to j+\Delta j) = 4\pi(2j+2\Delta j+1)$$
$$\sum_{\lambda\mu} (2\lambda+1)^{-2} \begin{pmatrix} j & j+\Delta j & \lambda \\ 0 & 0 & 0 \end{pmatrix}^2 N^2_{\lambda\mu}(V_{\Delta j\mu}/\hbar)^2 .$$

The connection between these two results may be established using the large j limit of a 3-j symbol:

$$(2j'+1)\begin{pmatrix} j & j' & \lambda \\ 0 & 0 & 0 \end{pmatrix}^2 = 4\pi(2\lambda+1)^{-1}\, N^2_{\lambda\Delta j}$$
$$\left[1 + \Delta j/(2j+1) + O(2j+1)^{-2}\right], \quad \Delta j = j' - j.$$

Including the j-dependent part of the detailed-balance correction (5.18), find the ratio of the first-order SCCP and quantal results for a $0 \to 1$ transition due to a P_1 potential and a $0 \to 2$ transition due to a P_2 potential.

$$\left(N^2_{11} = 3/8\pi \; ; \quad N^2_{22} = 15/32\pi \; ; \quad \begin{pmatrix} 0 & j & j \\ 0 & 0 & 0 \end{pmatrix}^2 = (2j+1)^{-1}\right)$$

4. Fairly convenient comparisons between SCCP and quantal results can be made for transition probabilities in sudden collisions when only a P_1 anisotropy is present. Show that A in (5.7) can be written:

$$A = C\, V \cos(\gamma + \delta)$$

where $V \equiv V^1_{01}$, $C = (1 - \cos^2\alpha \sin^2\beta)^{\frac{1}{2}}$ and δ is a phase depending only on α and β. Now use (5.15) to verify that

$$|\bar{a}(j \to j+\Delta j)|^2 = J^2_{\Delta j}(CV).$$

To do the average over α and β in (5.14) it is convenient to choose new coordinates (θ,ϕ) for the rotor axis relative to Ox as polar axis, i.e. the new z' axis points along the line of closest approach. Proceeding as in Q.2, show $\cos\theta = \cos\alpha \sin\beta$ and hence prove

$$\mathcal{P}(j \rightarrow j + \Delta j) = \int_0^{\pi/2} \sin\theta\, d\theta\, J^2_{\Delta j}(V \sin\theta). \qquad (1)$$

No closed-form expression for this integral is known, though it may be expressed in terms of the sum of an infinite series of Bessel functions of argument (2V) (see ref. 39, eq. (4.20)).

The quantal result for the $0 \rightarrow j$ probability is particularly simple. Following Q.2, show that, for curved orbits,

$$A = V\, \hat{\underline{r}} \,.\, \hat{\underline{x}} \equiv V \cos\theta$$

in the new coordinates. Now use (3.19) and the standard expansion of a plane wave in Legendre polynomials to derive

$$\begin{aligned} a^S(00 \rightarrow j0) &= i^j (2j+1)^{\frac{1}{2}}\, j_j(V), \\ P^S(0 \rightarrow j) &= (2j+1)\, j_j^2(V), \end{aligned} \qquad (2)$$

where j is a spherical Bessel function.

Compare the results from (1) and (2) for the $0 \rightarrow 1$ transition to 4th order in V, including the detailed-balance correction (5.18).

$$(j_1(x) = x/3\ (1 - x^2/10 + \dots);$$

$$J_1(x) = x/2\ (1 - x^2/8 + \dots)\ )$$

It is interesting to compare the SCCP result and the quantal result for larger values of V and watch the SCCP result get steadily out of phase with the quantal result for $V \gtrsim 2$. This is because the SCCP for matrix elements is being employed when the quantity involved is changing more rapidly than the wavefunctions.

The new coordinates introduced here are those employed in the Body-Fixed and M-Conserving approximations (see §5.4).

REFERENCES

(1) Gryzinski, M.: 1957, Phys. Rev. 107, pp. 1471-1475.
(2) Bates, D.R.: 1978, Phys. Reports 35, pp. 305-372.
(3) Percival, I.C.: 1978, Electronic and Atomic Collisions, Invited Papers, 10th ICPEAC, ed. G. Watel, pp. 569-578, (North Holland, Amsterdam).
(4) Percival, I.C. and Richards, D.: 1975, Adv. Atom. Molec. Phys., 11, pp. 1-82.
(5) Clark, A.P., Dickinson, A.S. and Richards, D.: 1977, Adv. Chem. Phys. 36, pp. 63-139.
(6) Miller, W.H.: 1974, Adv. Chem. Phys. 25, pp. 69-177.
(7) Miller, W.H.: 1975, Adv. Chem. Phys. 30, pp. 77-136.
(8) Fröman, N.: 1979, this volume.
(9) Percival, I.C.: 1977, Adv. Chem. Phys. 36, pp. 1-61.
(10) Handy, N.C.: 1979, this volume.
(11) Goldstein, H.: 1959, Classical Mechanics, ch. 9, (Addison-Wesley, New York).
(12) Born, M.: 1927, Mechanics of the Atom, ch. 2 (Bell, London).
(13) Landau, L.D. and Lifshitz, E.M.: 1958, Quantum Mechanics, pp. 432-435.
(14) Burhop, E.H.S.: 1961, Quantum Theory I, ch. 9, ed. D.R. Bates (Academic, New York).
(15) Dickinson A.S.: 1979, Computer Physics Communications, 17, pp. 51-80.
(16) Child, M.S.: 1974, Molecular Collision Theory, ch. 7 (Academic, London).
(17) Percival, I.C.: 1974, Computer Physics Communications, 6, pp. 347-357.
(18) Balint-Kurti, G.G.: 1975, Int. Rev. Sci., Physical Chemistry, Series 2, Vol. 1, pp. 285-326.
(19) Presynakov, L.P. and Urnov, A.M.: 1970, J. Phys. B, 3, pp. 1267-1271.
(20) Percival, I.C. and Richards, D.: 1970, J. Phys. B, 3, pp. 1035-1045.
(21) Richards, D.: 1972, J. Phys. B 5, pp. L53-55.
(22) Child, M.S.: 1979, this volume.
(23) Child, M.S.: 1976, Dynamics of Molecular Collisions, Part B, ed. W.H. Miller, ch. 4 (Plenum, New York).
(24) Miller, W.H. and Smith, F.T.: 1978, Phys. Rev. A17, pp. 939-953.
(25a) Secrest, D. and Johnson, B.R.: 1966, J. Chem. Phys. 45, pp. 4556-4570.
(25b) Clark, A.P. and Dickinson, A.S.: 1973, J. Phys. B 6, pp. 164-180.
(26) Parker, G.A. and Pack, R.T.: 1978, J. Chem. Phys. 68, pp. 1585-1601.
(27) Kerner, E.H.: 1958, Can. J. Phys. 36, pp. 371-377.
(28) Abramowitz, M. and Stegun, I.A.: 1965, Handbook of Mathematical Functions (Dover, New York).

(29) Clark, A.P. and Dickinson, A.S.: 1971, J. Phys. B 4, pp. L112-116.
(30) Naccache, P.F.: 1972, J. Phys. B 5, pp. 1308-1319.
(31) Pechukas, P. and Child, M.S.: 1976, Mol. Phys. 31, pp. 973-987.
(32) Child, M.S. and Hunt, P.M.: 1977, Mol. Phys. 34, pp. 261-272.
(33) Thomas, L.D., Kraemer, W.P., Diercksen, G.H.F. and McGuire, P.: 1978, Chem. Phys. 27, pp. 237-243.
(34) Böttner, R., Ross, U. and Toennies, J.P.: 1976, J. Chem. Phys. 65, pp. 733-746.
(35) Arthurs, A.M. and Dalgarno, A.: 1960, Proc. Roy. Soc. A256, pp. 540-551.
(36) Kistemaker, P.G. and de Vries, A.E.: 1975, Chem. Phys. 7, pp. 371-382.
(36a) Cohen, A.O. and Marcus, R.A.: 1970, J. Chem. Phys. 52, pp. 3140-51.
(37) Edmonds, A.R.: 1974, Angular Momentum in Quantum Mechanics, (Princeton University Press, Princeton).
(38) Mukherjee, D. and Smith, F.T.: 1978, Phys. Rev. A17, pp. 954-967.
(39) Dickinson, A.S. and Richards, D.: 1974, J. Phys. B 7, pp. 1916-1936.
(40) Dickinson, A.S. and Richards, D.: 1977, J. Phys. B 10, pp. 323-343.
(41) Dickinson, A.S. and Richards, D.: 1976, J. Phys. B 9, pp. 515-533.
(42) Pack, R.T.: 1975, J. Chem. Phys. 62, pp. 3143-3148.
(43) McGuire, P. and Kouri, D.J.: 1974, J. Chem. Phys. 60, pp. 2488-99.
(44) McGuire, P.: 1976, Chem. Phys. 13, pp. 81-94.
(45) Dickinson, A.S. and Richards, D.: 1978, J. Phys. B 11, pp. 3513-3528.
(46) Lin, C.S. and Secrest D.: 1979, J. Chem. Phys. 70, pp. 199-209.
(47) Schaefer, J. and Lester, W.A.: 1975, J. Chem. Phys. 62, pp. 1913-1924.
(48) Clark, A.P.: 1977, J. Phys. B 10, pp. L389-394.
(49) Gordon, M.D. and Secrest, D.: 1970, J. Chem. Phys. 52, pp. 120-131.
(50) Dickinson, A.S. and Munoz, J.: 1979, work in progress.
(51) De Pristo, A.E. and Alexander, M.H.: 1977, J. Chem. Phys. 66, pp. 1334-1342.
(52) Bhattacharyya, S.S., Dickinson, A.S. and Richards, D.: 1979, to be submitted to J. Phys. B.
(53) Schröder, M.: 1978, Bericht 9/1978, Max-Planck-Institut für Strömungsforschung, Göttingen, W. Germany.
(54) Norcliffe, A.: 1973, Case Studies in Atomic Physics, 4, pp. 1-55.
(55) Gentry, W.R.: 1974, J. Chem. Phys. 60, pp. 2547-2553.
(56) Dickinson, A.S.: 1977, J. Phys. B 10, pp. 967-981.

THE DETERMINATION OF BOUND STATES BY SEMICLASSICAL METHODS

N. C. Handy

University Chemical Laboratory, Lensfield Road, Cambridge

ABSTRACT

The form of classical trajectories for many dimensional bound state problems is discussed. The distinction is made between regular and irregular trajectories.

There are several semiclassical methods in use for the determination of low lying bound quantal states of small systems. Each is reviewed. Results are discussed and compared with quantum mechanical results. No method yet exists for the determination of semiclassical eigenvalues which lie in the irregular region.

The implications of classical irregular motions on quantum mechanical wavefunctions is discussed.

It is shown how these methods may be extended to potentials with many minima, in particular showing how tunneling effects may be included.

I. THE NATURE OF CLASSICAL TRAJECTORIES

All workers have based their investigations on Hamiltonians of the form

$$H = \frac{1}{2}\sum_i (p_i^2/m_i + m_i\omega_i^2 x_i^2) + \varepsilon V(\underline{x}) \tag{1.1}$$

where $\varepsilon<1$, usually indicating a small perturbation from the separable Hamiltonian. In particular the Henon-Heiles (1) and the Barbanis (2) Hamiltonians have been used:

M. S. Child (ed.), Semiclassical Methods in Molecular Scattering and Spectroscopy, 297-321.

Henon-Heiles: $H = \frac{1}{2}(p_1^2+p_2^2 + \omega_1^2x_1^2 + \omega_2^2x_2^2) + ax_1^2x_2 + bx_2^3$ (1.2)

Barbanis: $H = \frac{1}{2}(p_1^2+p_2^2 + \omega_1^2x_1^2 + \omega_2^2x_2^2) + \varepsilon x_1x_2^2$ (1.3)

Classical trajectories for these Hamiltonians fall into two categories R and E

R: Motion does not fill the whole of the allowed phase space. In (x_1,x_2) cartesian space, motion lies inside well-defined caustics AB, BC, CD, DA. Trajectories return, after sufficient time, infinitesimally close to their starting position in phase space. At each point in (x_1,x_2) space, four values $\underline{p}^\gamma$, γ = I, II, III, IV exist for the momentum $\underline{p}$. See Figure 1. Poincaré surfaces of section are well-defined curves. See Figure 2. [A Poincaré surface of section are the values of (p_2,x_2) for which x_1=0 and p_1>0, for a given trajectory].

The above motions are understood in terms of invariant tori. There must be N constants of the motion in the 2N dimensional phase space. Motion therefore lies on N dimensional toroids in 2N dimensional phase space. Kolmogorov, Arnol'd and Moser (5) proved the existence of these quasi periodic motions for non-separable Hamiltonians. They showed that $(\underline{p},\underline{x})$ may be represented by convergent Fourier series e.g.,

$$\underline{x} = \sum_{\underline{k}} \underline{c}_k \exp(i\underline{k}\cdot\underline{\theta}) \quad (1.4)$$

where $\underline{\theta}$ are the action variables which obey

$$\underline{\theta} = \bar{\underline{\omega}}t + \underline{\delta} \quad (1.5)$$

Depending upon the values of $\underline{\omega}$ (eq^n (1.3)), the patterns of these trajectories may become exceedingly complex. See Figure 2.

All such trajectories are classified as REGULAR.

E: Motion does not lie within well-defined caustics, but it is inclined to wander all over the allowed (x_1,x_2) space. Poincaré surfaces of section show a shotgun pattern. There are many values of $\underline{p}$ for a given (x_1,x_2). See Figure 3.

Such trajectories are said to be ERGODIC (or Irregular or stochastic). In this article we shall use any of these terms, although precise definitions exist for each.

At the same energy, there may exist both E and R type trajectories, dependent upon initial conditions. See Figure 3. It is found that as the energy increases for Hamiltonians (1.2) and (1.3), the number of E trajectories increases. For the Barbanis potential, with $\omega_1^2 = 1.6$, $\omega_2^2 = 0.9$ and $\varepsilon = -0.08$, there are no E trajectories below an energy of 70%D, where D is the escape energy.

II. METHODS FOR DETERMINING SEMICLASSICAL EIGENVALUES

The problem of finding constants of the motion for Hamiltonians (1.1) received much attention at the turn of the century. See the classic text by Whittaker (7). The onset of the Old Quantum Theory, where classical actions $\underline{N}$ were quantized, increased attention by Born (8), and Einstein (9), in particular, on these methods of classical mechanics. However, the development of the New Quantum Theory withdrew attention from these other, now approximate methods, except one-dimensional problems, which could be solved exactly (e.g., JWKB methods). The problem was that many dimensional non separable problems were so difficult. However, the advancement of computational methods has now reopened this field of study, with Marcus (10) leading the way in 1973. Here four current methods for determining semiclassical eigenvalues are discussed in chronological order.

a) The Methods of Marcus and Co-workers

The fact that a trajectory is Regular implies the existence of action-angle variables for the system, $\underline{N}$, $\underline{Q}$, with $\underline{N}$ being constants of the motion. Let $F_2(\underline{N},x)$ be the generator to these variables from the original variables $\underline{p}$, $\underline{x}$. F_2, being a solution of the Hamilton-Jacobi equation

$$\frac{1}{2}\left[\left(\frac{\partial F_2}{\partial x_1}\right)^2 + \left(\frac{\partial F_2}{\partial x_2}\right)^2 + \omega_1^2 x_1^2 + \omega_2^2 x_2^2\right] + \varepsilon V = E \qquad (2.1)$$

can take the form

$$F_2^\gamma \equiv \int_A^x \underline{p}^\gamma \cdot \underline{dx} \qquad \gamma = \text{I, II, III, IV.} \qquad (2.2)$$

Here A is a corner of the caustic pattern (see Fig. 1), and γ refers to any of the four directions for $\underline{p}$ (see Fig. 1). The wavefunction will be a linear combination of $A^\gamma(\underline{x})\ \exp(\underline{i}F_2^\gamma)$, where $A^\gamma(\underline{x})$ is a suitable weight function. By extending the usual one-dimensional ideas to many dimensions, these functions may be 'connected' near the corner A to give the semiclassical wavefunction

$$\psi(\underline{x}) = \left|\frac{\partial^2 F_2^I}{\partial \underline{x}\partial \underline{N}}\right| \cos(F_2^I - \frac{1}{2}\pi) + \left|\frac{\partial^2 F^{III}}{\partial \underline{x}\partial \underline{N}}\right| \cos F_2^{III}. \tag{2.3}$$

If instead of A, one uses the corner D, a similar solution with A replaced by D, and $(F_2^I - \frac{1}{2}\pi, F_2^{III})$ replaced by $(F_2^I, F^{III} - \frac{1}{2}\pi)$ is obtained. The fact that these two solutions are the same enabled Marcus to show that

$$\begin{aligned} 2\pi(N_2 + \tfrac{1}{2}) &= \int_o^{o'} \underline{p}^I \cdot \underline{dx} + \int_{o'}^{o} p^{III} \cdot \underline{dx} = \oint_{C_2} \underline{p} \cdot \underline{dx} \\ 2\pi(N_1 + \tfrac{1}{2}) &= \int_p^{p'} \underline{p}^I \cdot \underline{dx} + \int_{p'}^{p} \underline{p}^{III} \cdot \underline{dx} = \oint_{C_1} \underline{p} \cdot \underline{dx} \end{aligned} \tag{2.4}$$

where the actions take the values (integer + $\frac{1}{2}$)2π. In Figure 4 it is shown how the paths pp', oo' are deformable into closed curves C_1, C_2.

Eastes and Marcus (11) first applied the method by letting the paths C_1, C_2 be the caustics AB and BC. By adjusting the energy and the initial condition, they were able to use an iterative method to solve equations (2.2), for any integer values N_1, N_2. They applied their method to the Henon-Heiles hamiltonian, and results are given in Table I.

From these results it is seen that there is very close agreement between these semiclassical eigenvalues and the exact quantum mechanical eigenvalues, obtained by diagonalizing the secular matrix. This type of agreement augurs well for the application of these methods to problems of real physical interest.

Noid and Marcus (12) improved on the above approach by introducing the Poincaré surface of section. The topological equivalence of the paths C_1, C_2 and the surface of section curves, see Figure 4, enabled them to write the quantum conditions as

$$\begin{aligned} 2\pi(N_1 + \tfrac{1}{2}) &= \oint_{C_1} \underline{p} \cdot \underline{dx} = \text{area of surface of section, } x_2 = 0,\ p_2 > 0 \\ 2\pi(N_2 + \tfrac{1}{2}) &= \oint_{C_2} \underline{p} \cdot \underline{dx} = \text{area of surface of section, } x_1 = 0,\ p_1 > 0 \end{aligned} \tag{2.5}$$

Essentially the same results were obtained by this method.

This method has now been extended (13) to systems which have more complex shaped caustics at higher energies (see Sec. II(c)).

b) The Method of Chapman, Garrett and Miller

The philosophy behind this approach is to find a Fourier series solution, for the F_2 generator, from the Hamilton-Jacobi equation (2.1). In zeroth-order action-angle variables $\underline{n},\underline{q}$ introduced by

$$x_i = \sqrt{\frac{2n_i}{\omega_i}} \cos q_i, \quad p_i = \sqrt{2n_i\omega_i} \sin q_i \tag{2.6}$$

the Hamiltonians (1.2) and (1.3) take the form

$$H = \underline{n}\cdot\underline{\omega} + \varepsilon V(\underline{n},\underline{q}) \tag{2.7}$$

Now introduce the F_2 generator from $\underline{n},\underline{q}$ to the good action-angle variables $\underline{N}$, $\underline{Q}$. It will have the form

$$F_2(\underline{N},\underline{q}) = \underline{N}\cdot\underline{q} + \varepsilon G(\underline{N},\underline{q}) \tag{2.8}$$

Because of the periodic nature of the motion, G must be periodic in $\underline{q}$. [CGM consider the semiclassical wavefunction $\exp(iF_2)$, and argue that G must be periodic for the wavefunction to be single valued]. Thus $G(\underline{N},\underline{q})$ is represented as

$$G(\underline{N},\underline{q}) = \sum_{\underline{k}} B_{\underline{k}} \exp(i\underline{k}\cdot\underline{q}) \tag{2.9}$$

Using the relation

$$\underline{n} = \frac{\partial F_2}{\partial \underline{q}} \tag{2.10}$$

The Hamilton-Jacobi equation becomes

$$E(\underline{N}) = (\underline{N}+\varepsilon \frac{\partial G}{\partial \underline{q}})\cdot\underline{\omega} + \varepsilon V(\underline{N}+\varepsilon \frac{\partial G}{\partial \underline{q}},\underline{q}) \tag{2.11}$$

The coefficients $B_{\underline{k}}$ in (2.9) may be obtained by equating to zero all the Fourier components of (2.11). In effect this is achieved by multiplying (2.11) by $\exp(-i\underline{k}\cdot\underline{q})$ and integrating. The resulting equations can be solved using an iterative technique.

Chapman, Garrett and Miller applied their method to the Henon-Heiles system, and within numerical error, they obtained the same results as Marcus and co-workers, as they should.

Handy, Colwell and Miller (15) used a real representation for $G(\underline{N},q)$

$$G(\underline{N},\underline{q}) = \sum_{\underline{k}} B_{\underline{k}} \begin{Bmatrix} \sin k_1 q_1 \\ \cos k_1 q_1 \end{Bmatrix} \begin{Bmatrix} \sin k_2 q_2 \\ \cos k_2 q_2 \end{Bmatrix} \quad \begin{Bmatrix} \sin k_t q_t \\ \cos k_t q_t \end{Bmatrix} \tag{2.12}$$

and then applied the method to the determination of vibrational energy levels of triatomic molecules.

The form of the classical hamiltonian for this problem has been determined by Wilson, Decius and Cross (16):

$$H = \frac{1}{2}(P_1^2+P_2^2+P_3^2) + \frac{1}{2} \sum_{\alpha,\beta}^{3} \pi_\alpha \, \mu_{\alpha\beta} \, \pi_\beta + V \tag{2.13}$$

where $\underline{P}$, $\underline{Q}$ are conjugate normal coordinates and momenta, $\underline{\pi}$ is the so-called vibrational angular momentum,

$$\pi_\alpha = \sum_{k,\ell} \zeta_{k\ell}^{\alpha} \, Q_k \, P_\ell , \tag{2.14}$$

$\zeta_{k\ell}^{\alpha}$ are coriolis coupling coefficients

$$\zeta_{k\ell}^{\alpha} = \varepsilon_{\alpha\beta\gamma} \sum_{i}^{3} \ell_{\beta i,k} \, \ell_{\gamma i,\ell} \tag{2.15}$$

The $\ell_{\beta i,k}$ are transformation coefficients from mass-weighted cartesian displacement coordinates to the normal coordinates

$$Q_k = \sum \ell_{\alpha i,k} \, m_i^{1/2} (r_{\alpha i} - r_{\alpha i}^0) \tag{2.16}$$

$\underline{r}_i = (r_{xi}, r_{yi}, r_{zi})$ is the position of the ith nucleus in molecule fixed coordinates, r_i^o is the corresponding equilibrium position and m_i is the mass of the ith nucleus. $\underline{\mu}$ is closely related to the inverse of the moment of inertia matrix.

The form of the potential V used in these calculations was

$$V = \sum_{ijk} K_{ijk} (\Delta r_{12})^i (\Delta r_{13})^j (\Delta\theta)^k \tag{2.17}$$

In H_2O, for example, Δr_{12} is the change in the OH_1 bond length, Δr_{13} is the change in the OH_2 bond length and $\Delta\theta$ is the change in the HOH bond angle from its equilibrium value. The normal coordinates Q_k are found by expanding (2.17) in cartesian displacement coordinates up to quadratic powers. If this truncated V is denoted V_o, then in normal coordinates it has the form

$$V_0 = \frac{1}{2}(\omega_1^2 Q_1^2+\omega_2^2 Q_2^2+\omega_3^2 Q_3^2). \quad (2.18)$$

Thus

$$H_0 = \sum_k^3 \frac{1}{2}(P_k^2+\omega_k^2 Q_k^2), \quad (2.19)$$

and

$$H_1 = \sum_{\alpha,\beta} \frac{1}{2} \pi_\alpha \mu_{\alpha\beta} \pi_\beta + V-V_0, \quad (2.20)$$

where H_1 will be a small perturbation to H_0 in most cases.

The forms of the potential used were

: SO_2. That of Kutchitsu and Morino (17). Fundamental frequencies are 1171, 525 and 1378 cm^{-1}.
: H_2O. That of Hoy, Mills and Strey (18). Fundamental frequencies are 3382, 1648 and 3942 cm^{-1}.

In Table 2, results are given for low-lying vibrational energy levels for these molecules. They are compared with results obtained by Whitehead and Handy (19), who solved the quantum mechanical secular equations

$$\langle \Phi_r | H_{QM}-E | \Sigma Z_s \Phi_s \rangle = 0, \quad (2.21)$$

where Φ_s are appropriate vibrational expansion functions, written as products of Hermite polynomials of the normal coordinates. It is recalled that (20)

$$H_{QM} = H - \frac{1}{8} \sum_{\alpha=1}^{3} \mu_{\alpha\alpha} \quad (2.22)$$

In Table 2, the semiclassical results were obtained using all the Fourier expansion functions in (2.12), with $k_i \leq 3$. It is interesting to observe that approximately the same number of expansion functions were needed by Whitehead and Handy to solve the quantum mechanical problem.

For SO_2, it is seen that the absolute semiclassical energies are in error by 0.5 cm^{-1}. For H_2O the error is more significant at approximately 6 cm^{-1}. Greater difficulty was encountered with

H_2O because of difficulty in solving the iteration equations. This is primarily due to the lightness of the protons. Indeed it is probable that insufficient Fourier functions were used.

c) The Method of Sorbie and Handy

Realizing that these quasi periodic trajectories closed, to any desired accuracy, in phase space after sufficient time, Sorbie and Handy (3,4,6,21) examined an approximate (13) semi-classical quantization condition

$$\int_\Gamma p_i \, dx_i = 2\pi \, M_i (N_i + \tfrac{1}{2}). \tag{2.23}$$

Γ is the closed trajectory and M_i is the number of complete circuits the trajectory makes when projected into the p_i, x_i plane. Equation (2.23) is an obvious extension of the one-dimensional quantization condition.

The accuracy of eq^n (2.23) is now examined. It can easily be shown that

$$\int_\Gamma p_i \, dx_i = \int_\Gamma n_i \, dq_i \tag{2.24}$$

where $\underline{n}, \underline{q}$ are zeroth-order action-angle variables. Introducing the F_2 generator, eq^n (2.8), and expanding $G(\underline{N},\underline{q})$ by

$$G = \sum_{\underline{k}}{}' (C_{\underline{k}} + \varepsilon \, C_{1\underline{k}} + \varepsilon^2 \, C_{2\underline{k}} + ..) \cos(\underline{k}\cdot\underline{q}) + \sum_{\underline{k}} (S_{\underline{k}} + \varepsilon \, S_{1\underline{k}} + \varepsilon^2 \, S_{2\underline{k}} + ..) \sin(\underline{k}\cdot\underline{q}) \tag{2.25}$$

$\underline{n}$ may be evaluated through

$$\underline{n} = \underline{N} + \varepsilon \frac{\partial G}{\partial \underline{q}} \tag{2.26}$$

Substitution into (2.24) gives

$$\int_\Gamma n_i \, dq_i = \int_\Gamma N_i \, dq_i + \varepsilon \int_\Gamma \frac{\partial G}{\partial q_i} dq_i \tag{2.27}$$

The first term is the r.h.s. of (2.23) as required. Using (2.25) and

$$\underline{q} = \underline{\omega} t + O(\varepsilon) \tag{2.28}$$

it is not difficult to show that

$$\int_{\Gamma} \frac{\partial G}{\partial q_i}\, dq_i = O(\varepsilon) \qquad (2.29)$$

Hence eq^n (2.23) holds for the exact action variables $\underline{N}$, correct to order ε^2. By expanding the energy in a Taylor series in $\bar{N}$-N, where $\bar{N}$ is the approximate action variable

$$\bar{N}-N = \frac{\varepsilon}{2\pi M_i} \int_{\Gamma} \frac{\partial G}{\partial q_i}\, dq_i \qquad (2.30)$$

it may be shown that

$$E(N) = E(\bar{N}) + O(\varepsilon^5) \qquad (2.31)$$

Hence the error in the energies obtained by using (2.23) as the quantization condition is of $O(\varepsilon^5)$, and it is therefore exceedingly unlikely that this error will exceed the difference between quantum mechanical and semiclassical energies.

The method of Sorbie and Handy has been much used. For example it has been used to obtain the 14 lowest lying vibrational energies of SO_2, and also low-lying states of H_2O. Results (21) are almost identical with those obtained by the CGM method (b) above.

The method has been successfully applied to problems with a zeroth order degeneracy, eg. $\omega_1^2 = \omega_2^2$ in the Barbanis Hamiltonian (1.3). It was this problem which gave such enormous difficulty to those workers (7,8) who applied classical perturbation theory 60 years ago. This is the problem of small divisors. When $\omega_1=\omega_2$, the trajectories and their caustics take weird shapes. See Figure 2. However, it was still possible to apply this method provided the axes intersect opposite caustics. Typical results for the Barbanis potential are given in Table 3. Notice how the small values of $|M_1-M_2|$ reflect the almost equal harmonic frequencies.

The method was also applied to higher eigenvalues (4) of the Barbanis Hamiltonian ($\omega_x^2 = 1.6$, $\omega_y^2 = 0.9$, $\varepsilon = -0.08$). The escape energy is 25.3125. For energies greater than 19.0, there are ergodic trajectories and it was not found possible to use this method for certain states above this energy. The method was straightforwardly applicable, but difficulties occur as the caustics become more weird. A selection of results is given in Table 4. Discussion on those states which can be classified as ergodic is given in Section III.

Noid and Marcus (13) have more recently extended their methods to deal with more complex cases such as when there is a zeroth order degeneracy. They examined the Henon-Heiles system. They found it appropriate to introduce polar coordinates r, θ with quantum conditions

$$\oint p_r \, dr = 2\pi(N_r + \tfrac{1}{2})$$
$$\oint p_\theta \, d\theta = 2\pi \, N_\theta \qquad \text{(no caustic intersection)} \tag{2.32}$$

With these coordinates they were able to examine higher eigenvalues and indeed were not able to find some because the corresponding trajectories were ergodic. Their paper shows that the methods of Marcus, and Sorbie and Handy, which are both trajectory based methods, are applicable to the same type of problem, each suffering from increasing difficulty as the caustics become increasingly weird in shape.

d) The Method of Percival and Pomphrey

In this method (22,23) the invariant toroid is represented as a Fourier series in the good angle variables $\underline{Q}$.

The mean values of the energy, over the toroid, is defined by

$$\langle E \rangle = \int d\underline{Q} \; H(\underline{x}(\underline{Q}), \; \underline{p}(\underline{Q})) \tag{2.33}$$

where the integral limits are 0 and 2π for each $\underline{Q}$. Likewise mean values of the action integrals are introduced

$$\langle N_k \rangle = \int d\underline{Q} \; \underline{p} \cdot \partial\underline{x}/\partial Q_k \tag{2.34}$$

Following the parallel with quantum mechanical variation methods, stationary values of E are found, subject to $\langle N_k \rangle$ being constant. Thus Φ is minimized, where

$$\Phi = \int d\underline{Q}[H - \Sigma \; \bar{\omega}_k \underline{p} \cdot \partial\underline{x}/\partial Q_k], \tag{2.35}$$

with respect to variations $\Delta\underline{p}$, $\Delta\underline{x}$. The resulting equations are

$$\sum_k \bar{\omega}_k \frac{\partial \underline{x}}{\partial Q_k} = \frac{\partial H}{\partial \underline{p}}; \quad \sum_k \bar{\omega}_k \frac{\partial \underline{p}}{\partial Q_k} = -\frac{\partial H}{\partial \underline{x}} \tag{2.36}$$

These are the same as Hamilton's equations if the identity

$$\frac{d}{dt} = \sum \bar{\omega}_k \frac{\partial}{\partial Q_k} \tag{2.37}$$

is made. Relation (2.37) implies that

$$Q_k = \bar{\omega}_k t + \delta_k \tag{2.38}$$

and thus the Lagrange multipliers become the frequencies of the system.

In detail, for a two dimensional problem, $\underline{r}$ and $\underline{p}$ are expanded in a Fourier series in $\underline{Q}$. Thus

$$\begin{aligned} \underline{r}(Q_1,Q_2) &= \sum \underline{r}_{s_1,s_2} \exp[i(s_1Q_1+s_2Q_2)] \\ \underline{p}(Q_1,Q_2) &= \sum i(s_1\bar{\omega}_1+s_2\bar{\omega}_2)\, \underline{r}_{s_1,s_2} \exp[i(s_1Q_1+s_2Q_2)] \end{aligned} \tag{2.39}$$

In terms of these expansion coefficients $\underline{r}_{s_1,s_2}$, Φ is

$$\begin{aligned} \Phi = \frac{1}{2} \sum (s_1\omega_1+s_2\omega_2)^2 (x_{s_1,s_2}\, x_{s_1,-s_2} + y_{s_1,s_2}\, y_{s_1,-s_2}) + \int d\underline{Q}\; V(\underline{r}(Q)) \\ -\bar{\omega}_1\langle N_1\rangle - \bar{\omega}_2\langle N_2\rangle \end{aligned} \tag{2.40}$$

(where $\underline{r} = (\underline{x},\underline{y})$). Differentiating Φ with respect to $\underline{r}_{-s_1,-s_2}$, one obtains

$$(s_1\bar{\omega}_1+s_2\bar{\omega}_2)^2\, \underline{r}_{s_1,s_2} = -\underline{F}_{s_1,s_2} \tag{2.41}$$

where $\underline{F}_{s_1,s_2}$ are the Fourier components of the force. Introducing the anharmonic parts of the force as $\underline{F}^{anh}$, one can rewrite (2.40) as

$$\begin{aligned} \{\omega_x^2 - (\bar{\omega}_1 s_1+\bar{\omega}_2 s_2)^2\}\, x_{s_1,s_2} &= F^{anh}_{xs_1,s_2} \\ \{\omega_y^2 - (\bar{\omega}_1 s_1+\bar{\omega}_2 s_2)^2\}\, y_{s_1,s_2} &= F^{anh}_{ys_1s_2} \end{aligned} \tag{2.42}$$

Equations (2.42) and (2.34) are then iterated together, continuing until the value of $\langle E\rangle$ obtained from eq^n (2.33) has converged. In practice this is obtained by noting that

$$N_1 = 2(x_{1,0}^2+y_{1,0}^2)\bar{\omega}_1 + \sum_{s_1s_2\neq 0} s_1^2(x_{s_1,s_2}^2+y_{s_1,s_2}^2)\bar{\omega}_1$$

$$+ \sum_{s_1,s_2} s_1s_2(x_{s_1,s_2}^2+y_{s_1,s_2}^2)\bar{\omega}_2 \qquad (2.43)$$

Initially, from (2.43), $x_{1,0}$ and $y_{0,1}$ are obtained by

$$x_{1,0} = x_{-1,0} = (N_1/2\omega_x)^{1/2}$$
$$y_{0,1} = y_{0,-1} = (N_2/2\omega_y)^{1/2} \qquad (2.44)$$

Then eq^n (2.42) is used with s_1=1, s_2=0 and s_1=0, s_2=1 to obtain values for $\bar{\omega}_1{}^2$ and $\bar{\omega}_2{}^2$. The other Fourier components are then obtained from eq^n (2.42), from the previously computed values for $\underline{F}^{anh}$. Equation (2.43) is then used to obtain updated values for $\bar{x}_{1,0}$ and similarly for $y_{0,1}$.

All indications are that this is probably the best of the four procedures discussed here. It is very fast, provided that the forces can be evaluated in some simple straightforward way. It has the distinct advantage that fast Fourier transform techniques can be used, and it is thus an order of magnitude faster than methods (a),(c) discussed. Its first application was to the Henon-Heiles Hamiltonian; M^2 Fourier terms were used, with $M \leq 16$.

More recently Percival and Pomphrey (24) have applied their method for obtaining vibrational energy levels of OCS, using a potential due to Foord et al (25), the latter being expanded in normal coordinates. The vibrational hamiltonian for a linear molecule has a form similar to (2.13). Percival and Pomphrey treated the terms involving π^2 as a perturbation, just calculating their average value at the end of the calculation. They reported obtaining 50 low lying levels of OCS using 145 seconds of CDC 7600. Difficulties arise in this method when small divisors are present, as can occur when, for some s_1,s_2, the ℓ.h.s bracket of eq^n (2.42) is nearly zero. Thus occurred with 17 of the first 67 levels of OCS. In these cases the trajectory methods appear preferable.

III. THE QUANTUM EFFECTS OF CLASSICAL ERGODICITY

Percival, in a recent review, has given an excellent account of the distinction between regular and irregular regions. [Ref. 26 pps 36-49.] Here we briefly summarize that review, concentrating on the portions relevant to semiclassical applications, in the following paragraph.

It has now been found, for the non-separable Henon-Heiles system, that even for low energies, there are irregular regions, albeit very small. They are associated with the small divisor problem, for large values of s_1, s_2 (eq^n 2.42). So the nature of full problem is very complex, even in the regular region. If such small irregular regions are ignored, then it is possible to say approximately, for the Henon-Heiles and like systems, that

(a) For low energies, the classical trajectories are regular. Semiclassical methods may be used by quantizing the actions. Such corresponding quantum states may thus be given quantum numbers.

(b) For higher energies, there will be both regular and irregular trajectories. There will also be some whose category is difficult to decide. There will therefore be some regular trajectories corresponding to integer actions, as in (a). There will also be an increasing number of integer actions for which it is not possible to find regular trajectories. No quantum numbers can therefore be assigned to the corresponding quantum eigenvalues. Percival anticipates therefore that the quantum energy levels for these irregular states will take on a random distribution. He thus expects these levels to be more sensitive to a slowly changing perturbation than the regular states. Pomphrey (27) examined this last prediction for the Henon-Heiles system ($\omega_x^2=\omega_y^2=1.0$, $a=\varepsilon$, $b=-\frac{1}{3}\varepsilon$), computing the eigenvalues and noting their change with the parameter. Using $0.086 \leq \varepsilon \leq 0.090$, and $E < 16.0$ (~74% Dissociation energy), he found all $\partial^2 E_i/\partial\varepsilon^2$ to be small. However, for $E > 16.0$, he found a few eigenvalues E_i whose $\partial^2 E/\partial\varepsilon^2$ was an order of magnitude larger. This thus indicates that there is an apparent noticeable difference in the quantum eigensolutions which correspond to irregular trajectories.

Nordholm and Rice (28) attempted to go further by analyzing the eigenfunctions of this Hamiltonian, in particular examining whether global extent of some eigenfunctions was significantly greater than that expected from zeroth order estimates. They indeed did find eigenfunctions which fell in this category, and therefore classified them as ergodic, but it has recently (13) been shown that these results were open to question because they used too small a quantum mechanical basis set.

More recently McDonald and Kaufman (29) have examined the eigenfunctions of the 'stadium" problem. The classical trajectories for this problem can be all shown to be ergodic. They find that the nodal pattern of the wavefunctions, has no apparent structure. This of course is in complete contradiction to separable harmonic oscillators, where the nodal lines form a rectangular grid, all lines intersecting. In this separable case, the quantum numbers (the action integers) correspond to the number of nodal lines in each direction.

Stratt, Handy and Miller (30) have also examined the nodal pattern for the Barbanis Hamiltonian. They found that for any $\varepsilon \neq 0$, the nodal lines do not intersect. However for all low lying states, and many higher ones, a discernible pattern is clear, and there is no ambiguity in counting nodal lines, and thus assigning quantum numbers. It is found that there is a 1-1 correspondence between these numbers of nodal lines, and the action integers from classical trajectories. However as the energy increases beyond 19, it becomes more difficult to note a discernible structure for the nodal patterns for some eigenvalues, and thus impossible to give quantum numbers. These states correspond to those for which Sorbie and Handy (6) were not able to find semiclassical eigenvalues. Some details are given in Table 4.

In conclusion it is apparent that the existence of classical ergodic trajectories does have some parallel effects on the nature of the quantum mechanical eigensolutions. It seems probable that the whole nature of the infrared vibrational spectrum will change in these ergodic regions. Ergodicity may have an important role in intramolecular energy transfer: how does the energy, originally in a local mode of vibration, become distributed about the whole molecule? No certain experimental observation of an ergodic state has yet been reported, probably because they are too high up in energy and also because one really does not know what to look for.

IV. PROBLEMS WHICH MUST INVOLVE TUNNELING

In one dimension, the simplest problem is the potential well. Thus far we have considered the many dimensional equivalent of this. It is more difficult in one dimension to deal with problems with more than one minimum, such as a double well potential. Such problems must involve quantum mechanical tunneling. Semiclassical approaches for this problem have been described elsewhere in this volume. A related important problem with chemical applications is the semiclassical determination of the energy levels of a rigid asymmetric rotor - solved recently by Colwell, Handy and Miller (31).

We now discuss many dimensional problems which have barriers. The most interesting problem, from the chemists point of view, is the one in which there is a simple barrier in one variable s, and simple minima in all the other variables $\underline{Q}$. If s is the 'reaction coordinate', the $\underline{Q}$ are vibrational coordinates, the related problem is a bimolecular reaction process.

The Hamiltonian discussed here has a double minimum $V_{DM}(x)$ potential in the x coordinate, a simple minimum $V_{SM}(y)$ in the y coordinate located at y=0, together with a non-separable perturbation $\varepsilon V(x,y)$,

$$H = \frac{p_x{}^2}{2} + \frac{p_y{}^2}{2} + V_{DM}(x) + V_{SM}(y) + \varepsilon V(x,y) \tag{4.1}$$

The barrier in $V_{DM}(x)$ is located round x=0. The problem splits into 3 regions, for energies below the barrier height

a) $x < 0$ and $E > V$ Region I

b) $V > E$ Region II

c) $x > 0$ and $E > V$ Region III

Regions I and III may be treated as two dimensional minima problems. In region I, where suppose V_{DM} has a minima at $x = -x_0$, the Hamiltonian may be written

$$H = H_0 + H_1 \tag{4.2}$$

where

$$H_0 = \frac{p_x{}^2}{2} + \frac{p_y{}^2}{2} + \frac{1}{2}\,\omega_x^2(x+x_0)^2 + \frac{1}{2}\,\omega_y^2 y^2 \tag{4.3}$$

$$H_1 = H - H_0$$

For any given values N_1, N_2 for the action variables, a value for E may be found by using any of the previously defined methods. The same applies for region III.

For region II, the appropriate zeroth order problem is for

$$H_0 = \frac{p_x{}^2}{2} + \frac{p_y{}^2}{2} - \frac{1}{2}\,\omega_x^2 x^2 + \frac{1}{2}\,\omega_y^2 y^2 \tag{4.4}$$

Appropriate zeroth-order action angle variables $\underline{n}$, $\underline{q}$ can be introduced by

$$x = \sqrt{\frac{in_x}{i\omega_x}}\cos q_x, \quad p_x = \sqrt{(in_x)(i\omega_x)}\sin q_x \tag{4.5}$$

$$y = \sqrt{\frac{n_y}{\omega_y}}\cos q_y, \quad p_y = \sqrt{n_y\omega_y}\sin q_y \tag{4.6}$$

In (4.5), the formal step of introducing pure imaginary action and frequency is made. In $\underline{n}$, $\underline{q}$ variables the Hamiltonian is

$$\begin{aligned} H &= H_0 + H_1(x,y) \\ &= -n_x\omega_x + n_y\omega_y + H_1\left(\sqrt{\frac{n_x}{\omega_x}}\cos q_x, \sqrt{\frac{n_y}{\omega_y}}\cos q_y\right) \end{aligned} \tag{4.7}$$

The problem is thus transformed to a purely real problem! It may be solved by any of the methods introduced previously. If the method of Sorbie and Handy is used, the actions are defined through

$$\begin{aligned} M_x N_x &= \int_\Gamma n_x \, dq_x \\ M_y N_y &= \int_\Gamma n_y \, dq_y \end{aligned} \tag{4.8}$$

The quantization conditions are

(i) $N_y = (\text{integer} + \frac{1}{2})\, 2\pi$

(ii) If N_1^γ, γ = I, II, III are action values from the three regions, the condition is taken from the corresponding one-dimensional double minimum problem. For a symmetrical double minimum it is (33)

$$(N^{I} + \Delta) \pm \tan^{-1}(\exp(-N^{II})) = (\text{integer})\pi \tag{4.9}$$

where

$$\Delta = \arg\Gamma\left(\frac{1}{2} + \frac{iN^{II}}{\pi}\right) - \frac{N^{II}}{\pi}\log\left(\frac{|N^{II}|}{e\pi}\right) \tag{4.10}$$

and Γ is the gamma function $\Gamma(z)$.

This method has been applied by Cerjan, Handy and Miller (32) to a symmetric double well potential. This last piece of theory has opened up an entirely new area of interest for the application of semiclassical methods. Research continues in this field.

Table 1. First results, obtained by Eastes and Marcus (11), for the semiclassical eigenvalues of a non-separable system. The Henon-Heiles Hamiltonian (1.2) was used with $\omega_x^2 = 0.29375$, $\omega_y^2 = 2.12581$, $a = -0.1166$, $b = 0.08414$

N_1	N_2	E_{SC}	E_{QM}	E_0
0	0	0.9920	0.9916	1.000
1	0	1.5164	1.5159	1.542
2	0	2.0313	2.0308	2.084
0	1	2.4196	2.4188	2.458

E_0 are the zeroth order eigenvalues.

E_{SC} are the semiclassical eigenvalues.

E_{QM} are the quantum mechanical eigenvalues.

Table 3. Semiclassical E_{SC} and Quantum E_{QM} Eigenvalues for the Barbanis Hamiltonian ($\omega_x^2 = \omega_y^2 = 0.5$, $\varepsilon = -0.025$)

State	E_{SC}	E_{QM}	M_1	M_2
00	0.4974	0.4979	68	69
10	0.9829	0.9856	142	145
01	0.9967	0.9942	87	89
20	1.4585	1.4602	137	140
11	1.4752	1.4739	659	691
02	1.4972	1.4945	124	127

Table 2. Semiclassical E_{SC} and quantum mechanical E_{QM} energies compared for SO_2 and for H_2O [Energies in cm^{-1}]

	State	E_{SC}	E_{QM}	δ^a
SO_2	000	1529.12	1529.60	7.7
	010	2685.11	2685.63	22.8
	020	2045.56	2045.81	16.3
	100	2555.86	2556.21	30.5
	001	2889.07	2889.53	25.6
H_2O	000	4645.11	4651.98	65
	010	6242.55	6249.33	117
	020	7804.80	7811.53	198
	100	8360.90	8369.29	178
	001	8466.58	8472.75	185

[a] δ is the difference between the zeroth order energy and E_{SC}.

Table 4. Semiclassical E_{SC} and Quantum E_{QM} eigenvalues for the Barbanis Hamiltonian ($\omega_x^2 = 1.6$, $\omega_y^2 = 0.9$, $\varepsilon = -0.08$)

Eigenvalue No.	State[a]	E_{SC}	E_{QM}	M_1	M_2	Type[b]
1	0 0	1.105	1.106	3	4	R
57	3 12	15.754	15.752	277	218	R
58	6 8	15.812	15.814	82	59	R
59	0 16	15.882	15.885	53	43	R
60	9 4	15.998	15.999	47	33	R
61	12 0	16.252	16.252	33	23	R
76	5 12	?	18.079	-	-	U
84	10 6	18.945	18.946	13	9	R
96	4 16	?	20.345	-	-	E
100	10 8	20.621	20.625	86	73	R
131	7?16?	?	23.625	-	-	E
135	10 12	23.865	23.892	53	36	R

[a] Action assignment (Sec II) and nodal assignment (Sec III).

[b] Regular, Uncertain or Ergodic (Sec III).

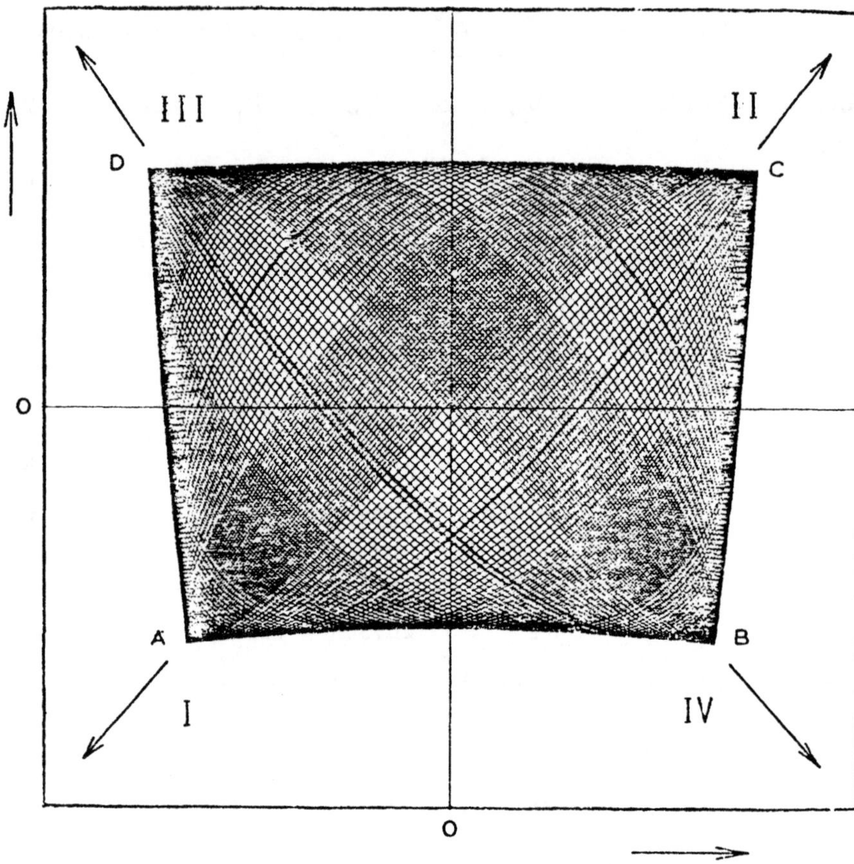

Figure 1. (From Ref. 3). A Regular classical trajectory for the Barbanis Hamiltonian with $\omega_x^2 = 1.6$, $\omega_y^2 = 0.9$, $\varepsilon = -0.08$ and $E = 1.1051$. The boundary of allowed phase space is an approximate ellipse passing through A, B, C, D.

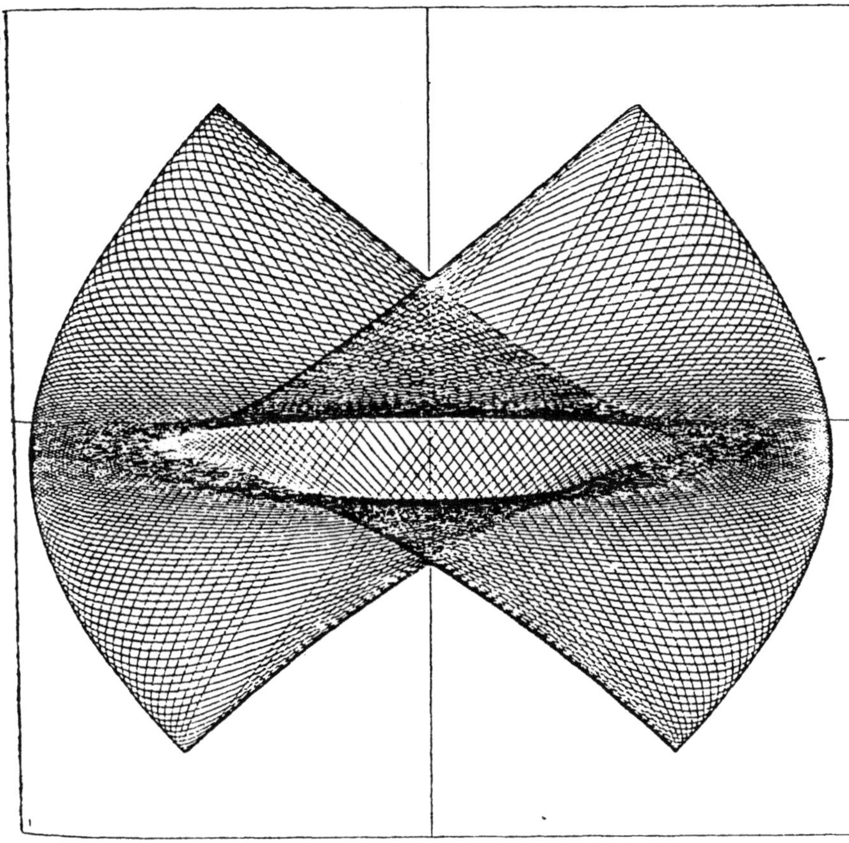

Figure 2. (From Ref. 4). A Regular trajectory for the Barbanis Hamiltonian with $\omega_x^2 = \omega_y^2 = 0.25$ and $\varepsilon = -0.0125$, with $E = 0.9959$.

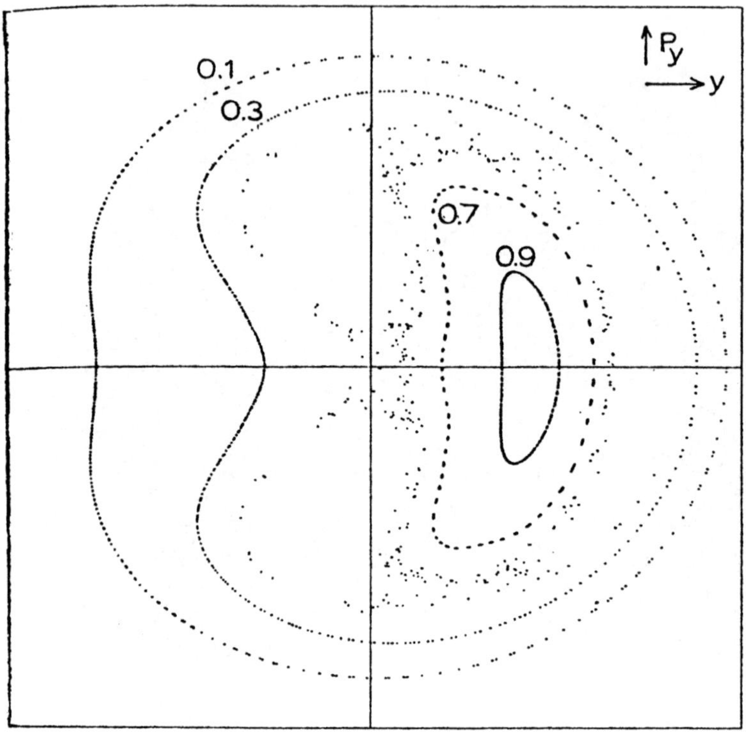

Figure 3. (From Ref. 4). Poincaré surfaces of section for Regular and Ergodic trajectories, for the same Hamiltonian as Figure 1. Here E = 20.0, and trajectories corresponding to different starting conditions are shown.

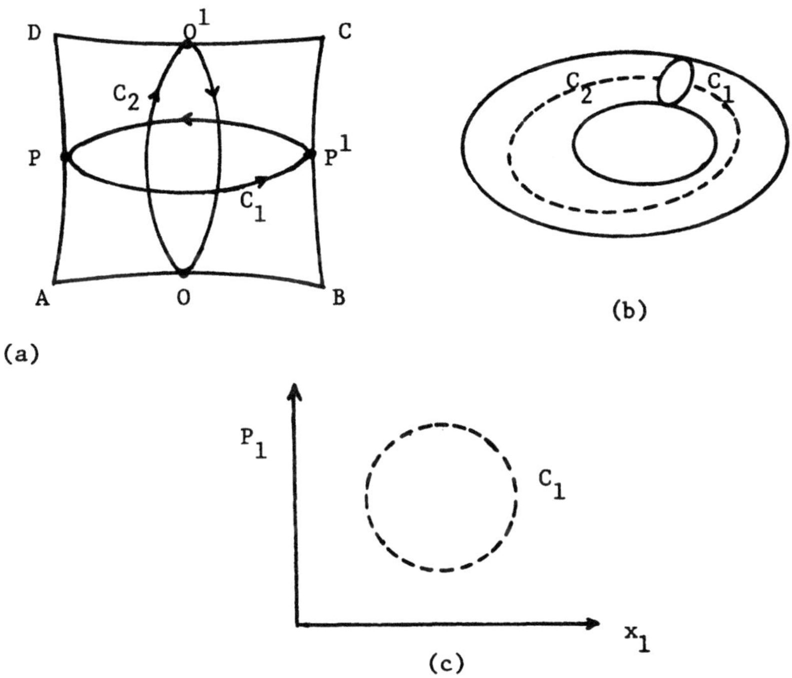

Figure 4. Equivalent topological paths C_1, C_2

(a) Joining opposite caustics is (x_1, x_2) space

(b) On the invariant toroid

(c) A Poincaré surface of section.

REFERENCES

1. Henon, M. and Heiles, C.: 1964, Astron. J. 69, p. 73.
2. Barbanis, B.: 1966, Astron. J. 71, p. 415.
3. Sorbie, K. S.: 1976, Molec. Phys. 32, p. 1577.
4. Sorbie, K. S. and Handy, N. C.: 1976, Molec. Phys. 32, p. 1327.
5. See C. L. Siegel and J. K. Moser "Lectures on Celestial Mechanics" (Springer-Verlag, Berlin, 1971), and references therein.
6. Sorbie, K. S. and Handy, N. C.: 1977, Molec. Phys. 33, p. 1319.
7. Whittaker, E. T.: 1904 'Analytical Dynamics', (Cambridge University Press).
8. Born, M.: 1960 "Mechanics of the Atom" (Ungar).
9. Einstein, A.: 1917, Verhand. Deut. Phys. Ges. 19, p. 82.
10. Marcus, R. A.: 1973, Chem. Soc. Far. Discuss. 55, p. 34.
11. Eastes, W. and Marcus, R. A.: 1974, J. Chem. Phys. 61, p. 4301.
12. Noid, D. W. and Marcus, R. A.: 1975, J. Chem. Phys. 62, p. 2119.
13. Noid, D. W. and Marcus, R. A.: 1977, J. Chem. Phys. 67, p. 559.
14. Chapman, S., Garrett, B. C., and Miller, W. H.: 1976, J. Chem. Phys. 64, p. 502.
15. Handy, N. C., Colwell, S. M., and Miller, W. H.: 1977, Chem. Soc. Far. Discuss. 62, p. 29.
16. Wilson, E. B., Decius, J. C., and Cross, P. C.: 1955, "Molecular Vibrations" (McGraw Hill).
17. Kutchitsu, K. and Morino, Y.: 1965, J. Chem. Soc. Japan, 38, p. 814.

18. Hoy, A. R., Mills, I. M., and Strey, G.: 1972, Molec. Phys. 24, p. 1265.

19. Whitehead, R. J. and Handy, N. C.: 1975, J. Molec. Spectrosc. 55, p. 356.

20. Watson, J. K. G.: 1968, Molec. Phys. 15, p. 479.

21. Colwell, S. M. and Handy, N. C.: 1978, Molec. Phys. 35, 1183.

22. Percival, I. C. and Pomphrey, N.: 1976, Molec. Phys. 31, p. 97.

23. Percival, I. C. and Pomphrey, N.: 1976, J. Phys. B, 9, p. 3131.

24. Percival, I. C. and Pomphrey, N.: 1978, Molec. Phys. 35, p. 649.

25. Foord, A., Smith, J. G., and Whiffen, D. H.: 1975, Molec. Phys. 29, p. 1685.

26. Percival, I. C.: 1977, Adv. Chem. Phys. 36, p. 1.

27. Pomphrey, N.: 1974, J. Phys. B, 7, p. 1909.

28. Nordholm, K. S. J. and Rice, S. A.: 1974, J. Chem. Phys. 61, 203; ibid 61, p. 768.

29. McDonald, S. W. and Kaufman, A. N.: 1979, Phys. Rev. Lett. 42, p. 1189.

30. Stratt, R., Handy, N. C., and Miller, W. H. to be published.

31. Colwell, S. M., Handy, N. C., and Miller, W. H.: 1978, J. Chem. Phys. 68, p. 745.

32. Cerjan, C., Handy, N. C., and Miller, W. H. to be published.

33. Child, M. S.: 1974, Mol. Spectrosc. 53, p. 280.

APPENDIX: THE AIRY FUNCTION

M.S. Child

Theoretical Chemistry Department
1 South Parks Road, Oxford.

ABSTRACT: Properties of the Airy Function and its first derivative are described and illustrated.

1. Properties

$$\mathrm{Ai}(z) = \frac{1}{2\pi}\int_{-\infty}^{\infty} \exp\,(iu^3/3 + iuz)du \quad (1)$$

$$= c_1\, f(z) - c_2 g(z) \quad (2)$$

$$\overset{z>1}{\sim}\ 0.5\ \pi^{-\frac{1}{2}}\ z^{-\frac{1}{4}}\ \exp\,(-z^{3/2}/1.5) \quad (3)$$

$$\overset{z<-1}{\sim}\ \pi^{-\frac{1}{2}}\ |z|^{-\frac{1}{4}}\ \sin\,(|z|^{3/2}/1.5 + \pi/4) \quad (4)$$

$$\mathrm{Ai}'(z) = \frac{i}{2\pi}\int_{-\infty}^{\infty} u \exp\,(iu^3/3 + iuz)du \quad (5)$$

$$= c_1 f'(z) - c_2 g'(z) \quad (6)$$

$$\overset{z>1}{\sim} -0.5\ \pi^{-\frac{1}{2}}\ z^{\frac{1}{4}}\ \exp\,(-z^{3/2}/1.5) \quad (7)$$

$$\overset{z<-1}{\sim}\ -\pi^{-\frac{1}{2}}\ |z|^{\frac{1}{4}}\ \cos\,(|z|^{3/2}/1.5 + \pi/4) \quad (8)$$

M. S. Child (ed.), Semiclassical Methods in Molecular Scattering and Spectroscopy, 323–324.

$$c_1 = 3^{-2/3}/\Gamma(2/3) = 0.355028 \tag{9}$$

$$c_2 = 3^{-1/3}/\Gamma(1/3) = 0.258819 \tag{10}$$

$$f(z) = 1 + z^3/3! + 1.4\, z^6/6! + 1.4.7 z^9/9! + \ldots \tag{11}$$

$$g(z) = z + 2z^4/4! + 2.5\, z^7/7! + 2.5.8.\, z^{10}/10! + .. \tag{12}$$

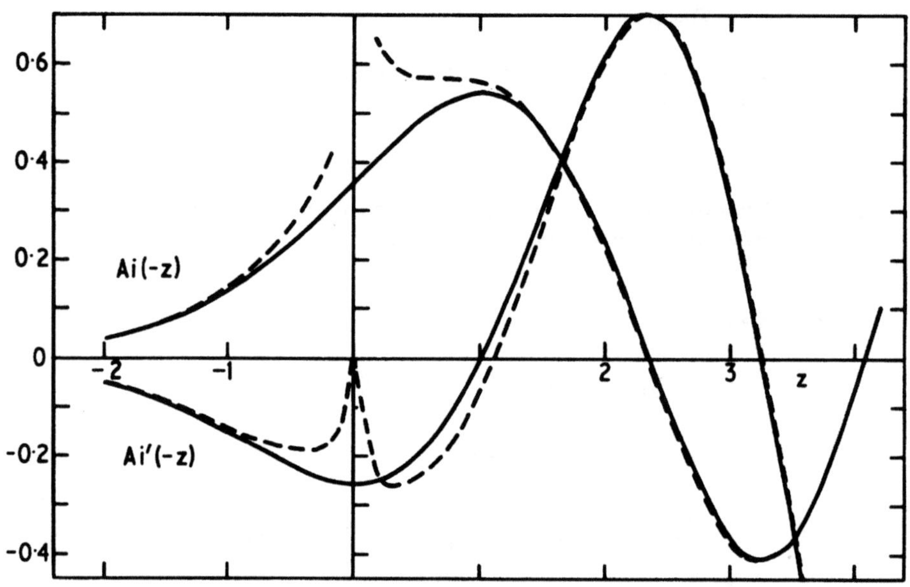

The functions Ai(-z) and Ai'(-z) (——) and asymptotic approximations (----)

2. References

Miller, J.C.P.: 1946, *The Airy Integral*, **B.A. Math Tables**, Part Vol.B (Cambridge Univ. Press).

Erdélyi, A.: 1953, *Higher Transcendental Functions*, vol 1,2 (McGraw Hill).

Abramowitz, M., and Stegun, I.A.: 1965, *Handbook of Mathematical Functions*, (Dover).

AUTHOR INDEX

SUBJECT INDEX

GPSR Compliance

The European Union's (EU) General Product Safety Regulation (GPSR) is a set of rules that requires consumer products to be safe and our obligations to ensure this.

If you have any concerns about our products, you can contact us on ProductSafety@springernature.com

In case Publisher is established outside the EU, the EU authorized representative is:

Springer Nature Customer Service Center GmbH
Europaplatz 3
69115 Heidelberg, Germany

Batch number: 10607220

Printed by Printforce, the Netherlands